Undergraduate Lecture Notes in Physics

For further volumes:
http://www.springer.com/series/8917

Undergraduate Lecture Notes in Physics (ULNP) publishes authoritative texts covering topics throughout pure and applied physics. Each title in the series is suitable as a basis for undergraduate instruction, typically containing practice problems, worked examples, chapter summaries, and suggestions for further reading.

ULNP titles must provide at least one of the following:

- An exceptionally clear and concise treatment of a standard undergraduate subject.
- A solid undergraduate-level introduction to a graduate, advanced, or non-standard subject.
- A novel perspective or an unusual approach to teaching a subject.

ULNP especially encourages new, original, and idiosyncratic approaches to physics teaching at the undergraduate level.

The purpose of ULNP is to provide intriguing, absorbing books that will continue to be the reader's preferred reference throughout their academic career.

Series Editors

Heinz Klose

E. G. Tsitsishvili

A. Komnik

Alexander O. Gogolin

Lectures on Complex Integration

Edited by

Elena G. Tsitsishvili
Andreas Komnik

Alexander O. Gogolin (1965–2011)

Editors
Elena G. Tsitsishvili
Institute for Cybernetics
Tbilisi Technical University
Tbilisi
Georgia

Andreas Komnik
Institut für Theoretische Physik
Universität Heidelberg
Heidelberg
Germany

ISSN 2192-4791 ISSN 2192-4805 (electronic)
ISBN 978-3-319-00211-8 ISBN 978-3-319-00212-5 (eBook)
DOI 10.1007/978-3-319-00212-5
Springer Cham Heidelberg New York Dordrecht London

Library of Congress Control Number: 2013940743

Printed on acid-free paper

Springer is part of Springer Science+Business Media (www.springer.com)

Preface

This is an amalgamation of lecture notes for various applied mathematics and mathematical physics courses I taught at the Mathematics Department, Imperial College London over the last couple of decades. I find that the theme of complex integration beautifully unites various, seemingly very different, topics in this field. While there are plenty of excellent textbooks and classical sources out there, I thought my particular teaching experience, selection of topics, and connections between them might be of interest to students and perhaps to a wider audience.

Assumed knowledge: real analysis, concepts such as limits, convergence, real integration, geometric series, and power series expansions of elementary functions. No knowledge about special functions is required, these are explained whenever the need arises.

Alexander O. Gogolin

The theory of complex functions is a strikingly beautiful and powerful area of mathematics. Some particularly fascinating examples are seemingly complicated integrals which are effortlessly computed after reshaping them into integrals along contours, as well as apparently difficult differential and integral equations, which can be elegantly solved using similar methods. The author was most proficient in this field and the purpose of this book is to summarize his vast knowledge.

We were confronted with the finalization of this book after Alexander tragically passed away in April 2011. The parts of the manuscript available were written in a very concise and clear style, that we have endeavored to emulate.

As a book written as lecture notes, in some places the reader may find it not going into much detail. For more in-depth understanding we therefore recommend a perusal of the classics, such as Refs. [1–3], alongside these notes. We have made reference to the specialized literature wherever possible, and have included a large number of examples and problems with detailed solutions.

We would like to thank O. V. Gogolin and S. Gogolina for help and support during our work on the manuscript.

E. G. Tsitsishvili
A. Komnik

Contents

Chapter 1
Basics

1.1 Basic Definitions

1.1.1 Complex Numbers

The set of complex numbers is given by

$$z = x + iy,$$

where x and y are real numbers and $i^2 = -1$. Here $x = \mathrm{Re}z$ and $y = \mathrm{Im}z$ are the real and imaginary parts of the complex number z. For complex conjugated variable the notation

$$\bar{z} = x - iy$$

will be used. Another useful representation is the polar one:

$$z = re^{i\theta},$$

where r is the modulus of z, $r = |z|$, or the distance between point z and the origin in the complex plane (this construction is called Argand diagram), see the left panel of Fig. 1.1, and $\theta = \arg z$ is the argument of z, or the polar angle. By convention, θ is always counted counterclockwise. It is usually - but not always - measured from the real axis, so that $-\pi < \theta \leq \pi$. In relation to the polar representation, note the seminal Euler formula:

$$e^{i\theta} = \cos\theta + i\sin\theta.$$

The importance of complex numbers and usefulness of complex analysis in applications in general is probably related to the fact that the set of complex numbers, unlike the set of real numbers, is closed under all algebraic operations. Also, a less

A. O. Gogolin (edited by E. G. Tsitsishvili and A. Komnik), *Lectures on Complex Integration*,
Undergraduate Lecture Notes in Physics, DOI: 10.1007/978-3-319-00212-5_1,

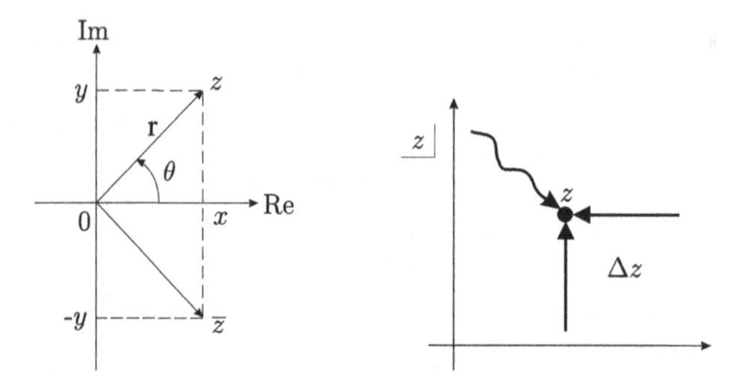

Fig. 1.1 *Left*: Argand diagram; *right*: the derivative of the analytic function is independent on the direction the limit $\delta z \to 0$ is taken

frequently acknowledged fact is the redundance freedom, like in gauge theory of physics.

1.1.2 Key Concept: Analyticity

We first need to learn how to differentiate functions of complex variable. The function $f(z)$ is differentiable at the point z if the limit—the derivative—

$$f'(z) = \lim_{\Delta z \to 0} \frac{f(z + \Delta z) - f(z)}{\Delta z}$$

exists irrespectively of the direction along which the limit $\Delta z \to 0$ is taken, as is schematically shown in the right panel of Fig. 1.1.

Now if the function $f(z)$ is differentiable at all points of an area (domain) D then $f(z)$ is said to be *analytic* in D. The concept of a domain—a continuum region in complex plane, simply connected or otherwise, is important here: as we shall see shortly non-trivial functions that are analytic in the whole complex plane simply do not exist. The boundary of a finite domain, sometimes denoted ∂D, is a closed curve (or a collection of closed curves), see Fig. 1.2.

Note that a formal discussion of properties of continuum sets can get very complicated very quickly. In tune with the applied nature of these notes we shall not dwell on these matters but instead rely on geometric intuition. For example, we shall

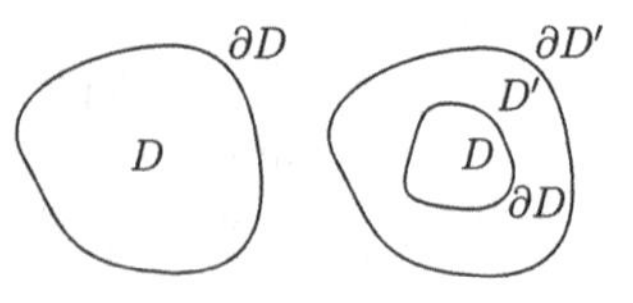

Fig. 1.2 *Left* a finite domain D with its boundary ∂D; *right* a domain D' has ∂D and $\partial D'$ as its boudaries

regard statements like 'a smooth closed curve separates the complex plane into two distinct regions: curve's interior and exterior' as obvious.

Cauchy–Riemann equations. One of the consequences of analyticity are the Cauchy–Riemann equations. Let

$$f(z) = u(x, y) + iv(x, y),$$

that is $u = \mathrm{Re} f$ and $v = \mathrm{Im} f$. The following equations then hold:

$$\frac{\partial u}{\partial x} = \frac{\partial v}{\partial y}, \quad \frac{\partial u}{\partial y} = -\frac{\partial v}{\partial x}.$$

Indeed, if we take the limit $\Delta z = \Delta x + i\Delta y \to 0$ 'horizontally', that is setting $\Delta y = 0$ first and computing the limit $\Delta x \to 0$, see Fig. 1.1, then from the definition of the derivative we obtain:

$$f'(z) = \lim_{\Delta x \to 0} \frac{u(x + \Delta x, y) + iv(x + \Delta x, y) - u(x, y) - iv(x, y)}{\Delta x} = \frac{\partial u}{\partial x} + i\frac{\partial v}{\partial x}.$$

On the other hand taking the limit 'vertically' one finds

$$f'(z) = \lim_{\Delta y \to 0} \frac{u(x, y + \Delta y) + iv(x, y + \Delta y) - u(x, y) - iv(x, y)}{i\Delta y} = \frac{\partial v}{\partial y} - i\frac{\partial u}{\partial y}.$$

For an analytic function the two expressions for the derivative must coincide, hence the Cauchy–Riemann equations.

It is easy to show that the converse is also true: if u and v satisfy Cauchy–Riemann equations in the domain D then the function $u + iv$ is analytic in that domain.

1.1.3 Contour Integrals

We take it for granted that the reader is familiar with basic properties of the real calculus integration

$$\int_a^b f(x)dx,$$

such as

$$\int_a^b f(x)dx = -\int_b^a f(x)dx,$$

or the integration by parts formula

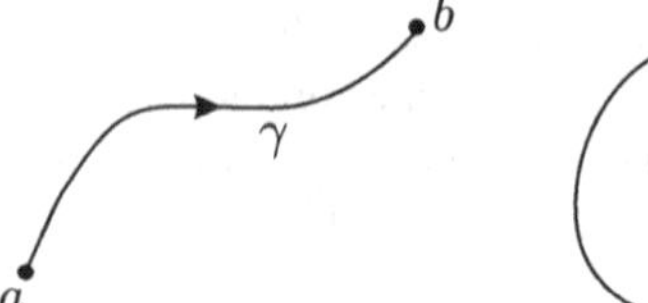

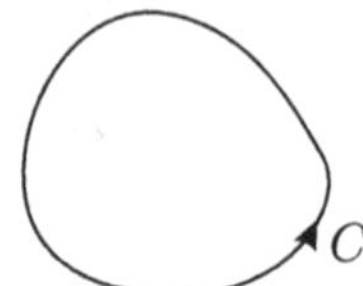

Fig. 1.3 *Left* an open path γ between points a and b; *right* a closed contour C for the case $a = b$

$$\int_a^b g(x)dh(x) = gh\Big|_a^b - \int_a^b h(x)dg(x).$$

We are now interested in generalising this integral to the case of functions of complex variable

$$\int_\gamma f(z)dz,$$

where γ is a smooth path (a contour) in complex plane starting at the point a and ending at the point b (now complex). It may be that $a = b$, then the contour is closed and usually denoted as C, see Fig. 1.3.

To this end we adopt a parametric representation of the contour: $x(t) + iy(t) = z(t) \in \gamma$, where the real parameter t takes values $t_0 \leq t \leq t_1$, such that $z(t_0) = a$ and $z(t_1) = b$. We then define

$$\int_\gamma f(z)dz = \int_{t_0}^{t_1} f(z(t))\frac{dz}{dt}dt,$$

thereby reducing the contour integral to a real integral (to be precise to four real integrals $\int u\dot{x}dt$, $\int u\dot{y}dt$, $\int v\dot{x}dt$, and $\int v\dot{y}dt$).

For example, if γ is a part of the real axis we have simply

$$\int_\gamma f(z)dz = \int f(x)dx,$$

with appropriate limits. If γ is a part of the imaginary axis then

$$\int_\gamma f(z)dz = \int f(iy)idy.$$

In a more interesting example consider a circle C_R with a radius R centered at the point $z = a$, see Fig. 1.4. Then

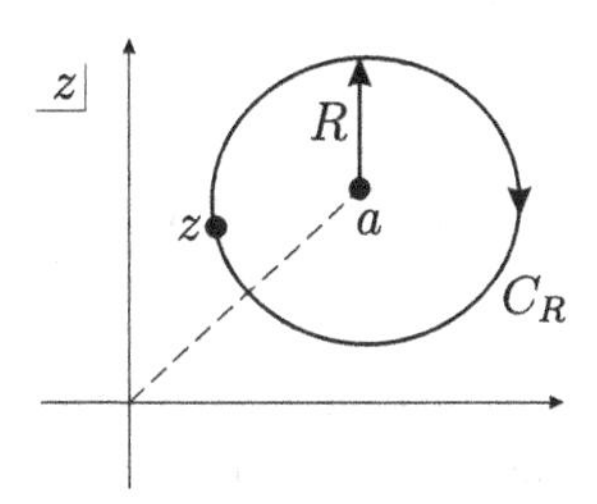

Fig. 1.4 Circular contour C_R with radius R around the point a

$$z|_{\text{on circle}} = a + Re^{i\theta},$$

with a and R fixed and θ varying in e.g. $[-\pi, \pi]$, so that $dz = iRe^{i\theta}d\theta$ and we have

$$\int_{C_R} f(z)dz = \int_{-\pi}^{\pi} f\left(a + Re^{i\theta}\right) iRe^{i\theta}d\theta.$$

1.2 Cauchy Theorem

1.2.1 Formulation and Proof

This theorem states that if C is a smooth closed contour and the function $f(z)$ is analytic on and inside C then

$$\int_C f(z)dz = 0. \tag{1.1}$$

It is truly wondrous how so many sophisticated results in applied mathematics directly follow from this simple and elegant theorem about zero. The literature contains many different proofs of Cauchy theorem. May be this is because this theorem is nearly obvious. Indeed, given a good natured character (analyticity) of the function $f(z)$ it is reasonable to assume that it has a primitive: a single valued differentiable (and so continuous) function $F(z)$ such that $F'(z) = f(z)$. Then moving z around C and returning it to the same point will not change the value of $F(z)$, so the integral must be zero. Yet the Cauchy theorem is too important to make assumptions, so a more rigorous proof is given below.

Define the object [4]

$$G(\lambda) = \lambda \int_C f(\lambda z)dz,$$

where the real parameter $\lambda \in [0, 1]$. By construction $G(0) = 0$ and the Cauchy theorem states that $G(1) = 0$. To see how the two limits are connected we compute the derivative

$$\frac{dG}{d\lambda} = \int_C f(\lambda z)dz + \lambda \int_C zf'(\lambda z)dz = \int_C f(\lambda z)dz + \int_C zdf(\lambda z).$$

Note that the derivative inside the second integral exists under the conditions of Cauchy's theorem: $f(z)$ is an analytic function inside C. Clearly the analog of the real integration by parts formula

$$\int_\gamma g(z)dh(z) = g(z)\, h(z)\Big|_a^b - \int_\gamma h(z)dg(z)$$

holds for contour integrals (indeed substituting here the parametric definition of the contour integral readily verifies this). It follows that

$$\frac{dG}{d\lambda} = \int_C f(\lambda z)dz + z\, f(\lambda z)\Big|_a^a - \int_C f(\lambda z)dz = 0$$

for all λ (a is an arbitrary point on C). As a result $G(\lambda) =$ const. and since $G(0) = 0$, it is zero for all λ including $\lambda = 1$, hence the Cauchy theorem.

1.2.2 Deformation of Contours

An immediate and important consequence of the Cauchy theorem is this: the value of a contour integral does not change if the contour is moved (deformed) within the analyticity domain of the integrand.

Indeed assume that we have two contours γ and γ' as shown in Fig. 1.5. We then have

$$\int_\gamma f(z)dz - \int_{\gamma'} f(z)dz = \int_{\gamma_{ab}+\gamma'_{ba}} f(z)dz = 0.$$

Indeed the second integral can be interpreted as integrating backward on γ'

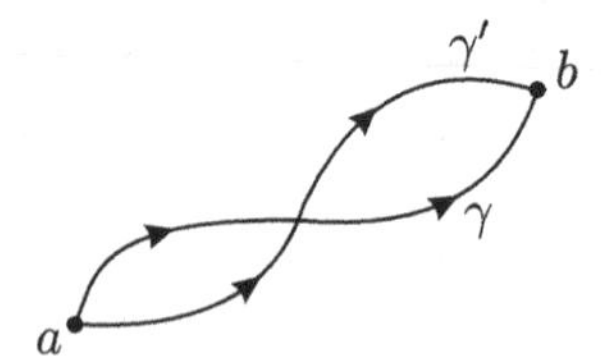

Fig. 1.5 Two different paths between the points a and b form a closed contour

$$-\int_{\gamma'_{ab}} f(z)dz = \int_{\gamma'_{ba}} f(z)dz$$

making the combined contour $\gamma_{ab} + \gamma'_{ba}$ closed. Because in addition $f(z)$ is assumed to be analytic inside this contour - the zero result quoted above follows from Cauchy theorem. Note: the end points can not be moved (rather they can be but that would introduce additional integration contours).

1.2.3 Cauchy Integral Formula

This is another direct consequence of the Cauchy theorem:

$$\frac{1}{2\pi i} \int_C \frac{f(z)dz}{z-a} = \begin{cases} 0, & a \text{ outside } C, \\ f(a), & a \text{ inside } C, \end{cases} \tag{1.2}$$

where the function $f(z)$ is again required to be analytic on and inside the closed contour C.

The first line in the above is obvious: if a is outside C then the function $f(z)/(z-a)$ is analytic on and inside C and so fulfills the conditions of the Cauchy theorem.

To see that the second line is also correct, we first deform the contour C into a circle of (small compared to the size of C) radius ρ (see Fig. 1.6) and continuously decrease this radius by taking the limit $\rho \to 0$. On the circle $z = a + \rho e^{i\theta}$ so we have

$$\frac{1}{2\pi i} \lim_{\rho \to 0} \int_{-\pi}^{\pi} \frac{f(a+\rho e^{i\theta}) i\rho e^{i\theta} d\theta}{\rho e^{i\theta}} = f(a).$$

This is a residue type calculation in spirit, as we shall see shortly. Here the function $f(z)/(z-a)$ is obviously not analytic inside C but has a singularity at $z = a$ called *simple pole*.

The next obvious question is about whether we can evaluate the integral (1.2) when the point a lies exactly on C. This is an important and interesting issue which we shall investigate in detail in the next chapters.

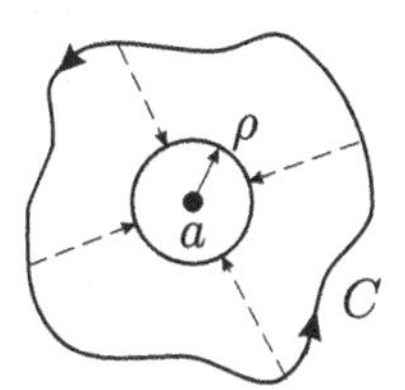

Fig. 1.6 A closed contour C can be collapsed onto a small circular one provided the integrand remains analytical in the area between them

Interestingly, if we want to define an analytic function using Cauchy type integral formula, the contour does not need to be closed. Let us consider $f(z)$ which is defined on γ, that is $z \in \gamma$, and which is continuous there.[1] Let us associate with $f(z)$ the following function $F(\xi)$, defined in whole z-plane:

$$F(\xi) = \frac{1}{2\pi i} \int_\gamma \frac{dz}{z-\xi} f(z).$$

The continuity of $f(z)$ guarantees convergence of the above integral as well as of the derivatives $F^{(n)}(\xi)$,

$$F^{(n)}(\xi) = \frac{n!}{2\pi i} \int_\gamma \frac{dz}{(z-\xi)^{n+1}} f(z)\,, \quad n = 1, 2, \ldots.$$

for all ξ away from γ. Therefore $F(\xi)$ is analytic for all ξ apart from $\xi \in \gamma$. When ξ approaches γ the integral becomes singular, we shall study this later in Sect. 1.3.3 and in more detail in Sect. 3.3.2.

Another 'difficult' case is when the endpoints of γ go to infinity, so that the whole contour γ is effectively closed on infinity. In particular γ can coincide with the real axis. It follows that the integral

$$F(\xi) = \frac{1}{2\pi i} \int_{-\infty}^{+\infty} \frac{dx}{x-\xi} f(x)$$

defines a function analytic everywhere (upper and lower half-planes) apart from the real axis γ. When ξ crosses the real axis the function $F(\xi)$ is discontinuous, we shall study this later in Sect. 3.3.2.

1.2.4 *Taylor and Laurent Expansions*

To arrive at the Taylor expansion (familiar from the real analysis), consider again a function $f(z)$ which is analytic on and inside a closed contour C. Let a and $a+h$ be two points in the complex plane both inside the contour C. We want to expand the function $f(a+h)$ around the point a in powers of h. To this end we write

$$f(a+h) = \frac{1}{2\pi i} \int_C \frac{f(z)dz}{z-a-h} = \frac{1}{2\pi i} \int_C \frac{f(z)dz}{z-a} \frac{1}{1-h/(z-a)}.$$

[1] Although this condition can be relaxed.

Making use of the geometric series in the integrand

$$\frac{1}{1 - h/(z-a)} = \sum_{n=0}^{\infty} \frac{h^n}{(z-a)^n}$$

we obtain the Taylor series in the form

$$f(a+h) = \sum_{n=0}^{\infty} a_n h^n, \tag{1.3}$$

with coefficients a_n given by the integral

$$a_n = \frac{1}{2\pi i} \int_C \frac{f(z)dz}{(z-a)^{n+1}}. \tag{1.4}$$

Note that re-expanding $f(z)$ around a in the integrand above and using the Cauchy integral formula is perfectly consistent with the familiarly looking expression $a_n = f^{(n)}(a)/n!$ (n^{th} derivative); singularities of higher order than the simple pole do not contribute as we shall see soon.

If the function $f(z)$ is not analytic at point a but has some kind of singularity at this point (see below for a classification) then the Taylor expansion is not applicable. Yet another kind of expansion, called Laurent expansion, could be found in this situation which converges in a ring outside point a.

In order to develop the Laurent expansion we first need to generalise the Cauchy theorem and the Cauchy integral formula to the case of non-simply connected domains. In fact, it will be sufficient to consider the simplest of such domains: a ring-shaped region between closed contours C and C' as shown in Fig. 1.7a. We assume that $f(z)$ is analytic in between C and C' but it is not necessarily analytic inside C'. To proceed we deform these two contours into a single one by building a small bridge $\gamma_\epsilon^\pm$ between C_ϵ and C'_ϵ as shown in Fig. 1.7b. The standard Cauchy theorem (1.1) then takes the form

$$\int_{C_\epsilon} f(z)dz + \int_{C'_\epsilon} f(z)dz + \sum_{\pm} \int_{\gamma_\epsilon^\pm} f(z)dz = 0.$$

In the limit $\epsilon \to 0^+$ Fig. 1.7b essentially reverts to Fig. 1.7a, the contributions from γ_ϵ^+ and γ_ϵ^- cancel each other out, and, keeping in mind the reversal of the direction of integration on C', we obtain the desired extension of the Cauchy theorem in the form

$$\int_C f(z)dz - \int_{C'} f(z)dz = 0.$$

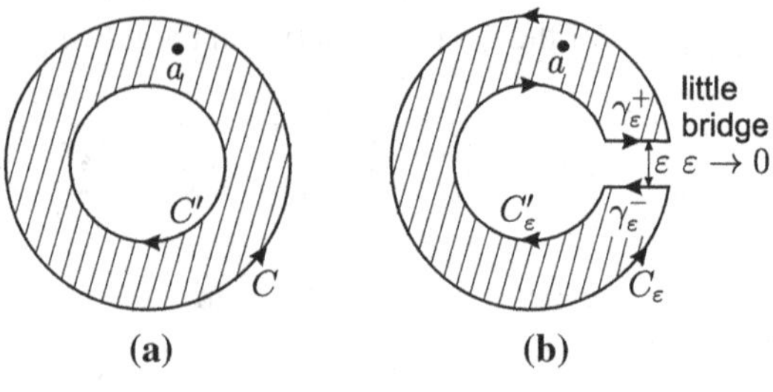

Fig. 1.7 An expansion in a non-simply connected annulus (**a**), can be constructed using the Cauchy formula for simply connected domains (**b**)

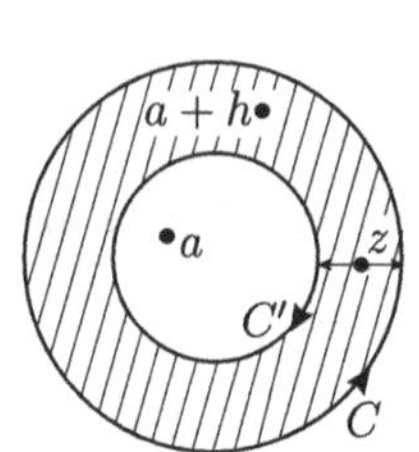

Fig. 1.8 The Laurent expansion is convergent in the annulus between the circles C and C'

Same 'narrow bridge' argument leads to the extension of the Cauchy integral formula as

$$f(a) = \int\limits_C \frac{f(z)dz}{z-a} - \int\limits_{C'} \frac{f(z)dz}{z-a},$$

where the point $z = a$ is in between the two contours.

Now we take a inside C' [that is generally out of the analyticity domain of the function $f(z)$], but the point $a + h$ between C' and C, see Fig. 1.8. The Cauchy integral formula for the latter point (obviously there is no Cauchy formula for the former) now reads:

$$f(a+h) = \int\limits_C \frac{f(z)dz}{z-a-h} - \int\limits_{C'} \frac{f(z)dz}{z-a-h}.$$

For $z \in C$ we use the geometric series expansion as in Taylor series:

$$\frac{1}{z-a-h} = \frac{1}{z-a}\frac{1}{1-h/(z-a)} = \frac{1}{z-a}\sum_{n=0}^{\infty}\frac{h^n}{(z-a)^n},$$

while for $z \in C'$

$$\frac{1}{z-a-h} = -\frac{1}{h}\frac{1}{1-(z-a)/h} = -\frac{1}{h}\sum_{n=0}^{\infty}\frac{(z-a)^n}{h^n} = \frac{1}{z-a}\sum_{n=-\infty}^{-1}\frac{h^n}{(z-a)^n},$$

leading to the Laurent expansion

$$f(a+h) = \sum_{n=-\infty}^{\infty} a_n h^n$$

with the coefficients

$$a_n = \frac{1}{2\pi i} \int_C \frac{f(z)dz}{(z-a)^{n+1}} \tag{1.5}$$

formally given by the same expression as in Taylor series.[2]

Example 1.1
We would like to expand the function $f(z) = 1/[z(1-z)]$ into a Laurent series in the annulus $\delta < |z| < 1$ with finite $\delta < 1$. The coefficients of the expansion are found with the help of (1.5),

$$a_n = \frac{1}{2\pi i} \int_C dz \, \frac{1}{z^{n+1}\, z(1-z)}.$$

Since $|z| < 1$ we can use the Taylor expansion for $1/(1-z)$. We chose the contour C to be a circle with radius $\delta < \epsilon < 1$ around the coordinate origin. It can be parametrized by the angle $0 < \phi < 2\pi$ via $z = \epsilon e^{i\phi}$. Then we obtain

$$a_n = \frac{1}{2\pi} \sum_{k=0}^{\infty} \epsilon^{k-n-1} \int_0^{2\pi} d\phi \, e^{i(k-n-1)\phi}.$$

The last integral is obviously finite and given by 2π only for $n = k-1$. Therefore the expansion starts with $n = -1$ and all coefficients are equal: $a_{-1} = a_0 = a_1 = \cdots = 1$. This series could, of course, be produced by a direct expansion as well

$$f(z) = \frac{1}{z(1-z)} = \frac{1}{z}\left(1 + z + z^2 + \ldots\right) = \frac{1}{z} + 1 + z + z^2 + \ldots$$

Example 1.2
As a less trivial application we try to expand

$$f(z) = e^{\zeta(z-1/z)/2} = \sum_{n=-\infty}^{\infty} J_n(\zeta)\, z^n \tag{1.6}$$

---

[2] Clearly the function $f(z)/(z-a)^{n+1}$ is analytic between C' and C for negative n, so the C' contour for these terms is deformed into C. The assumption that the function $f(z)$ is analytic between C and C' means for the Laurent series that the terms with non-negative powers of h should have a radius of convergence (around a) that encircles C. On the other hand the terms with the negative powers of h converge outside a smaller radius that lies inside C'. One way to look at the Laurent expansion is then that it provides the means of splitting the function into two: the one analytic outside a smaller circle and the other analytic inside a bigger circle.

around $z = 0$. The coefficients are again given by

$$J_n(\zeta) = \frac{1}{2\pi i}\int_C dz\, \frac{e^{\zeta(z-1/z)/2}}{z^{n+1}}, \tag{1.7}$$

where this time C can be a circle of any radius. We choose $z = e^{i\phi}$ and write

$$J_n(\zeta) = \frac{1}{2\pi}\int_{-\pi}^{\pi} d\phi\, e^{i\phi}\, e^{-i\phi(n+1)+i\zeta\sin\phi} = \frac{1}{\pi}\int_0^{\pi} d\phi\, \cos(n\phi - \zeta\sin\phi). \tag{1.8}$$

In the last step we divided the integral in half and exchanged ϕ to $-\phi$ in the first part. On the other hand after the substitution $z \to 2z/\zeta$ in (1.7) we obtain

$$J_n(\zeta) = \frac{1}{2\pi i}\left(\frac{\zeta}{2}\right)^n \int_C dz\, \frac{e^{z-\zeta^2/4z}}{z^{n+1}} = \frac{1}{2\pi i}\sum_{k=0}^{\infty}\frac{(-1)^k}{k!}\left(\frac{\zeta}{2}\right)^{n+2k}\int_C dz\, z^{-(n+k+1)}\, e^z.$$

The last integral can be computed by repeated partial integration on the unit circle and is equal to $2\pi i/[(n+k)!]$, so that we obtain the following series representation of the coefficient

$$J_n(\zeta) = \sum_{k=0}^{\infty}\frac{(-1)^k\,\zeta^{n+2k}}{2^{n+2k}\,k!\,(n+k)!}. \tag{1.9}$$

$J_n(\zeta)$ is referred to as *Bessel function of the first kind* and (1.7) is its representation via a contour integral.

1.2.5 Liouville Theorem

This theorem is crucial for solving integral equations in following chapters and is a direct consequence of the Cauchy integral formula (and the Taylor expansion). The classic Liouville theorem states that if a function $f(z)$ is analytic for all finite z and is bounded on infinity then $f(z)$ is identically a constant for all z.

The proof is immediate. If we take two points z and z' inside a circle C_R of radius R (see Fig. 1.9) then by the Cauchy integral formula the analyticity of $f(z)$ means

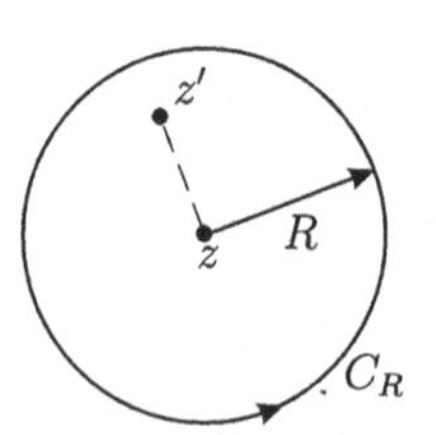

Fig. 1.9 Circular contour for the proof of Liouville theorem

that

$$f(z) = \frac{1}{2\pi i} \int_{C_R} \frac{f(\xi)d\xi}{\xi - z},$$

and similarly for $f(z')$. Now take the limit $R \to \infty$ whilst keeping the difference of $z - z'$ finite. We can then estimate the difference in the values of the function as

$$|f(z) - f(z')| = \frac{1}{2\pi} \left| \int_{C_R} \frac{f(\xi)d\xi}{\xi - z} - \int_{C_R} \frac{f(\xi)d\xi}{\xi - z'} \right|$$
$$= \frac{1}{2\pi} \left| \int_{C_R} \frac{(z - z')f(\xi)d\xi}{(\xi - z)(\xi - z')} \right| = O\left(\frac{1}{R}\right).$$

This is zero in the limit $R \to \infty$ and so $f(z) = f(z')$ for all z and z'.

We shall actually need an extension of this theorem when the function $f(z)$ is allowed to grow as $z \to \infty$, but not faster than algebraically. That is $f(z)$ is bounded as

$$|f(z)| < A|z|^N,$$

where A is a positive constant and N is a non-negative integer power. In that case the function $f(z)$ is a polynomial degree N. To see that this statement is correct it is sufficient to realise that, as $f(z)$ is analytic for all z, it can be Taylor expanded at any point and, in particular, at the origin, $f(z) = \sum_{n=0}^{\infty} a_n z^n$. The only way to satisfy the above bound now is to require that the Taylor expansion terminates at the Nth term, that is $f(z) = \sum_{n=0}^{N} a_n z^n$—manifestly a polynomial of degree N (provided $a_N \neq 0$).

1.2.6 Isolated Singular Points, Residues and Integrals

Under the conditions of Sect. 1.2.4, we call $z = a$ an isolated singular point if, while not analytic at $z = a$, $f(z)$ is still analytic at all points in the immediate vicinity of $z = a$. An example of such a function is provided by

$$e^{1/z} = \sum_{n=0}^{\infty} \frac{1}{n!z^n} = \sum_{n=-\infty}^{0} \frac{z^n}{(-n)!}.$$

This is a Laurent expansion that converges outside of an arbitrary small circle (that is for all $z \neq 0$) but has an infinite number of diverging powers as $z \to 0$: this is called an essential singularity.

A natural way of classifying isolated singular points now presents itself. If the Laurent expansion of $f(z)$ terminates at a finite negative N, i.e. if

$$f(z)=\frac{a_{-N}}{(z-a)^N}+\frac{a_{-N+1}}{(z-a)^{N-1}}+\ldots+\frac{a_{-1}}{z-a}+\varphi(z),$$

where $\varphi(z)$ is a function analytic around a, then $z=a$ is said to be an Nth degree pole of the function $f(z)$. The 1st degree pole is called a simple pole.

Now suppose we have a closed contour C running around $z=a$ in the positive direction. A reasonable question to ask then is: what is the value of the integral $\int_C f(z)dz$ under these circumstances? Clearly $\int_C \varphi(z)dz=0$ because of the Cauchy theorem. Next we consider the general diverging term. As before we deform C into a circle with radius ρ around a (see Fig. 1.6) and obtain

$$\int\limits_{C_\rho}\frac{dz}{(z-a)^r}\bigg|_{z=a+\rho e^{i\theta}}=\int\limits_0^{2\pi}\frac{i\rho e^{i\theta}d\theta}{\rho^r e^{ir\theta}}=i\rho^{1-r}\int\limits_0^{2\pi}e^{i(1-r)\theta}d\theta=\rho^{1-r}\left[\frac{e^{i(1-r)\theta}}{1-r}\right]\bigg|_0^{2\pi},$$

which is zero for $r\neq 1$ because of the angle integration (the primitive function is obviously periodic in θ). For $r=1$, on the other hand, we have

$$\int\limits_{C_\rho}\frac{dz}{z-a}=i\int\limits_0^{2\pi}d\theta=2\pi i,$$

and so the answer to our question is

$$\int\limits_C f(z)dz=2\pi i a_{-1}.$$

It is customary to call the number a_{-1}, the coefficient of the simple pole, a *residue* of the function $f(z)$ at the pole $z=a$ and to write

$$a_{-1}=\operatorname{Res}f(z)|_a.$$

Clearly if the function has another pole inside C it will contribute in an additive manner. This leads us to the following statement: if the function $f(z)$ is analytic inside the closed contour C, apart from a finite set of poles $\{a_p\}$, the integral over C is determined by the sum of residues as

$$\int\limits_C f(z)dz=2\pi i\sum_p\operatorname{Res}f(z)|_{a_p}. \tag{1.10}$$

This result is sometimes referred to as the *residue theorem*. However, with it being a direct consequence of the Cauchy theorem, we will continue to refer to these type of results as extended Cauchy theorem or simply Cauchy theorem in these notes.

We realise that while the integral of, for example $f(z) = 1/(z-a)^2$ around a is exactly zero, the integral of $f(z) = g(z)/(z-a)^2$, where $g(z)$ is a function analytic inside C is generally finite. Indeed, the calculation proceeds as follows:

$$\int_C \frac{g(z)dz}{(z-a)^2} = \int_C \frac{dz}{(z-a)^2}[g(a) + (z-a)g'(a) + O((z-a)^2)] = 2\pi i g'(a).$$

For a general N^{th} power in $f(z) = g(z)/(z-a)^N$, the term $g^{[N-1]}(a)/(N-1)!$ in the Taylor expansion of $g(z)$ will be responsible for the simple pole giving rise to the formula for the residue:

$$\text{Res} f(z)|_a = \frac{1}{(N-1)!}\frac{d^{N-1}}{d^{N-1}z}[(z-a)^N f(z)]|_{z=a}.$$

Sometimes it is expedient to use this formula rather than perform the Laurent expansion directly.

One useful class of integrals that can be evaluated by the residue technique is as follows (integrals over a period)

$$I = \int_0^{2\pi} R(\cos\theta, \sin\theta)d\theta, \tag{1.11}$$

where R is an arbitrary rational function (that is a ratio of polynomials) of its arguments. When substituting $z = e^{i\theta}$, the integration in z goes around the unit circle (in positive direction). We have

$$d\theta = \frac{dz}{iz}, \quad \cos\theta = \frac{1}{2}\left(z + \frac{1}{z}\right), \quad \sin\theta = \frac{1}{2i}\left(z - \frac{1}{z}\right),$$

and therefore

$$I = \int_{|z|=1} R(z)\frac{dz}{iz} = 2\pi \sum_p \text{Res}\left[\frac{R(z)}{z}\right]\Bigg|_{z_p,\, |z_p|<1}.$$

Example 1.3

$$\int_0^{2\pi} \cos\theta d\theta = \frac{1}{2i}\int_{|z|=1}\left(1 + \frac{1}{z^2}\right)dz = 0.$$

The integrand has no simple pole.

Example 1.4

$$\int_0^{2\pi} \sin^2 \theta d\theta = \frac{1}{4i} \int_{|z|=1} \left(-z - \frac{1}{z^3} + \frac{2}{z} \right) dz = \pi,$$

as expected. Indeed, when integrating over an integer number of periods, one can simply substitute $\cos^2 \theta \to 1/2$ or $\sin^2 \theta \to 1/2$.

Example 1.5

$$\int_0^{2\pi} \frac{d\theta}{1 - 2p\cos\theta + p^2} = \frac{1}{i} \int_{|z|=1} \frac{dz}{z - p(z^2+1) + p^2 z}$$
$$= \frac{1}{i} \int_{|z|=1} \frac{dz}{(z-p)(1-pz)} = \frac{2\pi}{1-p^2},$$

where $0 < p < 1$ is a real parameter.

Surprisingly many definite integrals on the real axis can be computed by the residue technique. Generally the integral is of the form

$$I = \int_a^b g(x) dx,$$

and the method proceeds in the following steps.

(i) Generate a domain D by choosing an appropriate contour C' connecting b with a in the complex plane, see Fig. 1.10.
(ii) Find an auxiliary function $f(z)$, analytic in D apart from a finite number of poles, and related to $g(x)$ when $z = x$ in a simple way, for example as $\mathrm{Re} f(x) = g(x)$.
(iii) Apply Cauchy theorem, which now takes the form:

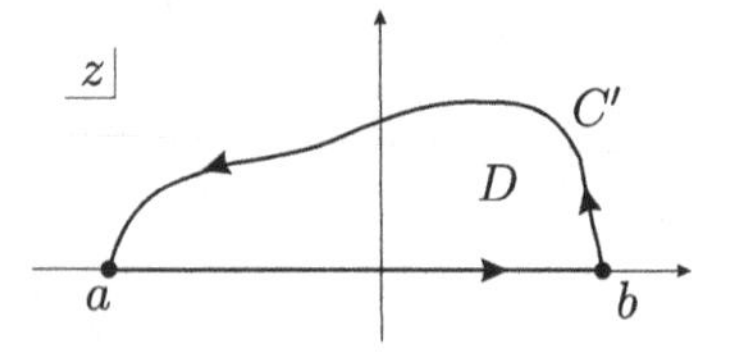

Fig. 1.10 Calculation of an integral along the real axis with the help of an auxiliary contour C'.

$$\int_a^b f(x)dx + \int_{C'} f(z)dz = 2\pi i \sum \mathrm{Res} f(z)\Big|_{\text{in } D}.$$

(iv) If the integral $\int_{C'} f(z)dz$ can be computed independently, or expressed in terms of the original integral I, then the problem of calculating I is solved.

Example 1.6

This is a Fourier integral transform of a Lorentzian, an integral of a type that commonly occurs in applications:

$$I(p) = \int_{-\infty}^{\infty} \frac{e^{ipx}dx}{x^2+1}, \tag{1.12}$$

where p is real. Here we simply take $f(z) = e^{ipz}/(z^2+1)$. We initially limit the x-integration to a segment $[-R, R]$ (the limit $R \to \infty$ to be taken later) and complete the analyticity domain by either an upper half-plane semicircle C_R^+ or the lower half-plane semicircle C_R^- as is shown in Fig. 1.11a,b, respectively. The poles are at $z = \pm i$ and the Cauchy theorem takes the form

$$\int_{-R}^{R} + \int_{C_R^+} = (2\pi i)\frac{e^{-p}}{2i} = \pi e^{-p},$$

or the form

$$\int_{-R}^{R} + \int_{C_R^-} = (-2\pi i)\frac{e^{p}}{(-2i)} = \pi e^{p},$$

depending on which semicircle is used. Note that in the second formula the minus sign of the residue is compensated by the minus sign coming from the angle integration (negative direction on C_R^-). Also note that the short hand notation where $\int_C$ stands

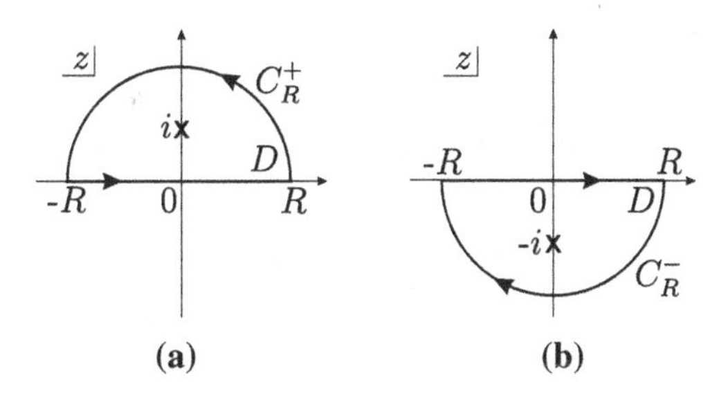

Fig. 1.11 The integration contour for Example 1.6: for positive p we close the contour in the upper half-plane (**a**), and for negative in the lower half-plane (**b**)

for $\int_C f(z)dz$ is used and will be used in similar calculations below. Now we take the limit $R \to \infty$. Clearly

$$\lim_{R\to\infty} \int_{-R}^{R} = I(p).$$

On the other hand $|e^{ipz}| = e^{-p\mathrm{Im}z}$. So if $p > 0$ then the integrand decays exponentially in the upper half-plane but diverges in the lower half-plane. In this case the contour C_R^+ should be used (C_R^- can not be used) as $\lim_{R\to\infty} \int_{C_R^+} = 0$ and $I(p) = \pi e^{-p}$. If $p < 0$, then the integrand decays exponentially in the lower half-plane but diverges in the upper half-plane. In this case, the contour C_R^- should be used (C_R^+ can not be used) as $\lim_{R\to\infty} \int_{C_R^-} = 0$ and $I(p) = \pi e^{p}$. Therefore the answer is

$$I(p) = \pi e^{-|p|}.$$

It is instructive to verify the inversion formula by elementary integration:

$$\frac{1}{2\pi}\int_{-\infty}^{\infty} e^{-ipx} I(p)dp = \frac{1}{2}\int_{0}^{\infty} e^{-ipx-p}dp + \frac{1}{2}\int_{-\infty}^{0} e^{-ipx+p}dp$$
$$= \frac{1}{2}\frac{1}{1+ix} + \frac{1}{2}\frac{1}{1-ix} = \frac{1}{x^2+1}.$$

Jordan Lemma. The helpful construction which we have used here can be applied for a large number of integrals of the type (1.12). It can be rigorously proven and is referred to as *Jordan lemma*. Usually, it is used to show that the C_R part of the integral goes to zero, so that one can equate the integral along the real axis to the one along the whole contour and subsequently use Cauchy's integral formula. The proposition is formulated as follows. Let $f(z)$ is analytic in the upper half–plane $\mathrm{Im}z \geq 0$, except a finite number of isolated points. Let also C_R be an arc of a semicircle $|z| = R$ in the upper half-plane. If for each z on C_R there is some constant μ_R such that $|f(z)| \leq \mu_R$ and $\mu_R \to 0$ as $R \to \infty$, then for $a > 0$

$$\lim_{R\to\infty} \int_{C_R} e^{iaz} f(z)dz = 0. \tag{1.13}$$

In order to prove this statement, set $z = Re^{i\theta}$ and use the following obvious relation: $\sin\theta \geq \frac{2}{\pi}\theta$ at $0 \leq \theta \leq \frac{\pi}{2}$. Then we have

$$\left|\int_{C_R} e^{iaz} f(z)dz\right| \le \mu_R R \int_0^{\pi} e^{-aR\sin(\theta)} d\theta \le \mu_R \frac{\pi}{a}(1 - e^{-aR}) \to 0, \text{ as } R \to \infty.$$

If $a < 0$ and $f(z)$ satisfies the conditions of Jordan lemma at $\text{Im}z \le 0$, then formula (1.13) is still valid but at the integration over the arc C_R in the lower half–plane. Similar statements take place at $a = \pm i\alpha$ ($\alpha > 0$) if the C_R–integration occurs in the right ($\text{Re}z \ge 0$) or left ($\text{Re}z \le 0$) half–plane, respectively.

Jordan lemma is a basis of a great number of integration methods based on taking appropriately chosen closed contours. Very often (and especially in physics literature) they are summarily referred to as *residue formulas*.

Example 1.7
For the calculation of this integral we use the periodic properties of a (complex) exponential:

$$I(a) = \int_{-\infty}^{\infty} \frac{e^{ax}dx}{e^x + 1},$$

where the real number $a \in [0, 1]$, otherwise the integral diverges. Again, the best candidate for the auxiliary function is simply $f(z) = e^{az}/(e^z + 1)$ and the useful contour is around a rectangular domain D, shown in Fig. 1.12. The poles are at z given by the solutions of $e^z = -1$, that is at $z = \pi i + 2\pi i n$ (n is an integer). There is only one pole in D at $z = \pi i$. To find the residue one best computes a direct Laurent expansion of $f(z)$. Put $z = \pi i + \delta$, then $e^z = e^{\pi i + \delta} = -e^{\delta} = -1 - \delta + O(\delta^2)$ and

$$f(\pi i + \delta) = \frac{e^{\pi a i}[1 + O(\delta)]}{-\delta} = \frac{-e^{\pi a i}}{\delta} + O(\delta^0)$$

and we have the residue equal to $-e^{\pi a i}$. The Cauchy theorem thus states that

$$\int_I + \int_{II} + \int_{III} + \int_{IV} = -2\pi i e^{\pi a i}.$$

Now it is time to take the limit $R \to \infty$. Clearly

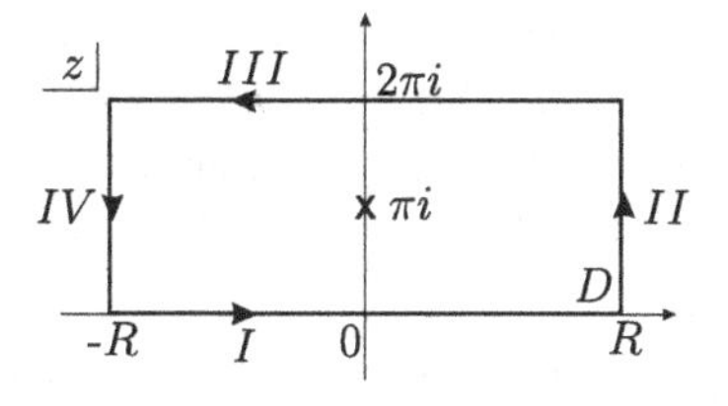

Fig. 1.12 Schematic representation of the contour for the calculation of the integral in Example 1.7

$$\lim_{R\to\infty} \int_{I} = I(a),$$

and also

$$\lim_{R\to\infty} \int_{III} = \int_{+\infty}^{-\infty} dx \, \frac{e^{a(x+2\pi i)}}{e^{x+2\pi i}+1} = -e^{2\pi a i} I(a).$$

On segments II and IV, $z = \mp R + iy$ with $0 < y < 2\pi$. Therefore the estimates for $R \to +\infty$ are

$$\int_{II} = O(e^{(a-1)R}), \quad \int_{IV} = O(e^{-aR}),$$

that is both integrals vanish when $0 < a < 1$. So we have

$$I(a) - e^{2\pi a i} I(a) = -2\pi i e^{\pi a i}$$

and the answer is

$$I(a) = \frac{2\pi i e^{\pi a i}}{e^{2\pi a i}-1} = \frac{2\pi i}{e^{\pi a i} - e^{-\pi a i}} = \frac{\pi}{\sin(\pi a)}.$$

1.3 Branch Cut Integration

1.3.1 Multivalued Functions

The simplest example for a multivalued function is the square root $f(z) = z^{1/2}$. Even on the real positive axis there are obviously two different values to the same argument: $\pm|z|^{1/2}$, which makes the function non-analytic. One can try to redefine the function in such a way that, for instance $(z = 1)^{1/2} = 1$. In order to find the value of the function at $z = -1$ we can then follow the semi-circled paths clockwise as well as anti-clockwise so that along each path the function remains analytic. Then we immediately realise that both values are simply $\pm i$. An easy remedy is to declare the negative real axis including the coordinate origin to a no-go area, or to 'cut' the complex plane $\mathbb{C}$. The coordinate origin is the point where the two options $\pm|z|^{1/2}$ come arbitrarily close to each other and is called *branching point* and the cut is naturally called *branch cut*.[3] After cutting $\mathbb{C}$ we still have two options to define an *analytic branch* of the square root and in most cases we are free to choose. The points $-1 \pm i\delta$ lie on the upper/lower shores of the cut and the procedure of relating their

[3] We would like to remark, that cutting along the negative real axis is not the only option. The function can be uniquely defined in any open simply connected set obtained from the whole complex plane $\mathbb{C}$ by cutting from the branching point to infinity.

values to each other is the simplest case of analytic continuation. Similar procedure can be applied to other non-integer powers $f(z) = z^a$, $a \neq n$, $n \in \mathbb{Z}$.

Yet another very important function with a branching point is the complex logarithm, which has an infinite number of analytic branches. Here the procedure of analytic branch extraction is very similar and one obtains an infinite number of branches differing by an offset of $i2\pi n$.

In many cases multivalued functions require complex plain cuts along finite segments. A typical example is the function $1/\sqrt{z^2-1}$, which needs a cut for $-1 < \mathrm{Re}\, z < 1$ (see Sect. 1.3.4).

The integrals involving functions with branching points are as a rule more complicated. On the other hand a door with new opportunities opens and even in the situations with branch cut segments, in which the analyticity domain of the function in question is not simply connected, many integrals can still be very conveniently evaluated.

1.3.2 Integrals of the Type $\int_0^\infty x^{a-1} Q(x)dx$

Here $Q(x)$ is assumed to be a rational function of x, without poles on the positive side of the real axis (for now), and such that $x^a Q(x) \to 0$ when $x \to 0$ or $x \to \infty$ (that is the integral converges).

Keeping in mind the standard definition of the argument, we take the auxiliary function as

$$f(z) = (-z)^{a-1} Q(z)$$

and integrate over the contour C shown in Fig. 1.13, which consists of C_ρ, C_R, and $C_\pm$. The Cauchy theorem takes the form

$$\int_C = 2\pi i \sum_p \mathrm{Res} f(z)\Big|_{z_p},$$

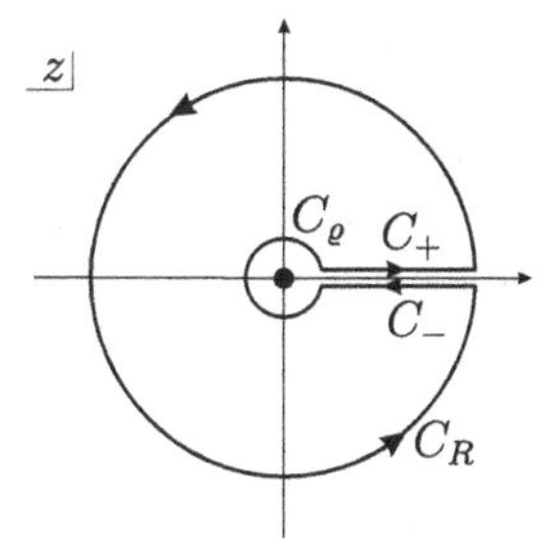

Fig. 1.13 Schematic representation of the contour for the calculation of the integrals from Sect. 1.3.2

where the sum is taken over all residues of $f(z)$. It is easy to see (by counting powers of ρ and R) that, under the conditions when the real integral $\int_0^\infty x^{a-1}Q(x)dx$ converges, we will also have

$$\lim_{\rho\to 0}\int_{C_\rho} = \lim_{R\to\infty}\int_{C_R} = 0.$$

To find the contributions from $C_\pm$ we write $-z = -x \mp i\delta = xe^{\mp i\pi}$ and so

$$f_+(x) = \lim_{\delta\to 0^+} f(x+i\delta) = e^{-i\pi(a-1)}x^{a-1}Q(x),$$

on the upper part of the branch cut and

$$f_-(x) = \lim_{\delta\to 0^+} f(x-i\delta) = e^{i\pi(a-1)}x^{a-1}Q(x),$$

on the lower one. Taking into account that along C_- we integrate in the negative direction we conclude that

$$\int_C = [e^{-i\pi(a-1)} - e^{i\pi(a-1)}]\int_0^\infty x^{a-1}Q(x)dx$$

and hence the general formula is:

$$\int_0^\infty x^{a-1}Q(x)dx = \frac{\pi}{\sin(\pi a)}\sum_p \mathrm{Res} f(z)\Big|_{z_p}. \tag{1.14}$$

Example 1.8
Calculate the integral

$$I(a) = \int_0^\infty \frac{x^{a-1}dx}{x+1}.$$

We need to limit the parameter a to the segment $(0, 1)$ for the integral to converge in both limits. We have $f(z) = (-z)^{a-1}/(z+1)$, which has a single pole at $z = -1$ with residue $a_{-1} = (+1)^{a-1} = 1$, so the above general formula (1.14), in this particular case, becomes

$$I(a) = \frac{\pi}{\sin(\pi a)}.$$

It is instructive to calculate the same integral in a different way. If we define the argument of z as $0 < \theta < 2\pi$ then we can use a different auxiliary function $f(z) = z^{a-1}/(z+1)$. According to the Cauchy theorem we then have

$$\sum_{\pm} \int_{C_\pm} = 2\pi i \mathrm{Res} f(z)|_{z=-1} = 2\pi i(-1)^{a-1} = 2\pi i e^{i\pi(a-1)},$$

where we use $-1 = e^{i\pi}$ because of the definition of the argument. On the other hand we now have

$$\frac{z^{a-1}}{z+1}\Big|_{z=x+i\delta} = \frac{x^{a-1}}{x+1} \quad \Rightarrow \int_{C_+} = I(a),$$

and

$$\frac{z^{a-1}}{z+1}\Big|_{z=x-i\delta} = \frac{e^{2\pi i(a-1)}x^{a-1}}{x+1} \quad \Rightarrow \int_{C_-} = -e^{2\pi a i} I(a),$$

leading to the same answer $I(a) = \pi/\sin(\pi a)$. Here we have explicitly used the relation between the function values $z_\pm$ on the upper/lower ($C_\pm$) shores of the branch cut:

$$z_-^{a-1} = z_+^{a-1}\, e^{2\pi(a-1)i}. \tag{1.15}$$

1.3.3 Principal Value Integrals

Let us suppose we are confronted with the integral of the previous section, but modified

$$I(a) = \int_0^\infty \frac{x^{a-1}dx}{x-1}$$

in such a way that $Q(x)$ now has a pole on the positive real axis. This integral, as such, is singular (divergent). However, the divergency is weak—logarithmic—and has opposite signs on the opposite sides of the singularity. This fact allows one to extend the definition of the integration so that a finite number is associated with integrals like the one written above.

This extended definition is called *principal value integral* and is as follows

$$\mathrm{P}\int_a^b \frac{f(x')dx'}{x'-x} = \lim_{\epsilon\to 0^+}\left[\int_a^{x-\epsilon} \frac{f(x')dx'}{x'-x} + \int_{x+\epsilon}^b \frac{f(x')dx'}{x'-x}\right], \tag{1.16}$$

where the function $f(x)$ is to be assumed differentiable on the segment $[a, b]$ (for now, we shall formulate more precise conditions in the next chapter).

We begin the discussion of this definition with calculating simplest possible principal value integral

$$\mathrm{P}\int_{-1}^{1} \frac{dx'}{x'-x} = \int_{-1}^{x-\epsilon} \frac{dx'}{x'-x} + \int_{x+\epsilon}^{1} \frac{dx'}{x'-x} = \ln|x'-x|_{-1}^{x-\epsilon} + \ln|x'-x|_{x+\epsilon}^{1}$$

$$= \ln\epsilon - \ln(1+x) + \ln(1-x) - \ln\epsilon = \ln\left(\frac{1-x}{1+x}\right).$$

Observe that it is really important that the singularity is approached at the same pace from left and right; indeed, a definition of the kind

$$\int_{-1}^{x-\epsilon_1} \frac{dx'}{x'-x} + \int_{x+\epsilon_2}^{1} \frac{dx'}{x'-x} = \ln\left(\frac{1-x}{1+x}\right) + \ln\left(\frac{\epsilon_1}{\epsilon_2}\right)$$

would produce an entirely different answer (infinite or finite depending on the behavior of ϵ_1/ϵ_2 in the limit). In what follows we shall always require that $\epsilon_1 = \epsilon_2$.

This simple calculation leads to an important conclusion that the principal value integral exists and is unique for any differentiable function $f(x)$. To see this we rewrite

$$\mathrm{P}\int_{a}^{b} \frac{f(x')dx'}{x'-x} = \mathrm{P}\int_{a}^{b} \frac{f(x') - f(x) + f(x)}{x'-x} dx'$$

$$= f(x)\ln\left(\frac{b-x}{x-a}\right) + \int_{a}^{b} \frac{f(x') - f(x)}{x'-x} dx',$$

where the symbol P in front of the second integral is removed as it is not singular anymore [indeed from differentiability $f(x') - f(x) = f'(x)(x'-x) + O((x'-x)^2)$ so that the singularity (pole) is removed].

Example 1.9
As an example we calculate the integral from which we started

$$I(a) = \mathrm{P}\int_{0}^{\infty} \frac{x^{a-1}dx}{x-1},$$

for $0 < a < 1$, now understood as a principal value integral. The suitable contour is shown in Fig. 1.14 and, as we are using the standard definition of the argument, we take $f(z) = (-z)^{a-1}/(z-1)$. For $0 < a < 1$ we have

$$\lim_{\rho\to 0}\int_{C_\rho} = \lim_{R\to 0}\int_{C_R} = 0$$

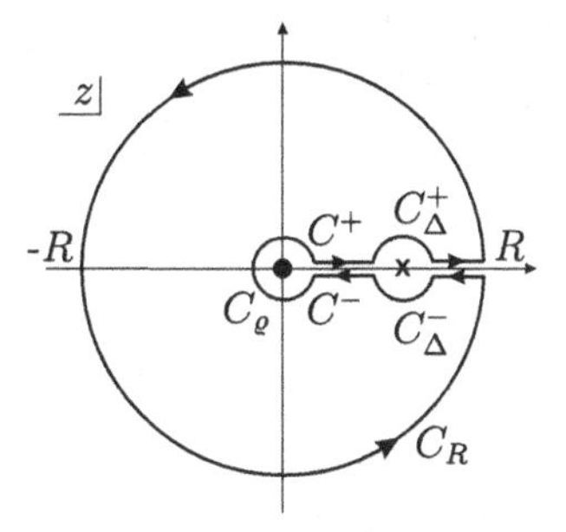

Fig. 1.14 Schematic representation of the contour for the calculation of the integrals from Sect. 1.3.3

as before, and the Cauchy theorem takes the form

$$\sum_{\pm}\int_{C^\pm} + \sum_{\pm}\int_{C_\Delta^\pm} = 0.$$

On $C^\pm$ we put $z = x \pm i\delta$ in the spirit of (1.15) and so

$$f(x \pm i\delta) = \frac{e^{\mp i\pi(a-1)} x^{a-1}}{x-1} = \frac{e^{\mp i\pi a} x^{a-1}}{1-x}$$

resulting in

$$\int_{C^+} = e^{-i\pi a} I(a), \quad \int_{C^-} = -e^{i\pi a} I(a).$$

It is understood that, strictly speaking here $C^\pm$ are segments with 2Δ around $x = 1$ cut out, so that the above equalities hold only in the limit $\Delta \to 0$. On $C_\Delta^\pm$ we must substitute $z = 1 + \Delta e^{i\theta}$. Following the sign of the imaginary part we conclude that in the limit

$$\begin{aligned}
\lim_{\Delta \to 0^+} (-z)^{a-1} &= \lim_{\Delta \to 0^+} (-1 - \Delta\cos\theta - i\Delta\sin\theta)^{a-1} \\
&= \begin{cases} e^{-i\pi(a-1)}, & 0 < \theta < \pi, \\ e^{i\pi(a-1)}, & -\pi < \theta < 0. \end{cases}
\end{aligned}$$

[It really can not be otherwise: as all points on both $C^\pm$ and $C_\Delta^\pm$ approach the real axis from the same direction (above or below), their arguments should reach the same value in the limit.] So around the pole at $z = 1$, we obtain:

$$\int_{C_\Delta^+} = \int_{\pi}^{0} \frac{e^{-i\pi(a-1)} i\Delta e^{i\theta} d\theta}{\Delta e^{i\theta}} = i\pi e^{-i\pi a}, \quad \int_{C_\Delta^-} = \int_{0}^{-\pi} \frac{e^{i\pi(a-1)} i\Delta e^{i\theta} d\theta}{\Delta e^{i\theta}} = i\pi e^{i\pi a}.$$

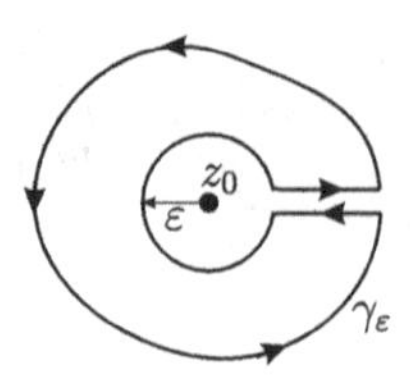

Fig. 1.15 Representation of the contour γ_ϵ, used in the definition (1.17)

Note that the θ–integral has the same (clockwise) direction on both C^+_Δ and C^-_Δ while the x–integration on $C^\pm$ is in opposite directions, hence an extra change of sign for $\int_{C^-}$. Finally, we collect the results to obtain:

$$(e^{-i\pi a} - e^{i\pi a})I(a) + i\pi(e^{-i\pi a} + e^{i\pi a}) = 0 \quad \Rightarrow \quad I(a) = \pi \cot(\pi a).$$

We shall encounter many interesting examples of principal value integrals later on. To conclude this section we give a straightforward generalisation of the principal value definition to the case when the integral is taken over a smooth (but otherwise) arbitrary contour γ in complex plane. The integral

$$F_p(z_0) = \frac{1}{2\pi i} \lim_{\epsilon \to 0} \int_{\gamma_\epsilon} \frac{f(\xi)d\xi}{\xi - z_0}, \tag{1.17}$$

where $z_0 \in \gamma$ and γ_ϵ is the contour γ with the section 2ϵ around z_0 cut out (see Fig. 1.15), is referred to as the Cauchy principal value integral. Clearly our discussion of the existence of this integral on the real axis is directly transferrable to the complex case.

1.3.4 Integrals Around Branch Cut Segments

We start with an elementary integral,

$$I = \int_{-1}^{1} \frac{dx}{\sqrt{1 - x^2}} = \int_{-\pi/2}^{\pi/2} \frac{d\sin\theta}{\cos\theta} = \int_{-\pi/2}^{\pi/2} d\theta = \pi$$

to illustrate a general idea. We integrate the function $f(z) = (z^2 - 1)^{-1/2}$ over circular contour C_R (see Fig. 1.16) with $R > 1$.

The method is very simple. We observe that the integral $\int_{C_R}$ can be computed in two ways: (i) send $R \to \infty$ and expect a simple answer (residue at infinity) or (ii) deform the contour C_R in such a way that it has collapsed onto the branch cut at $z = x \in [-1, 1]$, where it will be related to I. Then recover the value of I by

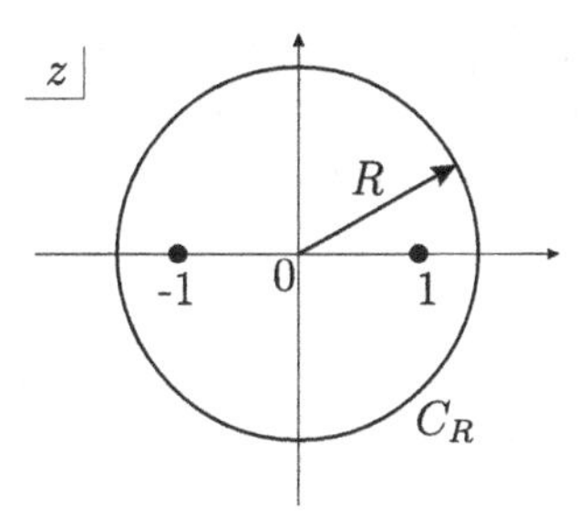

Fig. 1.16 Schematic representation of the contour C_R used in Sect. 1.3.4

comparing the two results. Next we follow these steps for the particular function $f(z) = (z^2 - 1)^{-1/2}$.

(i) Here we need the Laurent expansion of $f(z)$ for $z \to \infty$:

$$\frac{1}{\sqrt{z^2-1}} = \frac{1}{z\sqrt{1-1/z^2}} = \frac{1}{z(1-1/(2z^2)+O(z^{-4}))}$$
$$= \frac{1}{z}\left(1+\frac{1}{2z^2}+O(z^{-4})\right) = \frac{1}{z}+\frac{1}{2z^3}+O(z^{-5}).$$

Then, on C_R ($z = Re^{i\theta}$), we have

$$\int_{C_R} = \int_{-\pi}^{\pi} iRe^{i\theta}\left(\frac{1}{Re^{i\theta}}+\frac{1}{2R^3e^{3i\theta}}+O(R^{-5})\right)d\theta = 2\pi i.$$

We see that terms other then the residue vanish in the limit $R \to \infty$ and vanish even for finite R due to the angle integration, just as in the standard residue calculation previously.

(ii) Next we deform the contour C_R towards the branch cut as shown in Fig. 1.17. The integrals over little circles (with radius ρ) vanish in the limit $\rho \to 0$ (they are of the order of $\rho^{1/2}$). On $C_\pm$, with the standard definition of the argument, we obtain:

$$f(z)|_{z=x\pm i\delta} = \frac{1}{e^{\pm i\pi/2}\sqrt{1-x^2}} = \mp i\frac{1}{\sqrt{1-x^2}},$$

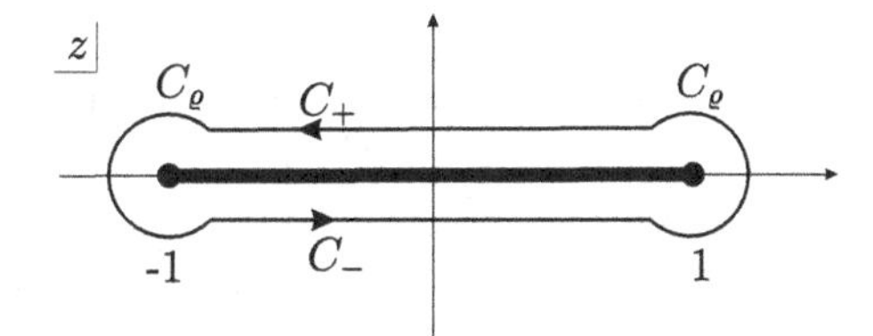

Fig. 1.17 Schematic representation of the contour used in the method (ii) of Sect. 1.3.4 for the function $f(z) = \sqrt{z^2-1}$

and therefore $\sum_{\pm} \int_{C_{\pm}} = 2iI$. Now equating the result of the calculation (i) with the result of the calculation (ii), $2\pi i = 2iI$, we recover the elementary result $I = \pi$.

Let us now consider another elementary integral,

$$I = \int_{-1}^{1} \sqrt{1-x^2}dx = \int_{-\pi/2}^{\pi/2} d\theta \cos^2\theta = \int_{-\pi/2}^{\pi/2} \left(\frac{1}{2} + \frac{1}{2}\cos(2\theta)\right) d\theta = \frac{\pi}{2},$$

to illustrate a separate point. We integrate the function $f(z) = (z^2 - 1)^{1/2}$ over the same contour C_R.

(i) The Laurent expansion is

$$f(z) = z - \frac{1}{2z} + O(z^{-3})$$

leading to

$$\int_{C_R} = \int_{-\pi}^{\pi} iRe^{i\theta}\left(Re^{i\theta} - \frac{1}{2Re^{i\theta}} + O(R^{-3})\right) d\theta = -\pi i.$$

The first term in the above equation may be a cause for concern as it formally diverges as $R \to \infty$. However there is no divergency here as $\int e^{2i\theta} d\theta$ over the period is clearly zero, for all finite R. Some authorities on the subject would perform a subtraction and work with the function $f(z) - z$ instead of function $f(z)$ (no added polynomial, being analytic across the branch cut, could produce a finite contribution on the collapsed contour, so the answer is the same either way around). But there is no need to do a subtraction and the lesson we are learning here is that, quite in analogy with residues at a finite z, for the residues at infinity if the function $f(z)$ expands as

$$f(z) = a_N z^N + a_{N-1} z^{N-1} + \ldots + a_1 z + a_0 + \frac{a_{-1}}{z} + O(z^{-2}),$$

for $z \to \infty$ (and some positive N), then

$$\int_{C_R} = 2\pi i a_{-1}.$$

Continuing with our example, for (ii), on $C_{\pm}$ we now have $f(x \pm i\delta) = \pm i\sqrt{1-x^2}$ resulting in $\sum_{\pm} \int_{C_{\pm}} = -2iI$ and thus $-\pi i = -2iI$ recovers the correct result $I = \pi/2$.

A more interesting example is the following integral:

$$I(a) = \int_{-1}^{1} \left(\frac{1-x}{1+x}\right)^a dx,$$

where $-1 < a < 1$ for convergence. We integrate the function $f(z) = (z+1)^{-a}(z-1)^a$ along C_R. (i) Expanding

$$\left(\frac{z-1}{z+1}\right)^a = \left(\frac{1-1/z}{1+1/z}\right)^a = \left[1 - \frac{a}{z} + O(z^{-2})\right]^2 = 1 - \frac{2a}{z} + O(z^{-2}),$$

we conclude that the residue at infinity, $a_{-1}(a) = -2a$, becomes a function of a and that $\int_{C_R} = -4\pi i a$. (ii) For a standard definition of the argument (see Sect. 1.1.1), the limiting values are

$$f(x \pm i\delta) = e^{\pm i\pi a} \left(\frac{1-x}{1+x}\right)^a,$$

so that

$$\int_{C_\pm} = -e^{i\pi a} I(a) + e^{-i\pi a} I(a) = -2i \sin(\pi a) I(a) \tag{1.18}$$

resulting in

$$I(a) = \frac{2\pi a}{\sin(\pi a)}.$$

Obviously we immediately recover the elementary limit $I(a \to 0) = 2$.[4]

The next logical step is to calculate

$$I(a) = \int_{-1}^{1} \left(\frac{1-x}{1+x}\right)^a R(x) dx,$$

where $R(x)$ is a rational function satisfying the conditions

$$\lim_{x \to 1} (1-x)^{1+a} R(x) = 0 \quad \text{and} \quad \lim_{x \to -1} (1+x)^{1-a} R(x) = 0,$$

---

[4] This integral can be reduced to an integral of the type studied in Sect. 1.3.2 by substitution

$$y = \frac{1-x}{1+x} \quad \Rightarrow I(a) = 2 \int_0^\infty \frac{y^a dy}{(y+1)^2},$$

the double pole here explains why the result is proportional to a.

for the integral to converge and, for the time being, we assume that $R(z)$ has no poles on the branch cut segment. Again we integrate the function $f(z) = (z+1)^{-a}(z-1)^a R(z)$ along C_R, but now we have a choice as to how to position C_R with respect to the singularities (poles) of $R(z)$. For simplicity we take the radius R large enough for C_R to encircle all the poles of $R(z)$.

(i) With such contour the poles of $R(z)$ do not contribute to this calculation and the answer is $\int_{C_R} = 2\pi i a_{-1}(a)$, where $a_{-1}(a)$ is a residue of the function $f(z)$ at infinity, as defined above (it is now a function of the parameter a).

(ii) When deforming C_R so as to collapse it onto the branch cut will leave counter-clockwise circles around all poles of $R(z)$ bringing the Cauchy theorem into the form

$$\int_{C_R} = \sum_{\pm} \int_{C_\pm} + 2\pi i \sum \operatorname{Res} f(z).$$

Clearly multiplying the integrand by a rational function is not going to affect the calculation of the limiting values $f(x \pm i\delta)$ [except trivially multiplying them both by $R(x)$] and therefore, unifying (i) and (ii) and using (1.18) we have

$$-2i\sin(\pi a)I(a) + 2\pi i \sum \operatorname{Res} f(z) = 2\pi i a_{-1}(a),$$

resulting in the general formula[5]:

$$I(a) = -\frac{\pi a_{-1}(a)}{\sin(\pi a)} + \frac{\pi}{\sin(\pi a)} \sum \operatorname{Res} f(z). \tag{1.19}$$

As an example calculate

$$I(a) = \int_{-1}^{1} \left(\frac{1-x}{1+x}\right)^a \frac{dx}{x^2+1}.$$

(i) We have

[5] It is instructive to investigate the $a \to 0$ limit of this formula. The $O(1/a)$ term must vanish telling us that

$$\sum \operatorname{Res} R(z)|_{\text{finite } z} = a_{-1}(0),$$

that is the sum of all finite residues is equal to the residue at infinity for an arbitrary rational function. For non-rational functions this is, generally, not true. Indeed the constant, $O(a^0)$ term produces an interesting expression:

$$\int_{-1}^{1} R(x)dx = \left[\sum \operatorname{Res}|_{\text{finite } z} - \operatorname{Res}|_{z=\infty}\right] \ln\left(\frac{z-1}{z+1}\right) R(z).$$

$$f(z) = \left(\frac{z-1}{z+1}\right)^a \frac{1}{z^2+1} = O\left(\frac{1}{z^2}\right)$$

for $z \to \infty$, so that $a_{-1}(a) = 0$.

(ii) With the standard definition of the argument (shown in Fig. 1.1), we have, for $z = i$ residue, $i - 1 = \sqrt{2}\, e^{i3\pi/4}$ and $i + 1 = \sqrt{2}\, e^{i\pi/4}$, so that $(i+1)^{-a}(i-1)^a = (\sqrt{2})^{-a} e^{-i\pi a/4} (\sqrt{2})^a e^{i3\pi a/4} = e^{i\pi a/2}$. For the other, $z = -i$, residue, $-i - 1 = \sqrt{2}\, e^{-i3\pi/4}$ and $-i + 1 = \sqrt{2}\, e^{-i\pi/4}$ [clearly $\theta_1 + \theta_2$ is equal to $\pi(-\pi)$ for points on the upper (lower) half of the unit circle] , so that $(-i+1)^{-a}(-i-1)^a = (\sqrt{2})^{-a} e^{i\pi a/4} (\sqrt{2})^a e^{-i3\pi a/4} = e^{-i\pi a/2}$, resulting in:

$$\sum \operatorname{Res} f(z)|_{z=\pm i} = \frac{e^{i\pi a/2}}{2i} + \frac{e^{-i\pi a/2}}{-2i} = \sin(\pi a/2),$$

leading to the answer

$$I(a) = \frac{\pi \sin(\pi a/2)}{\sin(\pi a)} = \frac{\pi}{2\cos(\pi a/2)}.$$

Note that the elementary limit $I(a=0) = \int_{-1}^{1} (x^2+1)^{-1} dx = \left[\tan^{-1} x\right]_{-1}^{1} = \pi/2$ checks out.

Finally consider the integral

$$I_a(z_0) = \int_{-1}^{1} \left(\frac{1-x}{1+x}\right)^a \frac{dx}{x - z_0},$$

where z_0 is an arbitrary point in the complex plane outside the branch cut segment. (i) Here

$$f(z) = \left(\frac{z-1}{z+1}\right)^a \frac{1}{z - z_0} = \frac{1}{z} + O\left(\frac{1}{z^2}\right)$$

for $z \to \infty$, so that $a_{-1}(a) = 1$. (ii) With θ_1 and θ_2 being arguments of $z_0 - 1$ and $z_0 + 1$, respectively, computation of the residua results in $\operatorname{Res} f(z)|_{z=z_0} = |z_0+1|^{-a} |z_0-1|^a e^{ia(\theta_1-\theta_2)}$. Plugging these results into the general formula (1.19) above we obtain

$$I_a(z_0) = -\frac{\pi}{\sin(\pi a)} + \frac{\pi}{\sin(\pi a)} \frac{|z_0-1|^a}{|z_0+1|^a} e^{ia(\theta_1-\theta_2)}.$$

Let $z_0 = x_0 > 1$ be on the positive real axis, then $\theta_1 = \theta_2 = 0$, and

$$I_a(x_0) = -\frac{\pi}{\sin(\pi a)} + \frac{\pi}{\sin(\pi a)} \left(\frac{x_0-1}{x_0+1}\right)^a.$$

The elementary limit

$$\lim_{a\to 0} I_a(x_0) = -\frac{1}{a} + O(a) + \left[\frac{1}{a} + O(a)\right]\left[1 + a\ln\left(\frac{x_0-1}{x_0+1}\right) + O(a)\right]$$
$$= \ln\left(\frac{x_0-1}{x_0+1}\right) + O(a)$$

again checks out and is quite revealing. When $z_0 = -x_0 < -1$ is on the negative real axis, then $\theta_1 = \theta_2 = \pi$ or $\theta_1 = \theta_2 = -\pi$ but $\theta_1 - \theta_2 = 0$ in either case, so

$$I_a(-x_0) = -\frac{\pi}{\sin(\pi a)} + \frac{\pi}{\sin(\pi a)}\left(\frac{x_0+1}{x_0-1}\right)^a.$$

The case when $x_0 \in [-1, 1]$ is a principal value case warranting special investigation given below.

1.3.5 Branch Cut Segments: Principal Values

We now concentrate on the important case when the rational function $R(x)$, entering the main integral of Sect. 1.3.4, has a pole on the branch cut segment $[-1, 1]$. It will be convenient to factor the pole out of $R(x)$ and call the remaining factor $R(x)$ again. So our next challenge is

$$I_a(x_0) = \mathrm{P}\int_{-1}^{1}\left(\frac{1-x}{1+x}\right)^a \frac{R(x)dx}{x-x_0},$$

where $x_0 \in [-1, 1]$ (when x_0 is outside then the integral is simpler—not a principal value—and the ideas of Sect. 1.3.4 apply).

(i) If we position C_R to encircle all the poles of $R(x)$ then $\int_{C_R} = 2\pi i a_{-1}(a; x_0)$, where the residue at infinity of the function

$$f(z) = \left(\frac{z-1}{z+1}\right)^a \frac{R(z)}{z-x_0}$$

enters (it will generally be a function of both a and x_0). (ii) Here we have a contribution from the poles of $R(z)$ similar to that of Sect. 1.3.4 and, in addition, the contour is bent in little circles $C_\Delta^\pm$ around the point x_0 (this situation is, in turn, similar to that in the example in Sect. 1.3.3), as is shown in Fig. 1.14. The Cauchy theorem states:

$$\int_{C_R} = \sum_\pm \left\{\int_{C^\pm} + \int_{C_\Delta^\pm}\right\} + 2\pi i \sum \mathrm{Res} f(z).$$

On $C^{\pm}$ our function has essentially the same arguments as before:

$$f(x \pm i\delta) = e^{\pm i\pi a} \left(\frac{1-x}{1+x}\right)^a \frac{R(x)}{x - x_0},$$

which, in the limit, produces the principal value integral as

$$\lim_{\Delta \to 0} \sum_{\pm} \int_{C^{\pm}} = -2i \sin(\pi a) I_a(x_0).$$

On the other hand on $C_{\Delta}^{\pm}$ we put $z = x_0 + \Delta e^{i\theta}$ to recover in the limit $\Delta \to 0$ (like in Sect. 1.3.3) the same arguments as on $C^{\pm}$, the θ–integration, being in the same direction (anti-clockwise), results in the factor $i\pi$ (half-residue) for each semi-circle:

$$\sum_{\pm} \int_{C_{\Delta}^{\pm}} = i\pi \sum_{\pm} e^{\pm i\pi a} \left(\frac{1-x_0}{1+x_0}\right)^a R(x_0) = 2\pi i \cos(\pi a) \left(\frac{1-x_0}{1+x_0}\right)^a R(x_0).$$

Collecting the results,

$$\begin{aligned} -2i \sin(\pi a) I_a(x_0) + 2\pi i \cos(\pi a) \left(\frac{1-x_0}{1+x_0}\right)^a R(x_0) + 2\pi i \sum \operatorname{Res} f(z) \\ = 2\pi i a_{-1}(a; x_0), \end{aligned}$$

we obtain the answer for the integral in question

$$I_a(x_0) = -\frac{\pi a_{-1}(a; x_0)}{\sin(\pi a)} + \pi \cot(\pi a) \left(\frac{1-x_0}{1+x_0}\right)^a R(x_0) + \frac{\pi}{\sin(\pi a)} \sum \operatorname{Res} f(z),$$

which is the main result of this section.

Example 1.10
We begin the examples with the simplest possible case when $R(x) = 1$. The function

$$f(z) = \left(\frac{z-1}{z+1}\right)^a \frac{1}{z - x_0} = \frac{1}{z} + O\left(\frac{1}{z^2}\right)$$

for $z \to \infty$ so that $a_{-1}(a; x) = 1$. There are no residues and we simply have:

$$I_a(x_0) = -\frac{\pi}{\sin(\pi a)} + \pi \cot(\pi a) \left(\frac{1-x_0}{1+x_0}\right)^a.$$

In particular, putting $a = \pm 1/2$, gives us two integrals

$$I_{1/2}(x_0) = \mathrm{P}\int_{-1}^{1} \sqrt{\frac{1-x}{1+x}} \frac{dx}{x-x_0} = -\pi$$

and

$$I_{-1/2}(x_0) = \mathrm{P}\int_{-1}^{1} \sqrt{\frac{1+x}{1-x}} \frac{dx}{x-x_0} = \pi$$

which we shall use in the future (interestingly, none of these integrals actually depends on x_0).

Example 1.11
It is instructive to take $R(x) = x + 1$ so that this time we have

$$I_a(x_0) = \mathrm{P}\int_{-1}^{1} (1+x)^{1-a}(1-x)^a \frac{dx}{x-x_0}.$$

The large z expansion now becomes

$$f(z) = \frac{(z+1)^{1-a}(z-1)^a}{z-x_0} = \frac{(1+1/z)^{1-a}(1-1/z)^a}{1-x_0/z} = \left(1 + \frac{1-a}{z} + \ldots\right)$$
$$\times \left(1 - \frac{a}{z} + \ldots\right)\left(1 + \frac{x_0}{z} + \ldots\right) = 1 + \frac{1-2a+x_0}{z} + O\left(\frac{1}{z^2}\right),$$

so that $a_{-1}(a; x_0) = 1 - 2a + x_0$. There are no residues and we obtain

$$I_a(x_0) = -\frac{\pi(1-2a+x_0)}{\sin(\pi a)} + \pi \cot(\pi a)(1+x_0)^{1-a}(1-x_0)^a.$$

In particular, for $a = 1/2$, we have

$$I_{1/2}(x_0) = \mathrm{P}\int_{-1}^{1} \frac{\sqrt{1-x^2}}{x-x_0} dx = -\pi x_0.$$

Example 1.12
Next we take $R(x) = 1/(1-x)$:

$$I_a(x_0) = \mathrm{P}\int_{-1}^{1} (1+x)^{-a}(1-x)^{a-1} \frac{dx}{x-x_0}$$

and restrict the allowed values of a to $0 < a < 1$. Now

$$f(z) = \frac{(z+1)^{-a}(z-1)^{a-1}}{z - x_0} = O\left(\frac{1}{z^2}\right)$$

so that $a_{-1}(a; x_0) = 0$. Again, there are no residues, and the answer is:

$$I_a(x_0) = \pi \cot(\pi a)(1 + x_0)^{-a}(1 - x_0)^{a-1}.$$

In particular, for $a = 1/2$, we have

$$I_{1/2}(x_0) = \mathrm{P}\int_{-1}^{1} \frac{1}{\sqrt{1-x^2}} \frac{dx}{x - x_0} = 0$$

for all $x_0 \in (-1, 1)$.

We conclude this section by discussing the case when there is a residue, $R(x) = 1/(x - z_0)$, where z_0 is an arbitrary point in the complex plane except on the branch cut:

$$I_a(x_0) = \mathrm{P}\int_{-1}^{1} \frac{(1+x)^{-a}(1-x)^{a}}{(x - x_0)(x - z_0)} dx.$$

For large z, $f(z) = O(1/z^2)$, hence $a_{-1}(a; x_0) = 0$. The arguments around the pole at z_0 has already been worked out in Sect. 1.3.4, so we read off from the main formula:

$$I_a(x_0) = \pi \cot(\pi a)\left(\frac{1 - x_0}{1 + x_0}\right)^a \frac{1}{x_0 - z_0} + \frac{\pi}{\sin(\pi a)} \frac{|z_0 - 1|^a}{|z_0 + 1|^a} \frac{e^{ia(\theta_1 - \theta_2)}}{x_0 - z_0}.$$

1.3.6 Analytic Continuation

This is an important theme to which we shall often return in subsequent chapters. In this section we open a discussion with a simple theorem, an integral, and a remark.

Theorem: Let D be a joint analyticity domain of two complex functions $f_1(z)$ and $f_2(z)$ and γ is a curve in D. If the two functions coincide on γ then they coincide everywhere in D.

In the way of proof we observe that an analytic function can be expanded in convergent Taylor series at any point in its analyticity domain. So, if we expand $f_{(1,2)}(\xi) = \sum_{n=0}^{\infty} a_n^{(1,2)} (\xi - z_0)^n$, with both $\xi \in \gamma$ and $z_0 \in \gamma$, then (as powers are linearly independent on a line) we are forced to conclude that $a_n^{(1)} = a_n^{(2)}$ for all n.

The two expansions coincide and represent one and the same function everywhere in the area of convergence, that is in D.

This theorem helps rationalize certain intuitively obvious assumptions about analytic expressions. Suppose a function $f(x)$ is given by an explicit expression on some segment of the real axis. We can formally substitute z for x to obtain a complex function $f(z)$, defined in some 'bigger' region including the segment. This complex function coincides with $f(x)$ for $z = x$ by definition and thereby provides a unique analytic continuation of $f(x)$ due to the theorem, which states that if there were another analytic continuation it would have coincided with $f(z)$ anyway. For example, suppose we are given the following function of real variable $f(x) = \sqrt{x-1}$ defined on $[1, \infty)$. We therefore know that an analytic continuation of this function to the complex plane is $f(z) = \sqrt{z-1}$. Moreover, assuming the standard definition of the argument, we also know that this function has a branch cut on $(-\infty, 1]$, with the limiting values being $f(x \pm i\delta) = \pm i\sqrt{1-x}$.

Integral. One often obtains a function, for instance when solving a differential equation, in the form of Taylor series $f(z) = \sum_{n=0}^{\infty} a_n z^n$, with some finite radius of convergence R. On the circle $z = Re^{i\theta}$ there will be at least one singularity of $f(z)$ (otherwise the circle could be enlarged) and indeed may be more then one singularity. If the singularities are not dense, then $f(z)$ can be analytically continued into some region outside the circle of convergence.

In many cases it is possible to write down a complex integral—a function of z—which coincides with $f(z)$ in $|z| < R$ but continues to converge in some region where $|z| > R$ thus providing an analytic continuation of $f(z)$ in that region. Here we discuss the simplest possible, indeed almost trivial, example of this nature: take $a_n = 1$ for all n, hence $R = 1$ and $f(z) = \sum_{n=0}^{\infty} z^n$ is a geometric series and write down the integral

$$F(z) = -\frac{1}{2\pi i} \int_{\sigma-i\infty}^{\sigma+i\infty} \frac{\pi}{\sin(\pi\xi)} (-z)^{\xi} d\xi,$$

where $-1 < \sigma < 0$ (the contour is shown in Fig. 1.18). The function $1/\sin(\pi\xi)$ is chosen for having poles at integer ξ and $(-z)^{\xi}$ (not z^{ξ}) is written because the residues of these poles alternate in sign,

$$\operatorname{Res} \frac{\pi}{\sin(\pi\xi)} (-z)^{\xi} \bigg|_{\xi=n} = (-1)^n (-z)^n = z^n.$$

We shall not dwell on convergence, which, in this case, is justified by the result, but only observe that $\ln|z| < 0$ for $|z| < 1$. Therefore one would expect to be able to complete the contour on the right for $|z| < 1$ (and on the left otherwise). So, for $|z| < 1$ we shift the contour to the extreme right (see Fig. 1.19a), to $\sigma \to \infty$, leaving integrals over 'bubbles' around the poles and recovering (integration around poles clockwise)

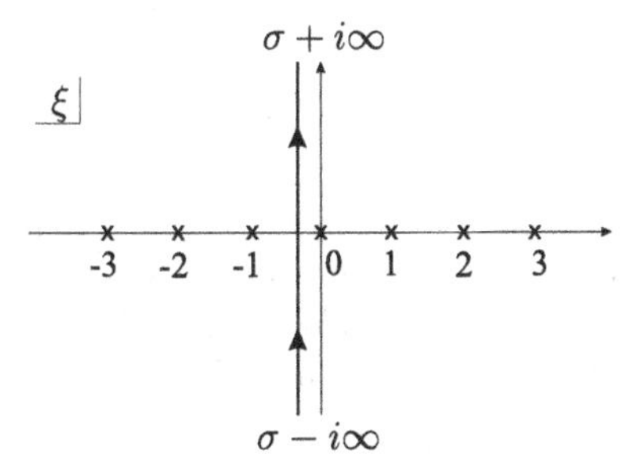

Fig. 1.18 Schematic representation of the contour used for the analytic continuation

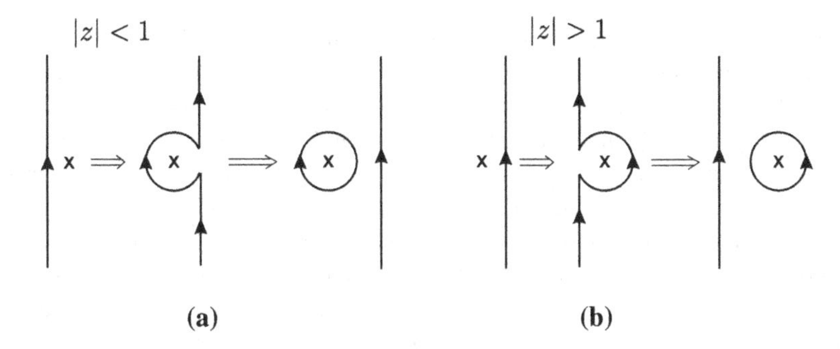

Fig. 1.19 In order to obtain $f(z)$ for $|z| < 1$ the contour is shifted to $+\infty$ (**a**), and for the opposite limit $|z| > 1$ to $-\infty$ (**b**)

$$F(z) = \sum_{n=0}^{\infty} \text{Res}\left[\frac{\pi}{\sin(\pi\xi)}(-z)^{\xi}\right]_{\xi=n} = \sum_{n=0}^{\infty} z^n,$$

the geometric series. For $|z| > 1$ we shift the contour to the extreme left, (see Fig. 1.19b), to $\sigma \to -\infty$ (integration anti-clockwise) to obtain a series in $1/z$ convergent for $|z| > 1$[6]:

$$F(z) = -\sum_{n=-1}^{-\infty} z^n = -\sum_{n=1}^{\infty}\left(\frac{1}{z}\right)^n = -\left(\frac{1}{1-1/z} - 1\right) = \frac{1}{1-z}.$$

Remark. Because we have an explicit and very simple expression for $f(z) = 1/(1-z)$, which will not happen in more complicated cases in following chapters, one may think that this example is completely trivial. Indeed while the integral trick is not really needed (as is given as an illustration in view of further non-trivial examples), there are lessons to be learnt here. Indeed, let us look again at the equality

[6] Alternatively, instead of shifting the contours we can think about closing the straight part with semicircles to the left/right. The convergence in both cases is given as $(-z)^{\xi} \to 0$ as $\text{Re}\,\xi \to -\infty$ in the former case and for $\text{Re}\,\xi \to \infty$ in the latter.

$$\frac{1}{1-z} = 1 + z + z^2 + z^3 + z^4 + \cdots .$$

The right-hand-side is the geometric series convergent for $|z| < 1$. The left-hand-side, on the other hand, is an algebraic formula valid for all z except one point $z = 1$. Now let us put $z = -1$ into this equation. While the left-hand-side is perfectly well defined, we can not really put $z = -1$ into the right-hand side, but let's do it anyway

$$\frac{1}{2} = 1 - 1 + 1 - 1 + 1 + \cdots .$$

The series $S = 1-1+1-1\ldots$ does not converge in the conventional sense (the partial sum oscillates between 0 and 1 and has no limit) and so is an example of divergent series. Yet the above calculation suggests that perhaps the standard definition of convergence could be extended so that a certain number could be ascribed to divergent series, like it happens in the case of principal value integrals. Here we come naturally to one such (somewhat loose) definition: a divergent series is—by definition—equal to the value taken by the analytic continuation of the appropriate function (and is said to be convergent even though it is divergent in conventional sense to this value under this definition). There are many more ways to define divergent series, see the seminal book by Hardy [5]. Different definitions often but not always result in the same value for given series (the S series above is actually convergent to $1/2$ for all known definitions).

1.4 Fourier Integral Transform

1.4.1 Definition, Inversion, and Dirac δ-Function

Let $f(x)$ be a continuous and differentiable function of $x \in (-\infty, \infty)$ except at finite number of points. For simplicity we assume that the integral

$$\int_{-\infty}^{\infty} |f(x)| dx$$

converges. The Fourier integral transform of $f(x)$ is then defined by the formula

$$F(y) = \int_{-\infty}^{\infty} e^{ixy} f(x) dx, \tag{1.20}$$

where both variables x and y are—for now—real. The inversion formula is known to be

$$f(x) = \frac{1}{2\pi} \int_{-\infty}^{\infty} e^{-ixy} F(y) dy. \tag{1.21}$$

Next we shall investigate the reasons why this formula is correct and introduce a useful concept in the process. Indeed, substituting the definition of $F(y)$ into the inversion formula in question one obtains

$$f(x) = \frac{1}{2\pi} \int_{-\infty}^{\infty} e^{-ixy} \int_{-\infty}^{\infty} e^{ix'y} f(x') dx' dy.$$

After re-arranging the integration

$$f(x) = \int_{-\infty}^{\infty} \left[\frac{1}{2\pi} \int_{-\infty}^{\infty} e^{i(x'-x)y} dy \right] f(x') dx'. \tag{1.22}$$

We see that the y-integral in this formula,

$$\frac{1}{2\pi} \int_{-\infty}^{\infty} e^{i(x'-x)y} dy,$$

is ill defined. However, as in the case of the principal part integration, the divergency is marginal: the regular oscillations of the integrand are likely to cancel out and produce a finite result given an additional definition.

There are at least two obvious suitable definitions. The procedure of making the limits of y finite, $y \in (-R, R)$ and then taking the limit $R \to \infty$ is employed in many textbooks. We follow a slightly different route here leading to the same result. We multiply the integrand by a decaying function, $e^{-\alpha|y|}$ is most convenient, with α a positive infinitesimal, and take the limit $\alpha \to 0$.

So with this definition we have a convergent y-integral, for which we write

$$\delta(x, \alpha) = \frac{1}{2\pi} \int_{-\infty}^{\infty} e^{ixy} e^{-\alpha|y|} dy = \frac{1}{\pi} \frac{\alpha}{x^2 + \alpha^2}$$

after performing an elementary integration. The function $\delta(x, \alpha)$ introduced here has an important property that

$$\int_{-\infty}^{\infty} \delta(x, \alpha) dx = 1$$

irrespectively of the value of α. By symmetry, it immediately follows that

$$\int_{-\infty}^{0} \delta(x,\alpha)dx = \int_{0}^{\infty} \delta(x,\alpha)dx = \frac{1}{2}.$$

The difficulty has now shifted to investigating the $\alpha \to 0$ limit of the following integral [compare to Eq. (1.22)]:

$$I(x,\alpha) = \int_{-\infty}^{\infty} \delta(x'-x,\alpha) f(x')dx' = \int_{-\infty}^{\infty} \delta(y,\alpha) f(x+y)dy.$$

At this point we allow the function $f(x)$ to be discontinuous at the point x, so that the limits $f(x+0)$ and $f(x-0)$ can be different [clearly, in the particular case when $f(x-0) = f(x+0)$, the continuity is restored]. It makes sense then to split the integral in two

$$I_{+}(x,\alpha) = \int_{0}^{\infty} \delta(y,\alpha) f(x+y)dy$$

and

$$I_{-}(x,\alpha) = \int_{-\infty}^{0} \delta(y,\alpha) f(x+y)dy.$$

We investigate $I_{+}(x,\alpha)$ first, the calculation for $I_{-}(x,\alpha)$ is completely analogous. The trick is to make a subtraction, that is to use the above properties of $\delta(x,\alpha)$ to rewrite the integral in the form

$$I_{+}(x,\alpha) = \frac{1}{2} f(x+0) + \int_{0}^{\infty} \delta(y,\alpha)[f(x+y) - f(x+0)]dy.$$

In order to separate the small y and large y properties of the integrand above, we need to split the integral further. We can do this at any finite y, so we choose to split it at $y = 1$ without loss of generality:

$$I_{+}(x,\alpha) = \frac{1}{2} f(x+0) + \int_{0}^{1} \delta(y,\alpha)[f(x+y) - f(x+0)]dy + \int_{1}^{\infty} \delta(y,\alpha)[f(x+y) - f(x+0)]dy. \quad (1.23)$$

For the latter integral we observe that the integral

$$\frac{1}{\pi}\int_{1}^{\infty}\frac{f(x+y)-f(x+0)}{y^2}dy$$

is convergent so that the last term of (1.23) is of the order $O(\alpha)$. Regarding the integral in the middle, we Taylor expand and write it as

$$\frac{\alpha}{\pi}\int_{0}^{1}\frac{dy}{y^2+\alpha^2}\left[f'(x+0)y+\frac{1}{2}f''(x+0)y^2+O(y^3)\right].$$

If we put $\alpha = 0$ then all integrals in this sum converge except the first one, which is

$$\frac{f'(x+0)}{\pi}\int_{0}^{1}\frac{ydy}{y^2+\alpha^2}=\frac{f'(x+0)}{2\pi}\ln\left(\frac{1+\alpha^2}{\alpha^2}\right).$$

We therefore come to the conclusion that the limiting form of $I_+(x,\alpha)$ is

$$I_+(x,\alpha)=\frac{1}{2}f(x+0)+\frac{f'(x+0)}{\pi}\alpha\ln\left(\frac{1}{\alpha}\right)+O(\alpha)$$

and in the limit we have simply

$$\lim_{\alpha\to 0}I_+(x,\alpha)=\frac{1}{2}f(x+0).$$

A completely analogous calculation results in

$$\lim_{\alpha\to 0}I_-(x,\alpha)=\frac{1}{2}f(x-0).$$

Therefore, in the framework of the 'δ regularization', the following result

$$\frac{1}{2\pi}\int_{-\infty}^{\infty}e^{-ixy}\int_{-\infty}^{\infty}e^{ix'y}f(x')dx'dy=\frac{1}{2}[f(x+0)+f(x-0)] \tag{1.24}$$

holds and is known as the *Fourier theorem*, which justifies the inversion formula (1.21).

Once the $\alpha \to 0$ limiting procedure is understood, it is customary, at least in the physics literature, to dispense with it entirely and introduce a simpler notation: the Dirac δ function with the defining property

$$\int_{-\infty}^{\infty} \delta(x'-x)f(x')dx' = f(x)$$

(assuming now that $f(x')$ is continuous around x). Note that to obtain $\delta(x)$ as an $\alpha \to 0$ limit of $\delta(x, \alpha)$ it does not have to be a Lorentzian as above but could be any smooth symmetric bell shaped function (or indeed a function with reasonable amount of oscillations).

Note that, via a partial fraction decomposition, the Dirac δ-function can be written as:

$$\delta(x) = \frac{1}{2\pi i} \lim_{\alpha \to 0} \left[\frac{1}{x - i\alpha} - \frac{1}{x + i\alpha} \right].$$

A natural question now is: do the individual limits for above fractions exist or only the limit for the difference? To answer this question we investigate the above sum, plugging it into the integral

$$\frac{1}{2\pi i} \int_{-\infty}^{\infty} \left[\frac{1}{y + i\alpha} + \frac{1}{y - i\alpha} \right] f(x+y)dy = \frac{1}{\pi i} \int_{-\infty}^{\infty} \frac{yf(x+y)dy}{y^2 + \alpha^2}.$$

If we formally put $\alpha = 0$ in this formula we end up with a divergent integral. It is similar to the one for which we have introduced the principal value definition. At finite α, the divergency is cut off at $y \sim \alpha$, so it is likely that the role of ϵ is played by α in the emerging principal value. We therefore subtract such a principal value integral

$$\frac{1}{\pi i} \int_{\alpha}^{\infty} dy \frac{f(x+y)}{y} + \frac{1}{\pi i} \int_{-\infty}^{-\alpha} dy \frac{f(x+y)}{y}$$

to end up with three integrals:

$$-\frac{\alpha^2}{\pi i} \left[\int_{-\infty}^{-\alpha} + \int_{\alpha}^{\infty} \right] \frac{f(x+y)dy}{y(y^2+\alpha^2)} + \frac{1}{\pi i} \int_{-\alpha}^{\alpha} \frac{yf(x+y)dy}{y^2+\alpha^2}.$$

The first two integrals converge for large y. So we can, in effect, Taylor expand $f(x)$ (strictly speaking one has to split them in two parts around $y \sim 1$ first and Taylor expand on finite segments only). We see that the main divergency $\sim f(x)/\alpha^2$ is, in fact, not there (odd function), and the lesser divergency $\sim f'(x)/\alpha$ leads to the overall order of magnitude $O(\alpha^2/\alpha) = O(\alpha)$ for the first two terms. In the third term we can Taylor expand without any caveats and we see that the main - logarithmic - divergency is equal to zero (odd function again) and the remaining sub-leading term is of the order $O(\alpha)$.

The conclusion is that

$$\frac{1}{2\pi i}\lim_{\alpha\to 0}\int_{-\infty}^{\infty}\left[\frac{1}{y-i\alpha}+\frac{1}{y+i\alpha}\right]f(x+y)dy=\frac{1}{\pi i}\lim_{\alpha\to 0}\left[\int_{-\infty}^{-\alpha}+\int_{\alpha}^{\infty}\right]\frac{f(x+y)}{y}dy$$

$$=\frac{1}{\pi i}\mathrm{P}\int_{-\infty}^{\infty}\frac{f(x+y)}{y}dy.$$

Combining this result with that for the δ–function (that is, adding and subtracting, then multiplying by πi), we see that

$$\lim_{\alpha\to 0}\int_{-\infty}^{\infty}\frac{f(x+y)dy}{y-i\alpha}=\mathrm{P}\int_{-\infty}^{\infty}\frac{f(x+y)}{y}dy+i\pi f(x),$$

and that

$$\lim_{\alpha\to 0}\int_{-\infty}^{\infty}\frac{f(x+y)dy}{y+i\alpha}=\mathrm{P}\int_{-\infty}^{\infty}\frac{f(x+y)}{y}dy-i\pi f(x).$$

Symbolically, we can write

$$\frac{1}{x\pm i0}=\mathrm{P}\frac{1}{x}\mp i\pi\delta(x). \tag{1.25}$$

These formulae are in fact a particular case of the *Plemelj formulae* which we shall study in the next chapter in a more general setting.

Alongside the Dirac δ–function, we introduce the Heaviside step function

$$\Theta(x)=\begin{cases}1 \text{ for } x>0,\\ 0 \text{ for } x<0,\end{cases} \tag{1.26}$$

and the sign function

$$\operatorname{sgn}(x)=\Theta(x)-\Theta(-x). \tag{1.27}$$

To see how these functions are related, we pause now to calculate their Fourier transforms. Obviously, we have

$$\int_{-\infty}^{\infty}e^{ixy}\delta(x)dx=1.$$

Next we regularize the integral as before to arrive at the Fourier transform of the step function:

$$\int_{-\infty}^{\infty} e^{ixy}\Theta(x)dx = \lim_{\alpha\to 0}\int_{0}^{\infty} e^{ixy-\alpha x}dx = \frac{1}{0-iy} = \frac{i}{y+i0},$$

and similarly (swapping the sign of y)

$$\int_{-\infty}^{\infty} e^{ixy}\Theta(-x)dx = -\frac{i}{y-i0}.$$

The inversion formula results in

$$\Theta(x) = \frac{i}{2\pi}\int_{-\infty}^{\infty}\frac{e^{-ixy}dy}{y+i0}.$$

Exchanging $x \leftrightarrow y$ and then $y \to -y$ and re-arranging, we obtain the result

$$\int_{-\infty}^{\infty}\frac{e^{ixy}dx}{x+i0} = -2\pi i\Theta(-y),$$

which is the Fourier integral transform of the function $1/(x+i0)$. Note that this result is easily understood in terms of complex integration. Indeed, making x a complex variable, we see that $|e^{ixy}| = e^{-y\mathrm{Im}x}$ and the only pole $x = -i0$ is situated in the lower half-plain. So for $y > 0$ we complete the contour in the upper half-plain and get a zero result, while for $y < 0$ we close the contour in the lower half-plane in the negative clockwise direction to obtain the above result. Now letting simultaneously $x \to -x$ and $y \to -y$, we find the related Fourier transform,

$$\int_{-\infty}^{\infty}\frac{e^{ixy}dx}{x-i0} = 2\pi i\Theta(y),$$

which can also easily be understood in terms of complex integration. Adding up these two results and using the previously derived property (1.25) we obtain the principal value Fourier transform:

$$\mathrm{P}\int_{-\infty}^{\infty} e^{ixy}\frac{dx}{x} = i\pi\mathrm{sgn}(y).$$

Finally, by either using the inversion formula for the principal value transform, or simply subtracting the transforms of $\Theta(x)$ and $\Theta(-x)$, which is equivalent, we obtain the Fourier transform of the sgn(x) function:

$$\int_{-\infty}^{\infty} e^{ixy}\,\mathrm{sgn}(x)dx = 2i\mathrm{P}\left(\frac{1}{y}\right).$$

We have now advanced sufficiently to define another important integral transform,

$$F_H(y) = \mathrm{P}\int_{-\infty}^{\infty} \frac{f(x)dx}{x-y}, \tag{1.28}$$

called the *Hilbert transform*. The basic question to ask about this (or indeed any other) integral transform is: what is the inversion formula?

To this end we Fourier transform the above expression (from now on we shall mostly use letters like k, p, q etc. as Fourier variable):

$$\int_{-\infty}^{\infty} e^{iky} F_H(y)dy = -\int_{-\infty}^{\infty} f(x)\left[\mathrm{P}\int_{-\infty}^{\infty} \frac{e^{iky}dy}{y-x}\right]dx = -i\pi\,\mathrm{sgn}(k)\,F(k),$$

where the Fourier transform of the principal value has been used. Therefore

$$F(k) = \frac{i}{\pi}\,\mathrm{sgn}(k)\int_{-\infty}^{\infty} e^{iky} F_H(y)dy,$$

so utilising the inversion formula for the Fourier transform

$$f(x) = \frac{i}{2\pi^2}\int_{-\infty}^{\infty}\left[\int_{-\infty}^{\infty} e^{ik(y-x)}\mathrm{sgn}(k)dk\right]F_H(y)dy$$

and the latter for the sign function we finally obtain

$$f(x) = -\frac{1}{\pi^2}\mathrm{P}\int_{-\infty}^{\infty} \frac{F_H(y)dy}{y-x},$$

which is the inversion formula for the Hilbert transform. Note that this formula implies the identity

$$\mathrm{P}\int_{-\infty}^{\infty}\left[\mathrm{P}\int_{-\infty}^{\infty} \frac{f(x')dx'}{x'-y}\right]\frac{dy}{y-x} = -\pi^2 f(x),$$

which is another incarnation of the Fourier theorem.

1.4.2 Analytic and Asymptotic Properties

Let us now make the Fourier variable in

$$F(k) = \int_{-\infty}^{\infty} e^{ikx} f(x)dx$$

complex: $k = k_1 + ik_2$. The integral is assumed to exist for all values of k_1 when $k_2 = 0$. When $k_2 \neq 0$ we have $|e^{ikx}| = e^{-k_2 x}$ so that the integrand decreases faster for $x > 0$ if $k_2 > 0$ and for $x < 0$ when $k_2 < 0$. However for $x > 0$ when $k_2 < 0$ and when $x < 0$ for $k_2 > 0$ the integrand acquires an exponentially growing factor and the convergency is in doubt. In fact unless $f(x)$ exhibits a pathologically fast, stretched exponential decay, the exponent will always win for sufficiently large $|k_2|$ and the integral will diverge. We see that the analytic properties of the contributions to the Fourier integral from large positive and large negative values of x are likely to compete.

For this reason alone it makes sense to separate these contributions and introduce the 'half-line' Fourier transforms

$$F_+(k) = \int_{0}^{\infty} e^{ikx} f(x)dx, \tag{1.29}$$

and

$$F_-(k) = \int_{-\infty}^{0} e^{ikx} f(x)dx. \tag{1.30}$$

Obviously

$$F(k) = F_+(k) + F_-(k)$$

so long as both integrals exist. It is advantageous to explicitly separate the $x > 0$ and $x < 0$ supported parts of the function $f(x)$:

$$f_+(x) = \begin{cases} f(x) & \text{for } x > 0 \\ 0 & \text{for } x < 0 \end{cases}, \quad f_-(x) = \begin{cases} 0 & \text{for } x > 0 \\ f(x) & \text{for } x < 0 \end{cases}. \tag{1.31}$$

Any function $f(x)$ can be written as $f(x) = f_+(x) + f_-(x)$ and the half-line Fourier transforms of $f(x)$ then become ordinary Fourier transforms of $f_\pm(x)$:

$$F_\pm(k) = \int_{-\infty}^{\infty} e^{ikx} f_\pm(x)dx,$$

again provided that the integrals in question converge.

As the imaginary part of the Fourier variable k_2 attaches an exponentially increasing (decaying) factor to the integrand, the convergence is best investigated in terms of exponential bounds of the functions $f_\pm(x)$. Namely we assume that

$$|f_+(x)| < A_+ e^{a_+ x} \quad \text{for} \quad x \to \infty, \tag{1.32}$$

and that

$$|f_-(x)| < A_- e^{a_- x} \quad \text{for} \quad x \to -\infty, \tag{1.33}$$

where the constants $A_\pm$ are positive and the *exponential bound indices* $a_\pm$ can have either sign. The precise values of the constants $A_\pm$ are usually not important. For a function $f(x)$ growing faster than an exponential for either $x \to \infty$ or $x \to -\infty$ the respective half-line Fourier transform simply does not exist. For a symmetric function like $f(x) = 1/\cosh x$, we have $a_+ = -1$ and $a_- = 1$. An asymmetric example is $f(x) = 1/(1+e^x)$, where $a_+ = -1$ while $a_- = 0$. Finally, consider an important example of an algebraic decay

$$|f_+(x)| < \frac{A}{|x|^\nu} \quad \text{for} \quad x \to \infty,$$

where $A > 0$ and $\nu > 0$. This is a decay faster than a constant (for which the exponential bound index is $a_+ = 0$) but slower than exponential decay (any $a_+ < 0$). We conclude that the algebraic decay, in terms of exponential bounds, is characterised by $a_+ = 0^-$ for $f_+(x)$, and by a similar argument $a_- = 0^+$ for $f_-(x)$.

Theorem: analyticity domains. Given that functions $f_\pm(x)$ are exponentially bounded, the half-line Fourier transform $F_+(k)$ is analytic in the region $k_2 > a_+$ and the transform $F_-(k)$ is analytic for $k_2 < a_-$.

Indeed, the integral

$$|F_+(k)| = |\int_0^\infty e^{ikx} f_+(x)dx| \leq \int_0^\infty |e^{ikx}||f_+(x)|dx < A_+ \int_0^\infty e^{-k_2 x} e^{a_+ x} dx = \frac{A_+}{k_2 - a_+}$$

converges absolutely for $k_2 > a_+$ thus guaranteeing existence of $F_+(k)$ and all its derivatives in this region. Similarly the expression

$$|F_-(k)| \leq \int_{-\infty}^0 |e^{ikx}||f_-(x)|dx < A_- \int_0^\infty e^{k_2 x} e^{-a_- x} dx = \frac{A_-}{a_- - k_2}$$

converges absolutely in the region $k_2 < a_-$.

Theorem: inversion formulae. For the right-handed Fourier transform we have

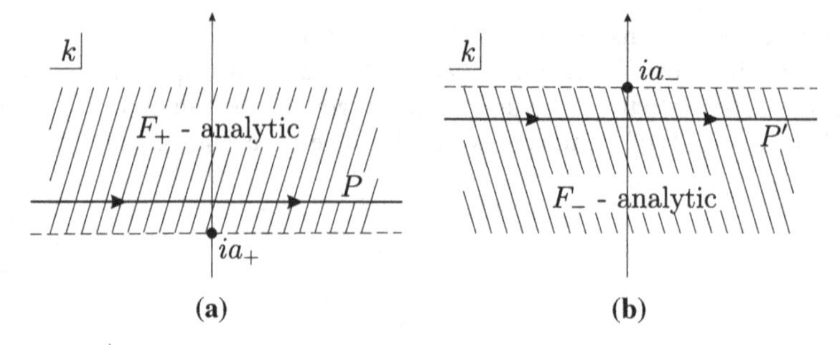

Fig. 1.20 Analyticity domain for the one-side Fourier transform $F_+(k)$ (**a**), and $F_-(k)$ (**b**)

$$f_+(x) = \frac{1}{2\pi} \int_P e^{-ikx} F_+(k) dk, \tag{1.34}$$

where P indicates integration along a straight horizontal line lying above all singularities of the function $F_+(k)$ see Fig. 1.20a. Similarly $f_-(x)$ is recovered via

$$f_-(x) = \frac{1}{2\pi} \int_{P'} e^{-ikx} F_-(k) dk, \tag{1.35}$$

where this time P' lies below all singularities of $F_-(k)$, see Fig. 1.20b.

Indeed, substituting $k = k_1 + ik_2$,

$$F_+(k) = \int_0^{+\infty} dx e^{ikx} f_+(x) = \int_{-\infty}^{+\infty} dx e^{ikx} f_+(x) = \int_{-\infty}^{+\infty} dx e^{ik_1x} [e^{-k_2x} f_+(x)].$$

Therefore the one-side Fourier transform $F_+(k)$ of the function $f_+(x)$ can also be regarded as an *ordinary* Fourier transform (with respect to k_1) of the function $f_+(x)e^{-k_2x}$, $\text{Im}k = k_2 > a_+$. To the latter the ordinary inversion formula is applicable:

$$f_+(x)e^{-k_2x} = \frac{1}{2\pi} \int_{-\infty}^{+\infty} dk_1 e^{-ik_1x} F_+(k_1 + ik_2),$$

where k_1 is now the Fourier variable and k_2 is a parameter. We multiply the result by e^{k_2x} and obtain

$$f_+(x) = \frac{1}{2\pi} \int_{-\infty}^{+\infty} dk_1 e^{-ik_1x+k_2x} F_+(k_1 + ik_2) = \frac{1}{2\pi} \int_{-\infty}^{+\infty} dk_1 e^{-i(k_1+ik_2)x} F_+(k_1 + ik_2).$$

That relation immediately leads us to the statement (1.34). The contour P lies above all singularities of $F_+(k)$ which is guaranteed by the requirement $k_2 > a_+$.

Note that we have $|e^{-ikx}| = e^{k_2 x} \to 0$ for $x < 0$ and $k_2 \to +\infty$. As additionally $F_+(k)$ is analytic for $k_2 > a_+$ we can shift the contour P upwards indefinitely. Hence

$$f_+(x) = \frac{1}{2\pi} \int_{ik_2-\infty}^{ik_2+\infty} dk\, e^{-ikx} F_+(k) = \frac{1}{2\pi} \lim_{k_2 \to +\infty} \int_{ik_2-\infty}^{ik_2+\infty} dk\, e^{-ikx} F_+(k) = 0$$

for any $x < 0$, as expected by construction of $f_+(x)$. On the contrary, for $x > 0$ we can not shift P upwards. We can shift it downwards but will hit singularities of $F_+(k)$ there. In the similar way one proves the inversion formula (1.35).

Therefore if in general the requirements (1.32) and (1.33) are satisfied by $f(x)$, then its (ordinary) Fourier transform

$$F(k) = \int_{-\infty}^{+\infty} dx e^{ikx} f(x) = \int_{-\infty}^{+\infty} dx e^{ikx} [f_+(x) + f_-(x)] = F_+(k) + F_-(k)$$

is analytic when $a_+ < \mathrm{Im} k < a_-$ (where both F_+ and F_- are analytic) in the strip where the upper half-plane ($\mathrm{Im} k > a_+$) and the lower half-plane ($\mathrm{Im} k < a_-$) overlap, see Fig. 1.21.

The inversion integral can be written as:

$$f(x) = \frac{1}{2\pi} \int_{i\sigma-\infty}^{i\sigma+\infty} dk e^{-ikx} F(k) \qquad \text{with} \quad a_+ < \sigma < a_-.$$

We would like to point out that the analyticity strip *only* exists for $a_+ < a_-$. This constitutes a restriction on the functions in question. For instance if $|f(x)| \leq A e^{-|x|}$ as $x \to \pm\infty$ $a_+ = -1$ and $a_- = +1$. The corresponding strip is shown in Fig. 1.21. It can happen though that $a_+ = a_-$, then the strip shrinks to a line. An important example is $|f(x)| \leq \frac{A}{|x|^\nu}$ as $x \to \pm\infty$, $\nu \geq 0$, when $a_+ = 0$ and $a_- = 0$, see the discussion above.

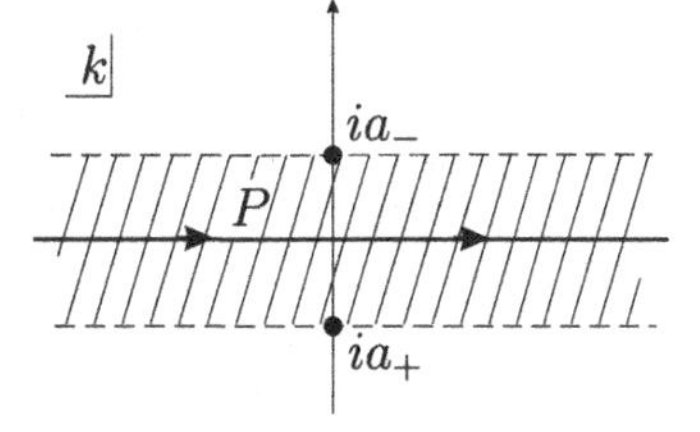

Fig. 1.21 The analyticity strip of the ordinary Fourier transform $F(k)$

Example 1.13
Ordinary differential equation (ODE)

$$\frac{d^2 f(x)}{dx^2} - f(x) = 0 \tag{1.36}$$

is defined on half-line $x > 0$ with $f(0) = 1$ and $f(x)$ bounded for $x \to \infty$. Clearly the solution is $f(x) = e^{-x}$, but can we find this using Fourier transform?

We use the definition (1.31) for $f_+(x)$. As it is required that $f_+(x)$ be bounded, hence $a_+ \leq 0$ and according to the above theorem, $F_+(k)$ is analytic in the upper half-plane $\mathrm{Im}k > 0$. Let us compute the Fourier transform of $f''_+(x)$:

$$\int_0^\infty dx f''_+(x) e^{ikx} = \int_0^\infty e^{ikx} d\left[\frac{df_+(x)}{dx}\right] = e^{ikx}\frac{df_+(x)}{dx}\bigg|_0^\infty - ik\int_0^\infty dx e^{ikx}\frac{df_+(x)}{dx}$$

$$= -f'_+(0) - ike^{ikx} f_+(x)\Big|_0^\infty + (ik)^2 \int_0^\infty dx e^{ikx} f_+(x).$$

Thus

$$\int_0^\infty dx f''_+(x)\, e^{ikx} = ik - k^2 F_+(k) - \alpha,$$

where $\alpha = f'_+(0)$ is an unknown constant.[7] Now the Fourier transform of the ODE (1.36) is an elementary algebraic equation

$$ik - k^2 F_+(k) - \alpha - F_+(k) = 0$$

solved by

$$F_+(k) = \frac{ik - \alpha}{k^2 + 1} = \frac{ik - \alpha}{(k + i)(k - i)}.$$

There is an apparent singularity at $k = i$. But we know that $F_+(k)$ is analytic in the upper half-plane, as a consequence of $f_+(x)$ being bounded. We must therefore fix it, that is remove the singularity. This is achieved by choosing α in such a way that the factor $(k - i)$ is canceled, therefore $\alpha = -1$. Thus

$$F_+(k) = \frac{i(k - i)}{(k + i)(k - i)} = \frac{i}{k + i},$$

[7] We would like to point out that the derivative with respect to x ($f'_+(x)$) *is not* equivalent to multiplying the Fourier transform by ik because of its 'one-sidedness', since the boundary contributions when integrating by part are non-trivial.

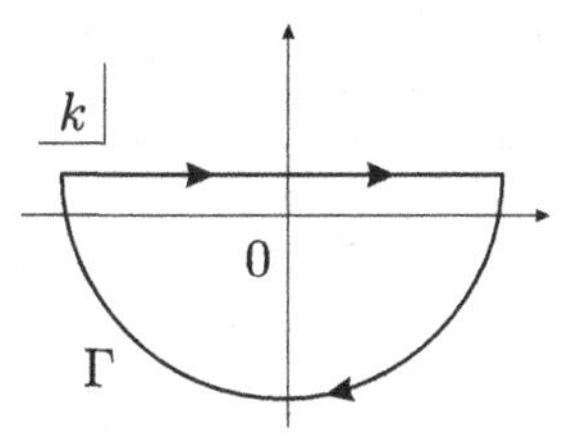

Fig. 1.22 The contour for the calculation of the function $f_+(x)$ in Example 1.13

which has a single pole at $k = -i$. To finally find $f_+(x)$ we use the inversion formula

$$f_+(x) = \frac{1}{2\pi} \int_P dk\, e^{-ikx} \frac{i}{k+i},$$

where P indicates integration along the straight horizontal line lying in the upper half-plane. If $x < 0$ then $|e^{-ikx}| = e^{(\mathrm{Im}k)x} \to 0$ as $k_2 \to \infty$ and closing the contour in the upper half-plane, we have $f_+(x < 0) = 0$. If $x > 0$, $|e^{-ikx}| = e^{(\mathrm{Im}k)x} \to 0$ as $k_2 \to -\infty$. Now we can only close the contour in the lower half-plane as is shown in Fig. 1.22. Then, by residue formula

$$f_+(x > 0) = \frac{1}{2\pi} \int_\Gamma dk e^{-ikx} \frac{i}{k+i} = \frac{1}{2\pi}(-2\pi i)\mathrm{Res}\frac{ie^{-ikx}}{k+i}\Big|_{k=-i}$$
$$= (-i)ie^{-i(-i)x} = e^{-x},$$

as expected.

Let us now discuss some asymptotic properties of Fourier transforms.

Consider an ordinary Fourier transform first

$$F(s) = \int_{-\infty}^{\infty} dx\, e^{isx} f(x).$$

Suppose that the function $f(x)$ with all its derivatives is continuous and bounded at $x \to \pm\infty$, that is analytic in a finite strip around the real axis. What can we say about the above integral for $s \to \infty$? In order to get the feeling what can happen we consider several examples.

Example 1.14

$$f(x) = \frac{1}{x^2 + a^2} = \frac{1}{(x+ia)(x-ia)}, \quad a > 0.$$

For $s > 0$ close the contour in the upper half-plane ($\mathrm{Im}x > 0$) [see Fig. 1.23a] and by residue formula

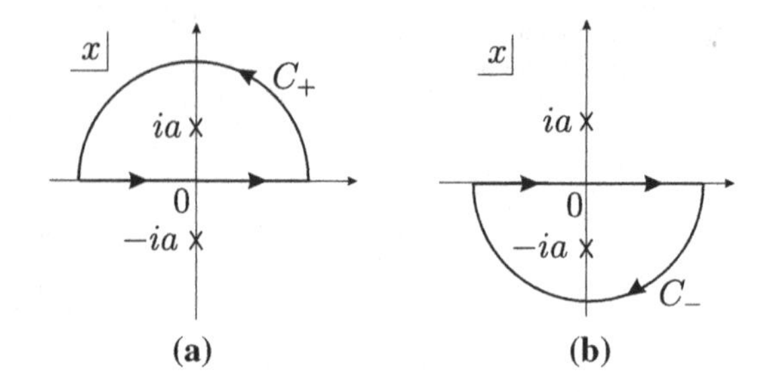

Fig. 1.23 For $s > 0$ we close the contour in the upper half-plane (**a**) while for $s < 0$ we close it in the lower half-plane (**b**)

$$\int_{C_+} dx \frac{e^{isx}}{x^2 + a^2} = 2\pi i \mathrm{Res}(...)\Big|_{x=ia} = 2\pi i \frac{e^{-as}}{2ia} = \frac{\pi}{a} e^{-as}.$$

For $s < 0$, on the other hand [see Fig. 1.23b]

$$\int_{C_-} dx \frac{e^{isx}}{x^2 + a^2} = (-2\pi i)\mathrm{Res}(...)\Big|_{x=-ia} = (-2\pi i)\frac{e^{as}}{-2ia} = \frac{\pi}{a} e^{as}.$$

Therefore

$$F(s) = \frac{\pi}{a} e^{-a|s|} \tag{1.37}$$

and thus the asymptotic behaviour is trivially

$$F(s) \sim e^{-a|s|} \quad \text{for} \quad s \to \infty.$$

Example 1.15

$$f(x) = \frac{1}{\sqrt{x^2 + 1}} = \frac{1}{\sqrt{(x - i)(x + i)}}.$$

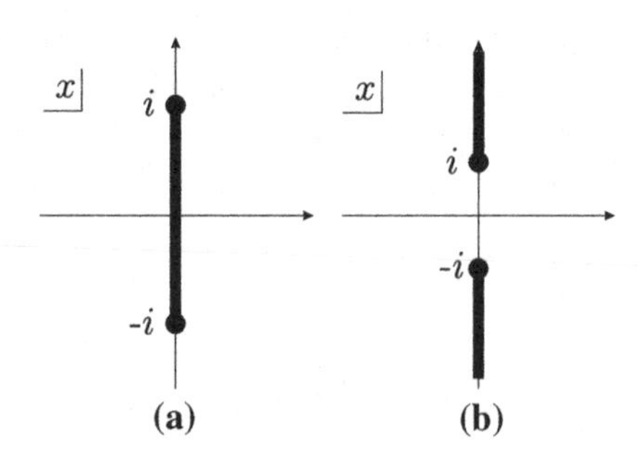

Fig. 1.24 Two different ways to define the cuts of the complex plane

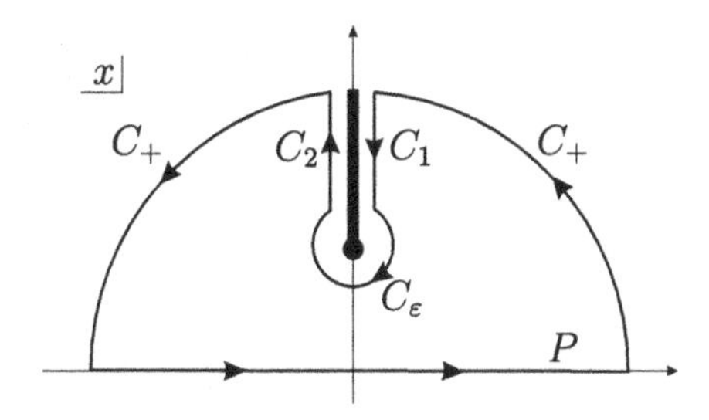

Fig. 1.25 Schematic representation of the contour for the calculation of the function *F(s)*

This function has a branch cut that can be chosen either as is shown in Fig. 1.24a or else as in Fig. 1.24b. The latter choice is more convenient here. To compute $F(s)$ for $s > 0$ observe that (see Fig. 1.25)

$$\int\limits_{C_+ + C_\epsilon + C_1 + C_2 + P} dx\, e^{isx} f(x) = 0,$$

and then

$$\int\limits_P dx\, e^{isx} f(x) = F(s).$$

Next we estimate the integral along the circle path C_ϵ around the upper branching point $z = i$, which we parametrise by $z = r\, e^{i\phi}$,

$$\int\limits_{C_\epsilon} dx\, e^{isx} f(x) \sim \int\limits_{C_\epsilon} \frac{dz}{\sqrt{z}} \sim \int\limits_0^{2\pi} \frac{r e^{i\phi} d\phi}{\sqrt{r}} \sim \sqrt{r}.$$

Therefore it vanishes for $r \to 0$. For $s > 0$

$$\int\limits_{C_+} dx\, e^{isx} f(x) \; = 0,$$

as C_+ can be pushed upwards. On C_1 we put $x = i(1 + \eta)$ and then obtain

$$\int\limits_{C_1} dx\, e^{isx} f(x) = \int\limits_\infty^0 \frac{i d\eta e^{-s-\eta s}}{i\sqrt{\eta(\eta+2)}} = -e^{-s} \int\limits_0^\infty \frac{d\eta}{\sqrt{\eta(\eta+2)}} e^{-\eta s} = +\int\limits_{C_2} dx\, e^{isx} f(x).$$

In this way we achieve a very convenient representation

$$F(s) = 2e^{-s} \int_0^\infty \frac{d\eta}{\sqrt{\eta(\eta+2)}} e^{-\eta s}.$$

The leading asymptotics of the remaining integral can be found by the following estimation

$$\int_0^\infty \frac{d\eta}{\sqrt{\eta(\eta+2)}} e^{-\eta s} \sim \int_0^{1/s} \frac{d\eta}{\sqrt{\eta}} \sim \sqrt{\eta}\Big|_0^{1/s} \sim \frac{1}{\sqrt{s}}.$$

So that in total we get a power-law correction to the leading exponential decay:

$$F(s) \sim \frac{1}{\sqrt{s}}\, e^{-s}.$$

In general, if a_0 is the location of the singularity of $f(x)$ in the upper (or lower) half-plane, which is closest to the real axis, then the large positive s asymptotic behaviour of the Fourier transform is given by

$$\int_{-\infty}^{\infty} e^{isx} f(x)\, dx \sim e^{-a_0|s|}.$$

If $f(x)$ is not continuous, but has a jump, the large s-decay is slower than exponential (usually a power law of some kind). Namely consider

$$f_+(x) = \begin{cases} f(x), & x > 0 \\ 0, & x < 0 \end{cases}$$

such that $f_+(0)$ is a finite constant and $f_+^{'}(0)$, $f_+^{''}(0)$, etc. are finite. Then the one-side Fourier transform can be expanded for large s in the following way,

$$F_+(s) = \int_0^\infty dx\, e^{isx} f_+(x) = \frac{1}{is} \int_0^\infty d(e^{isx}) f_+(x) = \frac{1}{is} e^{isx} f_+(x)\Big|_0^\infty$$
$$- \frac{1}{is} \int_0^\infty dx e^{isx} f_+^{'}(x) = -\frac{1}{is} f_+(0) + \frac{1}{(is)^2} f_+^{'}(0) + \cdots .$$

Therefore in this situation

$$F_+(s) \to \frac{if_+(0)}{s} - \frac{f_+^{'}(0)}{s^2} + \mathrm{O}(1/s^3) \to \frac{if_+(0)}{s} + \mathrm{O}(1/s^2).$$

It can also happen that $f_+(x)$ is singular at $x \to 0^+$. The following limiting form often occurs in practice:

$$f_+(x) \simeq a_0 x^\lambda \quad \text{as} \quad x \to 0^+$$

where $\lambda > -1$ (otherwise x-integral is diverging). Then as $s \to \infty$

$$F_+(s) \simeq \int\limits_0^\infty a_0 x^\lambda e^{isx} dx = \frac{a_0}{(-is)^{\lambda+1}} \int\limits_0^\infty t^\lambda e^{-t} dt = \frac{a_0 \Gamma(\lambda+1)}{(-is)^{\lambda+1}},$$

(Γ is the Gamma function, see Sect. 2.1) so that from

$$f_+(x)\Big|_{x\to 0^+} \sim x^\lambda \quad \text{follows} \quad F_+(s)\Big|_{s\to\infty} \sim \frac{1}{s^{\lambda+1}}.$$

Similarly if

$$f_-(x)\Big|_{x\to 0^-} \sim (-x)^\mu$$

than also

$$F_-(s)\Big|_{s\to\infty} \sim \frac{1}{s^{\mu+1}}.$$

1.4.3 Laplace Integral Transform

Laplace transform can be understood as a special case of the Fourier transform and shares a number of its properties. It is particularly useful for solving differential equations which in many cases reduce to algebraic ones for the images. This method is known under the name of *operational calculus* .

Let $f(t)$ be a function which is zero for $t < 0$ and exponential bound index of which is α (see Sect. 1.4.2). Then the Laplace transform, or the image $F(p)$ of the original $f(t)$ is defined by

$$F(p) = \int_0^\infty dt\, f(t)\, e^{-pt}, \tag{1.38}$$

where $\text{Re}\, p > \alpha$, otherwise the integral (1.38) diverges. It is linear just as the Fourier transform and has the scaling property $f(\beta t) \leftrightarrow F(p/\beta)/\beta$. We further agree that the function $f(t)$, unless otherwise stated, is viewed as the product $f(t)\Theta(t) \to f(t)$, where $\Theta(t)$ is the Heaviside function (1.26). Using the definition (1.38), let us find the images for some elementary functions.[8]

[8] In many practical applications both Fourier and Laplace transforms can become very complicated. Usually it is useful to consult the tables of integral transforms, see for example [6].

One readily obtains that for the exponential function

$$f(t) = e^{\alpha t} \quad \leftrightarrow \quad F(p) = \frac{1}{p - \alpha}. \tag{1.39}$$

Consequently, for example,

$$f(t) = \sin t \quad \leftrightarrow \quad F(p) = \frac{1}{p^2 + 1}. \tag{1.40}$$

In the same way we obtain

$$f(t) = \cos t \quad \leftrightarrow \quad F(p) = \frac{p}{p^2 + 1}.$$

For the power function t^n ($n = 0, 1, 2, \dots$), integrating n times by parts, we have:

$$f(t) = t^n \quad \leftrightarrow \quad F(p) = \frac{n!}{p^{n+1}}. \tag{1.41}$$

Using a similar procedure one can evaluate the image of the derivative,

$$\int_0^\infty dt\, f'(t)\, e^{-pt} = f'(t)e^{-pt}\Big|_0^\infty + p \int_0^\infty dt\, f(t)\, e^{-pt} = p\, F(p) - f(0). \tag{1.42}$$

For the image of the nth order derivative then follows:

$$f^{(n)}(t) \to p^n F(p) - p^{n-1} f(0) - p^{n-2} f'(0) - \cdots - p f^{(n-2)}(0) - f^{(n-1)}(0). \tag{1.43}$$

If all $f^{(0,\dots,n-1)}(0) = 0$ then $f^{(n)}(t) \to p^n F(p)$ which is probably the most important property of the Laplace transform.

Analogous relations can be obtained for the derivative of the image. Indeed,

$$F'(p) = \int_0^\infty dt\, \partial_p [f(t)\, e^{-pt}] = \int_0^\infty dt(-t) f(t)\, e^{-pt}. \tag{1.44}$$

By induction we obviously arrive at

$$F^{(n)}(p) \leftrightarrow (-t)^n\, f(t). \tag{1.45}$$

Interestingly, while a multiplication with the respective argument is equivalent to taking derivative of the image/original a division by the argument is translated to an integration. For example let $g'(t) = f(t)$ and $g(0) = 0$ then

$$f(t) = g'(t) \leftrightarrow pG(p) \quad \text{and therefore} \quad F(p) = pG(p) \tag{1.46}$$

and ultimately

$$F(p)/p \leftrightarrow \int_0^t d\tau\, f(\tau). \tag{1.47}$$

Inspired by the above calculation we can also show that

$$f(t)/t \leftrightarrow \int_p^\infty dq\, F(q). \tag{1.48}$$

Another very useful property of the Laplace transform is the transform of the convolution, defined as

$$h(t) = (f * g)(t) = \int_0^t ds\, f(s)\, g(t-s). \tag{1.49}$$

Then

$$H(p) = \int_0^\infty dt\, e^{-pt}\, h(t) = F(p)\, G(p). \tag{1.50}$$

This relation can be proven in the following way.

$$H(p) = \int_0^\infty dt\, e^{-pt} \int_0^t ds\, f(s)\, g(t-s) = \int_0^\infty ds \int_s^\infty dt\, e^{-pt}\, f(s)\, g(t-s),$$

where we have just changed the order of integrations, taking into account the geometry of the integration domain. In the next step we change variable by $q = t - s$ which leads to

$$= \int_0^\infty ds \int_0^\infty dq\, e^{-p(q+s)}\, f(s)\, g(q) = F(p)\, G(p).$$

Finally, let us construct the original function $f(t)$ for the given image $F(p)$. To this end we consider the auxiliary function $g(t) = f(t)e^{-bt}$, where $b > \alpha$ [recall that α is the exponential bound index of the original $f(t)$]. Then, remembering that $f(t) = 0$ at $t < 0$, the Laplace transform of $f(t)$

$$F(b+iu) = \int_0^\infty dt\, f(t)\, e^{-(b+iu)t} = \int_{-\infty}^\infty dt\, g(t)e^{-iut}$$

is the Fourier transform of $g(t)$. Its inversion

$$g(t) = \frac{1}{2\pi} \int_{-\infty}^\infty du\, F(b+iu)e^{iut}$$

should be understood in terms of principal value integral. Now we recover $f(t)$ and shift the integration contour,

$$f(t) = \frac{1}{2\pi} \int_{-\infty}^{\infty} du\, F(b+iu) e^{(b+iu)t} = \frac{1}{2\pi i} \int_{b-i\infty}^{b+i\infty} dp\, F(p) e^{pt}. \tag{1.51}$$

This is the inverse Laplace transform and is also referred to as the *Mellin formula.*

Example 1.16
Solve the initial value problem with $x(0) = x'(0) = 0$ for the differential equation

$$x''(t) - x(t) = t^2 + 2e^t. \tag{1.52}$$

Performing the Laplace transform of the equation we obtain an algebraic equation

$$p^2 X(p) - px(0) - x'(0) - X(p) = 2\left(\frac{1}{p-1} + \frac{1}{p^3}\right).$$

Therefore

$$X(p) = \frac{2}{p^2-1}\left(\frac{1}{p-1} + \frac{1}{p^3}\right).$$

The original function $x(t)$ can be found simply with a help of the convolution formula (1.50):

$$\frac{2}{p^2-1} \cdot \frac{1}{p-1} \quad \to \quad 2\int_0^t \sinh\tau\, e^{(t-\tau)} d\tau = te^t - \sinh t,$$

$$\frac{2}{p^2-1} \cdot \frac{1}{p^3} \quad \to \quad 2\int_0^t \sinh(t-\tau)\, \tau^2\, d\tau = -2 - t^2 + 2\cosh t.$$

The solution of the problem is thus

$$x(t) = te^t - t^2 - 2 + \frac{3}{2} e^{-t} + \frac{1}{2} e^t.$$

This result is easily checked by substituting into the original Eq. (1.52). We leave the reader to verify the obtained result using the Mellin formula (1.51). It is important to realize that in this calculation we must set $b > 1$ and close the integration along the (shifted) imaginary axis by the arc C_R, $|p - b| = R$ in the left half-plane; the C_R-integral tends to zero as $R \to \infty$ and the calculation reduces to the determination of residues.

1.5 Problems

Problem 1.1
Derive the Cauchy–Riemann equations in the polar co-ordinates. Hence show that the function $\ln z$ is analytic for all finite z except $z = 0$.

Problem 1.2
Calculate the integrals:

$$\textbf{(a)}: \quad \int_0^{2\pi} \frac{d\varphi}{(a+b\cos\varphi)^2}, \quad (a > b > 0),$$

$$\textbf{(b)}: \quad \int_0^{2\pi} \frac{\cos^2(3\varphi)d\varphi}{1-2p\cos(2\varphi)+p^2}, \quad (0 < p < 1),$$

$$\textbf{(c)}: \quad \int_0^{2\pi} \frac{(1+2\cos\varphi)^n \cos(n\varphi)d\varphi}{1-a-2a\cos\varphi}, \quad (0 < a < 1/3),$$

using the residue method.

Problem 1.3
The following integral equation

$$\left[\lambda - \frac{1}{\sqrt{1+3x^2/4}}\right] f(x) = \frac{2}{\pi} \int_{-\infty}^{\infty} \frac{f(x')dx'}{1+x^2+x'^2-xx'}$$

emerges in the study of bound states in a system of 3 bosons in one spatial dimension. Show that the function

$$f(x) = \frac{1}{1+x^2}$$

solves the above equation and determine the corresponding value of the spectral parameter λ.

Problem 1.4
Choose appropriate contours and apply Cauchy theorem to calculate the integrals:

$$\textbf{(a)}: \quad \int_0^{\infty} \frac{\sin x}{x} dx \quad \text{(Euler 1781)}.$$

Hint: complete the real axis with a small and large semi-circles;

$$\textbf{(b)}: \quad \int_0^{\infty} e^{-ax^2} \cos(bx)dx \quad (a > 0).$$

Hint: use a rectangular contour;

$$(\mathbf{c}): \quad \int_0^\infty \cos(x^2)dx, \quad \int_0^\infty \sin(x^2)dx.$$

Hint: use a triangular type contour.

Problem 1.5
Calculate the integrals:

$$(\mathbf{a}): \quad I = \int_0^\infty \frac{dx}{1+x^3},$$

$$(\mathbf{b}): \quad I(a,b) = \int_0^\infty \frac{\sin ax}{\sin bx} \frac{1}{1+x^2}\, dx, \quad |a| < |b|.$$

Problem 1.6
Using branch cut properties calculate the integrals:

$$(\mathbf{a}): \int_0^\infty \frac{x^{-p} dx}{1+2x\cos\lambda + x^2} \quad (-1 < p < 1, -\pi < \lambda < \pi),$$

$$(\mathbf{b}): \int_0^\infty \frac{\ln x\, dx}{(x^2+1)^2},$$

$$(\mathbf{c}): \int_{-1}^1 \frac{dx}{[(1-x)(1+x)^2]^{1/3}}.$$

Hint: in **(b)** use the contour of Problem 1.4 **(a)**.

Problem 1.7
Calculate Fourier transforms and identify the fundamental strips for the following functions:

$$(\mathbf{a}): \; g(\lambda) = -\frac{e^{-\lambda}}{1+e^{-2\lambda}},$$

$$(\mathbf{b}): \; k(\lambda) = -\frac{4U}{\pi} \frac{e^{-\lambda}}{(1+e^{-\lambda})^2 + U^2(1-e^{-\lambda})^2}.$$

Problem 1.8
Show that for the original $f(t)$ and its Laplace image $F(p)$ holds the following translation property:

$$f(t-\tau) \leftrightarrow F(p)\, e^{-p\tau}.$$

Using this relation calculate the image of $f(t) = |\sin t|$.

Problem 1.9
Determine the original function $f(t)$ for the given Laplace image:

$$(\mathbf{a}):\quad F(p) = \frac{1}{p^{\alpha+1}}, \quad -1 < \alpha < 0, \quad \mathrm{Re}\, p > 0\,,$$
$$(\mathbf{b}):\quad F(p) = \frac{1}{p}\, e^{-\alpha\sqrt{p}}, \qquad \alpha > 0, \quad \mathrm{Re}\, p > 0.$$

Answers:

Problem 1.2: **(a):** $2\pi a(a^2 - b^2)^{-3/2}$; **(b):** $\pi(1 - p + p^2)/(1 - p)$; **(c):**

$$\frac{2\pi}{\sqrt{1 - 2a - 3a^2}}\left(\frac{1 - a - \sqrt{1 - 2a - 3a^2}}{2a^2}\right)^n.$$

Problem 1.3: $\lambda = 2$.

Problem 1.4: **(a):** $\pi/2$; **(b):** $\sqrt{\pi/4a}\exp(-b^2/4a)$; **(c):** $\sqrt{\pi/8}$.

Problem 1.5: **(a):** $I = 2\pi/3\sqrt{3}$; **(b):** $I(a, b) = \frac{\pi}{2}\frac{\sinh a}{\sinh b}$.

Problem 1.6: **(a):**

$$\frac{\pi}{\sin(\pi p)}\frac{\sin(p\lambda)}{\sin\lambda},$$

(b): $-\pi/4$; **(c):** $2\pi/\sqrt{3}$.

Problem 1.7: **(a):** $G(k) = -\pi/[2\cosh(\pi k/2)]$; **(b):** $K(k) = -4\sinh(k\gamma)/\sinh(\pi k)$, where $\cos\gamma = (1 - U^2)/(1 + U^2)$.

Problem 1.8: $\coth(\pi p/2)/(1 + p^2)$.

Problem 1.9: **(a):** $t^\alpha/\Gamma(1 + \alpha)$; **(b):** $1 - \mathrm{erf}\big(\alpha/2\sqrt{t}\big)$; where $\mathrm{erf}(x)$ is the error function, defined in (2.62).

Chapter 2
Hypergeometric Series with Applications

Hypergeometric function arises in connection with solutions of the second order ordinary differential equations with several regular singular points. Although Gamma function does not satisfy any differential equation with algebraic coefficients, a detailed discussion would give us an opportunity to introduce all necessary notation as well as a number of very useful tools.

2.1 The Gamma Function

2.1.1 Definition and Recurrence Relation

One way to define the Gamma function is via the integral representation also referred to as *Euler integral of the second kind*:

$$\Gamma(z) = \int_0^\infty dt\, t^{z-1} e^{-t}. \tag{2.1}$$

By elementary integration we immediately find that

$$\Gamma(1) = \int_0^\infty dt\, e^{-t} = 1.$$

We observe that integral (2.1) is absolutely convergent for $\mathrm{Re} z > 0$ and therefore defines an analytic function in this domain.[1] For $\mathrm{Re} z < 0$, on the other hand, the

[1] Indeed the integral

$$\frac{d\Gamma(z)}{dz} = \int_0^\infty dt\ (\ln t) t^{z-1} e^{-t}$$

A. O. Gogolin (edited by E. G. Tsitsishvili and A. Komnik), *Lectures on Complex Integration*, 63
Undergraduate Lecture Notes in Physics, DOI: 10.1007/978-3-319-00212-5_2,

integral diverges at its lower limit. So we need to analytically continue the Gamma function to Re$z < 0$ region from Re$z > 0$ region. To this end, we integrate by parts,

$$\Gamma(z) = \frac{1}{z}\int_0^\infty d(t^z)e^{-t} = \frac{\Gamma(z+1)}{z},$$

to establish the *recurrence relation*:

$$\Gamma(z+1) = z\Gamma(z). \tag{2.2}$$

Obviously, for any integer n we then have a useful identity

$$\Gamma(n+1) = n!. \tag{2.3}$$

Therefore the Gamma function can be considered as a generalization of the factorial for non-integer arguments.

2.1.2 Analytic Properties

The identity (2.2) can be used to accomplish the analytic continuation. Indeed, the relation

$$\Gamma(z) = \frac{\Gamma(z+1)}{z}$$

defines $\Gamma(z)$ in the region $-1 < \text{Re}z < 0$ where Eq. (2.1) can still be used on the rhs of the above equation. Applying the recurrence relation repeatedly ($n+1$ times), one obtains

$$\begin{aligned}\Gamma(z) &= \frac{\Gamma(z+1)}{z} = \frac{\Gamma(z+2)}{(z+1)z} = \dots \\ &= \frac{\Gamma(z+n+1)}{(z+n)(z+n-1)\dots(z+1)z},\end{aligned} \tag{2.4}$$

which defines $\Gamma(z)$ in $-n-1 < \text{Re}z < -n$. As the above process can be continued indefinitely, we have the definition of $\Gamma(z)$ for all Re$z < 0$. In addition, we see that the Gamma function is analytic for all finite values of z except non-positive integers. To investigate the nature of the singularities of the Gamma function near $z = -n$ ($n = 0, 1, 2, \dots$), put

$$z = -n + \delta$$

(Footnote 1 continued)
is clearly convergent in Re$z > 0$ as are all higher derivatives.

in the above formula and expand in $\delta \to 0$:

$$\Gamma(-n+\delta) = \frac{\Gamma(1+\delta)}{\delta(1-\delta)\dots(n-\delta-1)(n-\delta)(-1)^n} = \frac{(-1)^n}{n!\,\delta} + O(\delta^0).$$

We thus discover that points $z = -n$ are simple poles of the Gamma function with residues

$$\mathrm{Res}\,\Gamma(z)|_{z=-n} = \frac{(-1)^n}{n!}.$$

These are the only singularities of the Gamma function for finite z.

Let us consider the integral

$$\frac{1}{2\pi i}\int_{-i\infty}^{i\infty} ds A(s)\Gamma(-s)(-z)^s,$$

where the integration contour is along the imaginary axis but is slightly deformed around the origin so as to leave the $s = 0$ pole on the right. Next assume that the function $A(s)$ is analytic in $\mathrm{Re}\, s > 0$ and is not too divergent as $\mathrm{Re}\, s \to +\infty$. Then for $|z| < 1$ we can shift the integration contour all the way to the right leaving zero result plus the contribution from the string of poles of the Gamma function in the integrand:

$$\frac{1}{2\pi i}\int_{-i\infty}^{i\infty} ds A(s)\Gamma(-s)(-z)^s = \sum_{n=0}^{\infty} \frac{A(n)}{n!} z^n. \qquad (2.5)$$

If the function $A(s)$ is chosen in such a way that the values $A(n)$ coincide with the coefficients of the Taylor expansion of a (different) function $f(z)$, that means $A(n) = f^{(n)}(0)$, then Eq. (2.5) provides a contour integral representation for the function $f(z)$ inside the unit circle.[2]

This provides a useful tool for the analytic continuation of Taylor series outside the unit circle as, depending on the properties of the function $A(s)$, it may happen that integral (2.5) still converges for $|z| > 1$ and can be computed by shifting the contour of integration to the left.

2.1.3 Complement Formula

For our subsequent studies of the hypergeometric series, we shall need another identity satisfied by the Gamma function for all complex z:

---

[2] In general, the Gamma function is not the only function that can be used to recover the Taylor series and other functions with similar pole structure, for example $1/\sin(\pi s)$ could be used as well see Sect. 1.3.6.

$$\Gamma(z)\Gamma(1-z) = \frac{\pi}{\sin(\pi z)}, \tag{2.6}$$

also referred to as *complement* or *reflection formula*. Indeed, from the pole structure of the Gamma function, it is clear that the expression $\sin(\pi z)\,\Gamma(z)\Gamma(1-z)$ has no poles at all on the real axis and is, in fact, analytic for all finite z and tends to π for $z \to 0$. We do not know, however, whether it is bounded and so the Liouville theorem cannot really be used here. We shall therefore prove identity (2.6) from the defining integral representation, Eq. (2.1), which we rewrite as ($t = x^2$)

$$\Gamma(z) = 2\int_0^\infty dx\, x^{2z-1}e^{-x^2}.$$

We shall for the moment restrict ourselves to the real values $z = \lambda$ belonging to the interval $\lambda \in (0, 1)$. One then finds

$$\Gamma(\lambda)\Gamma(1-\lambda) = 4\int_0^\infty\int_0^\infty dxdy x^{2\lambda-1}y^{-2\lambda+1}e^{-x^2-y^2} = 4\int_0^{\pi/2} d\varphi(\cot\varphi)^{2\lambda-1}\int_0^\infty d\rho\rho e^{-\rho^2},$$

where the polar co-ordinates $x = \rho\cos\varphi,\quad y = \rho\sin\varphi$ have been used.[3] The ρ–integral in the above is elementary and equal to $1/2$. To calculate the φ–integral substitute $x = \cot\varphi$:

$$\int_0^{\pi/2} d\varphi(\cot\varphi)^{2\lambda-1} = \int_0^\infty \frac{x^{2\lambda-1}dx}{x^2+1}.$$

This is the same type of branch–cut integral studied in Sect. 1.3.2 and so given by the formula (1.14) with $a = 2\lambda$:

$$\int_0^\infty \frac{x^{2\lambda-1}dx}{x^2+1} = \frac{\pi}{\sin(2\pi\lambda)}\sum_\pm \mathrm{Res}\frac{(-z)^{2\lambda-1}}{z^2+1}\Big|_{z=\pm i} = \frac{\pi}{\sin(2\pi\lambda)}\left[\frac{(-i)^{2\lambda-1}}{2i} + \frac{(i)^{2\lambda-1}}{-2i}\right]$$

$$= \frac{\pi}{2i\sin(2\pi\lambda)}\left(e^{-i\pi\lambda+i\pi/2} - e^{i\pi\lambda-i\pi/2}\right) = \frac{\pi\cos\pi\lambda}{\sin(2\pi\lambda)} = \frac{\pi}{2\sin\pi\lambda}.$$

We thus conclude that the equality

$$\Gamma(\lambda)\Gamma(1-\lambda) = \frac{\pi}{\sin\pi\lambda}$$

[3] This is a generally useful trick in mathematical physics, employed to obtain various product identities and for instance for the evaluation of the Gaussian integral.

holds on the interval $\lambda \in (0, 1)$. From the analytic continuation theorem of Sect. 1.3.6 it follows that Eq. (2.6) is valid for all finite z in the complex plane.

The value

$$\Gamma\left(\frac{1}{2}\right) = \sqrt{\pi} \tag{2.7}$$

immediately follows from (2.6). Setting $z \to \frac{1}{2} + z$, another complement identity follows from (2.6):

$$\Gamma\left(\frac{1}{2} + z\right)\Gamma\left(\frac{1}{2} - z\right) = \frac{\pi}{\cos \pi z}. \tag{2.8}$$

2.1.4 Stirling's Formula

In the following sections we shall need to estimate various integrals involving Gamma functions. Therefore the $|z| \to \infty$ asymptotic form is given here for future reference:

$$\Gamma(z) = \sqrt{2\pi}\, z^{z-1/2} e^{-z} \left[1 + \frac{1}{12z} + O\left(\frac{1}{z^2}\right)\right]. \tag{2.9}$$

This expression is known as the *Stirling's formula* and is valid for $|\arg z| < \pi - \delta$ (δ an infinitesimal). The complete proof of (2.9) is somewhat lengthy (the expansion in brackets [...] is in terms of Bernoulli numbers) and can be looked up in [3]. Equation (2.9) follows from (2.2) and the fact that $\Gamma(1) = 1$ and $\Gamma(n) = (n-1)!$. For positive integers $z = n$ Eq. (2.9) reduces to the Stirling's formula for $n!$ familiar from the elementary analysis.

2.1.5 Euler Integral of the First Kind

Assuming $\mathrm{Re} z > 0$ and $\mathrm{Re} w > 0$, we define the Beta function as

$$B(z, w) = \int_0^1 t^{z-1}(1-t)^{w-1} dt. \tag{2.10}$$

From the symmetry $t \leftrightarrow 1 - t$ it follows that $B(z, w) = B(w, z)$.

Next we prove that the following relation to Gamma functions holds:

$$B(z, w) = \frac{\Gamma(z)\Gamma(w)}{\Gamma(z + w)}. \tag{2.11}$$

Proceed by reducing the double integral as in Sect. 2.1.3

$$\Gamma(z)\Gamma(w) = 4\int_0^\infty\int_0^\infty dxdy\, x^{2z-1}y^{2w-1}e^{-x^2-y^2} = 4\int_0^\infty d\rho\,\rho^{2(z+w)-1}e^{-\rho^2}$$
$$\times \int_0^{\pi/2} d\varphi\,(\cos\varphi)^{2z-1}(\sin\varphi)^{2w-1}.$$

The ρ–integral is now equal to $\Gamma(z+w)/2$. In the φ–integral substitute $\sin^2\varphi = t$:

$$(\cos\varphi)^{2z-1}(\sin\varphi)^{2w-1}d\varphi = \frac{1}{2}(1-t)^{z-1}t^{w-1}dt,$$

and thus obtain the definition (2.10). Hence

$$\Gamma(z)\Gamma(w) = \Gamma(z+w)B(w,z) = \Gamma(z+w)B(z,w)$$

indeed holds.

2.1.6 Duplication Formula

One practical usage of the Euler integral is for the derivation of the duplication formula for the Gamma function. To see this set $z = w$ in the Euler integral

$$B(z,z) = \int_0^1 t^{z-1}(1-t)^{z-1}dt = \int_0^1 [t(1-t)]^{z-1}dt.$$

The integrand in the above is evidently symmetric about the point $t = 1/2$ (i.e. under the substitution $t \leftrightarrow 1-t$) and therefore

$$B(z,z) = 2\int_0^{1/2} [t(1-t)]^{z-1}dt.$$

Now substitute $t = (1-\sqrt{\tau})/2$. We have $dt = -d\tau/(4\sqrt{\tau})$, $t(1-t) = (1-\sqrt{\tau})(1+\sqrt{\tau})/4$ and therefore

$$B(z,z) = \frac{2}{4^z}\int_0^1 \tau^{-1/2}(1-\tau)^{z-1}d\tau = 2^{1-2z}B\left(\frac{1}{2},z\right).$$

This expression is known as *Legendre duplication formula*. Recalling that $\Gamma(1/2) = \sqrt{\pi}$ [see (2.7)] as well as using (2.11), we deduce the identity for the Gamma function

$$\Gamma(2z) = \frac{2^{2z-1}}{\sqrt{\pi}} \Gamma(z)\Gamma\left(z + \frac{1}{2}\right), \tag{2.12}$$

which is also known as the *duplication formula*. While the Euler integral only converges for $\text{Re}z > 0$, formula (2.12) is actually valid for all complex z in the spirit of the analytic continuation.

2.1.7 Infinite Product Representation

It is not difficult to show that for any b and c the following remarkable statement holds (see for example Sect. 1.1 of Ref. [7] for the proof),

$$\lim_{u\to\infty} \frac{u^b\,\Gamma(u+c)}{u^c\,\Gamma(u+b)} = 1. \tag{2.13}$$

One way to verify it is to use the Stirling formula (2.9). Let us now set $b = z + 1$ and $c = 1$ and restrict u to integers. Then

$$\lim_{n\to\infty} \frac{n^z\,\Gamma(n+1)}{\Gamma(z+n+1)} = 1.$$

Using (2.3) and the fact that $\Gamma(1) = 1$ we obtain

$$\lim_{n\to\infty} \frac{n^z\,n!}{z(z+1)\ldots(z+n)\,\Gamma(z)} = 1,$$

which immediately leads to the infinite product representation:

$$\Gamma(z) = \lim_{n\to\infty} \frac{n^z\,n!}{z\,(z+1)\cdots(z+n)}. \tag{2.14}$$

Obviously, this representation is valid for all complex numbers z, except the poles $z = 0, -1, -2, \ldots$. With its help some of the important relations can be recovered in a very convenient way. For example for the recurrence relation we have

$$\begin{aligned}
\Gamma(z+1) &= \lim_{n\to\infty} \frac{n^{z+1}\,n!}{(z+1)\cdots(z+n+1)} \\
&= \lim_{n\to\infty} \frac{nz}{z+n+1} \lim_{n\to\infty} \frac{n^z\,n!}{z(z+1)\cdots(z+n)} = z\Gamma(z).
\end{aligned}$$

There is also another way to show the compatibility of the representations (2.14) and (2.1). For this purpose consider the integral

$$G_n(z) = \int_0^n \left(1 - \frac{t}{n}\right)^n t^{z-1} dt,$$

for Re $z > 0$. As is known from the elementary analysis, the factor $(1 - t/n)^n$ in the integrand limits to the function e^{-t} at $n \to \infty$. One may expect therefore that at infinite growth of n the function $G_n(z)$ limits to the Gamma function (2.1). To prove it, put $t = n(1 - e^{-v})$ so that the above integral rewrites as

$$G_n(z) = n^z \int_0^\infty v^{z-1} e^{-nv} f(v) dv, \quad f(v) = \left(\frac{1 - e^{-v}}{v}\right)^{z-1} e^{-v},$$

where $f(v)$ is a continuous function of v and $f(0) = 1$. Hence as $n \to \infty$

$$G_n(z) = \Gamma(z)\Big[1 + O(n^{-1})\Big].$$

On the other hand we can verify that the function $G_n(z)$ satisfies (2.14). To do that we integrate by parts taking into account that n are positive integers. After the substitution $\tau = t/n$ one then finds

$$\begin{aligned} G_n(z) &= n^z \int_0^1 (1-\tau)^n \tau^{z-1} d\tau = n^z \Big[\underbrace{\frac{1}{z}\tau^z (1-\tau)^n \Big|_0^1}_{=0} \\ &+ \frac{n}{z}\int_0^1 (1-\tau)^{n-1}\tau^z d\tau\Big] = n^z \frac{n(n-1)}{z(z+1)} \int_0^1 (1-\tau)^{n-2}\tau^{z+1} d\tau = \dots \\ &= \frac{1 \cdot 2 \dots n}{z(z+1)\dots(z+n)}\, n^z. \end{aligned}$$

Taking the limit $n \to \infty$ we then recover (2.14).

Representation (2.14) is often used to establish new properties of $\Gamma(z)$. In particular, after some algebra[4] one obtains the Gamma function in the form of the Weierstrass infinite product

$$\Gamma(z) = \frac{e^{-\gamma z}}{z} \prod_{n=1}^{\infty} \Big[\left(1 + \frac{z}{n}\right)^{-1} e^{z/n}\Big]. \tag{2.15}$$

[4] The rhs of (2.14) rewrites as

Here γ is the Euler constant, which is defined by

$$\gamma = \lim_{n\to\infty}\left[\sum_{k=1}^{n}\frac{1}{k} - \ln(n)\right] = 0.5772157\ldots$$

Clearly, from the Weierstrass formula (2.15) it follows that only negative integers or zero are the poles of $\Gamma(z)$.

It could also be shown that the complement formula (2.6) follows directly from (2.14):

$$\frac{1}{\Gamma(z)\Gamma(-z)} = -\frac{z}{\Gamma(z)\Gamma(1-z)} = -z^2\prod_{k=1}^{\infty}\Big(1-\frac{z^2}{k^2}\Big) = -\frac{z\sin\pi z}{\pi}.$$

The same applies to the duplication formula (2.12).

Furthermore, the infinite product (2.14) allows to prove the following general formula (known as the Gauss multiplication formula):

$$\Gamma(z)\Gamma\Big(z+\frac{1}{m}\Big)\ldots\Gamma\Big(z+\frac{m-1}{m}\Big) = (2\pi)^{\frac{1}{2}(m-1)}m^{\frac{1}{2}-mz}\Gamma(mz), \qquad (2.16)$$

where $m = 2, 3\ldots$. To prove this multiplication rule, notice that $\Gamma(mz)$, by a slight modification of definition (2.14), is given by

$$\Gamma(mz) = \lim_{n\to\infty}\frac{(mn)!(mn)^{mz}}{\prod_{k=0}^{mn}(mz+k)}. \qquad (2.17)$$

Similarly, for a product of the Gamma functions in the lhs of (2.16), by using the same identity (2.14), we have

$$\left[\prod_{k=0}^{m-1}\Gamma\Big(z+\frac{k}{m}\Big)\right]^{-1} = \lim_{n\to\infty}\frac{1}{(n!)^m n^{mz+(m-1)/2}m^{mn+m}}\prod_{k=0}^{nm+m-1}(mz+k).$$

(Footnote 4 continued)

$$\lim_{n\to\infty}\frac{n!n^z}{z(z+1)\ldots(z+n)} = \lim_{n\to\infty}\frac{e^{z\ln n}}{z}\prod_{k=1}^{n}(1+\frac{z}{k})^{-1}$$

$$= \frac{e^{\overbrace{-\lim_{n\to\infty}(1+\frac{1}{2}+\ldots+\frac{1}{n}-\ln n)z}^{-\gamma z}}}{z}\prod_{k=1}^{\infty}(1+\frac{z}{k})^{-1}e^{z/k}.$$

Note that although it seems that here the Euler's constant γ is introduced artificially, we will see shortly that it is deeply related to the Gamma function.

Combining this expression with (2.17) we obtain,

$$\frac{(m)^{-mz}\Gamma(mz)}{\prod_{k=0}^{m-1}\Gamma(z+\frac{k}{m})} = \lim_{n\to\infty}\frac{(mn)!\,n^{(m-1)/2}}{(n!)^m\,m^{mn+m}}\prod_{k=nm+1}^{nm+m-1}[(mz+k)/n].$$

Now we see that in fact the rhs of the above relation is independent of z and is equal to[5]

$$m^{-\frac{1}{2}}(2\pi)^{-\frac{1}{2}(m-1)}.$$

Hence finally we have

$$\frac{\Gamma(mz)}{\prod_{k=0}^{m-1}\Gamma(z+\frac{k}{m})} = (2\pi)^{-\frac{1}{2}(m-1)}m^{-\frac{1}{2}+mz},$$

as required.

Another interesting identity is derived from the multiplication formula (2.16) by setting $z = 1/m$,

$$\Gamma\left(\frac{1}{m}\right)\Gamma\left(\frac{2}{m}\right)\ldots\Gamma\left(\frac{m-1}{m}\right) = \frac{(2\pi)^{\frac{1}{2}(m-1)}}{\sqrt{m}},\quad m = 2, 3, \ldots.$$

2.2 Mellin Integral Transform

Let us get back to the definition (2.1) and rescale the integration variable by a parameter p, $t \to pt$, then we obtain

$$\int_0^\infty dt\, t^{z-1}\, e^{-pt} = \Gamma(z)/p^z. \tag{2.19}$$

---

[5] Indeed, at infinite growth of n the involved product reduces to

$$\lim_{n\to\infty}\prod_{k=nm+1}^{nm+m-1}[(mz+k)/n] = m^{m-1}$$

and the pre–factor, by using the Stirling's formula for the factorial

$$n! \sim \sqrt{2\pi n}\,(n/e)^n, \tag{2.18}$$

is given by

$$\lim_{n\to\infty}\frac{(mn)!n^{(m-1)/2}}{(n!)^m m^{mn+m}} = (2\pi)^{-\frac{1}{2}(m-1)}m^{\frac{1}{2}-m}.$$

In this result we immediately recognize the Laplace transform of the power function $f(t) = t^{z-1}$. One can define another integral transform assuming t^{z-1} to be its kernel instead of the exponential e^{-pt}. It is referred to as *Mellin transform* and is formally defined as

$$F_M(s) = \int_0^\infty dx\, f(x)\, x^{s-1}, \tag{2.20}$$

as long as the integral is convergent. In general, the integral does exist for complex s in some strip depending on the function $f(x)$.

Let $f(x)$ be continuous on the positive real semiaxis. Similar to the exponential bound indices we can introduce power law bound indices α_0 and α_∞, defined as

$$\lim_{x\to 0} |f(x)| \le C_0\, x^{\alpha_0} \quad \text{and} \quad \lim_{x\to\infty} |f(x)| \le C_\infty\, x^{\alpha_\infty},$$

where $C_{0,\infty}$ are some positive constants. Then we can perform the following estimation

$$\begin{aligned} \left| \int_0^\infty dx\, f(x) x^{s-1} \right| &\le \int_0^1 dx\, |f(x)| x^{\mathrm{Re}(s)-1} + \int_1^\infty dx\, |f(x)| x^{\mathrm{Re}(s)-1} \\ &\le C_0 \int_0^1 dx\, x^{\mathrm{Re}(s)+\alpha_0-1} + C_\infty \int_1^\infty dx\, x^{\mathrm{Re}(s)+\alpha_\infty-1}. \end{aligned}$$

The last two integrals exist for $\mathrm{Re}(s) > -\alpha_0$ and $\mathrm{Re}(s) < -\alpha_\infty$, respectively. Thus the Mellin transform of $f(x)$ exists for any complex s in the *fundamental strip* $-\alpha_0 < \mathrm{Re}(s) < -\alpha_\infty$. For any polynomial $\alpha_\infty > \alpha_0$ and therefore the Mellin transform does not exist. For the exponential function the fundamental strip is $(0, \infty)$.

The Mellin transform is closely related to the Gamma function. This is obvious as Mellin transform of the exponential function e^{-t} $(t > 0)$ is the Gamma function $\Gamma(s)$. Another examples are:

- for the function $f(t) = e^{-1/t}/t$ we obtain

$$F_M(s) = \int_0^\infty dt\, \frac{e^{-1/t}}{t}\, t^{s-1} \underbrace{\Rightarrow}_{1/t=x} \int_0^\infty dx x^{-s} e^{-x} = \Gamma(1-s),$$

- for the function $f(t) = (1+t)^{-a}$, setting $t = x(1-x)^{-1}$ we get $(1+t)^{-1} = 1-x, dt = (1-x)^{-2} dx$ and thus:

$$\begin{aligned} F_M(s) &= \int_0^\infty t^{s-1}(1+t)^{-a} dt = \int_0^1 x^{s-1}(1-x)^{a-s-1} dx = B(s, a-s) \\ &= \frac{\Gamma(s)\Gamma(a-s)}{\Gamma(a)}, \end{aligned}$$

where in the last step we have used Eq. (2.10) for the Beta-function.

Mellin transforms have interesting properties, a number of which we shall use in the subsequent chapters. Probably the most important of them is the transform of a function with a scaled argument, $f(x) = g(ax)$, then

$$F_M(s) = \int_0^\infty dx\, g(ax)\, x^{s-1} = a^{-s} \int_0^\infty dx\, g(x)\, x^{s-1} = a^{-s} G_M(s). \tag{2.21}$$

It is particularly useful in solving functional equations of the type

$$f(x) = g(x) + f(ax), \tag{2.22}$$

which, after applying the Mellin transform take a form of a simple algebraic relationship

$$F_M(s) = G_M(s)/(1 - a^{-s}).$$

This property can be very useful when deriving interesting functional relations. The inversion formula for the Mellin transform is given by

$$f(x) = \frac{1}{2\pi i} \int\limits_{a-i\infty}^{a+i\infty} F_M(s) x^{-s} ds, \tag{2.23}$$

where the integration is along a vertical line through $\mathrm{Re} s = a$. To get this formula, we use definition (2.20) and introduce the new variables by the substitution $x = e^{-\xi}$ and $s = a + 2\pi i \beta$. The Mellin transform writes then as

$$F_M(s) = \int\limits_{-\infty}^{\infty} d\xi\, f(e^{-\xi}) e^{-a\xi} e^{-2\pi i \beta \xi}.$$

Hence for a given value $\mathrm{Re} s = a$ belonging to the fundamental strip, the Mellin transform of a function is expressed as a Fourier transform. Using Fourier's inversion theorem the original is:

$$f(e^{-\xi}) e^{-a\xi} = \int\limits_{-\infty}^{\infty} F_M(s) e^{2\pi i \beta \xi} d\beta.$$

Going now back to variables x and s, we obtain the result (2.23):

$$f(x) = x^{-a} \int\limits_{-\infty}^{\infty} F_M(s) x^{-2\pi i \beta} d\beta = \frac{1}{2\pi i} \int\limits_{a-i\infty}^{a+i\infty} F_M(s) x^{-s} ds.$$

Example 2.1
Let us consider the Jacobi's *elliptic* ϑ_3*-function*, which is defined as

$$\vartheta_3(0, e^{-\pi x}) = 1 + 2\sum_{n=1}^{\infty} e^{-\pi x n^2} = 1 + 2f(x).$$

We would like to show that

$$\sqrt{y}\,\vartheta_3(0, y) = \vartheta_3(0, 1/y) \quad \text{or} \quad 1 + 2f(x) = \frac{1}{\sqrt{x}} + \frac{1}{\sqrt{x}}\, f(1/x). \tag{2.24}$$

The Mellin transform of $f(x)$ can be found to be

$$F_M(s) = \frac{1}{\pi^s}\,\Gamma(s)\,\zeta(2s),$$

where

$$\zeta(s) = \sum_{n=1}^{\infty} \frac{1}{n^s} \tag{2.25}$$

is the Riemann *Zeta function*. The fundamental strip for the function $f(x)$ is $\mathrm{Re}s > 0$, so that the inverse transformation is

$$f(x) = \frac{1}{2\pi i} \int_{1-i\infty}^{1+i\infty} ds\, x^{-s}\, \frac{1}{\pi^s}\,\Gamma(s)\,\zeta(2s),$$

with the integration contour P being a straight line parallel to the imaginary axis, see Fig. 2.1. The integrand $F_M(s)$ has two poles at $s = 0$ and $s = 1/2$, the respective residua at which can be read off the following expansions,

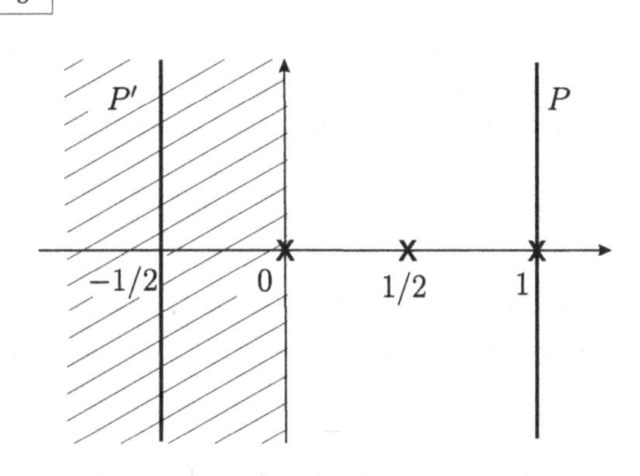

Fig. 2.1 Contour shift used in Example 2.1

$$F(s)\Big|_{s\to 0} = -\frac{1}{2s} + \dots, \quad F(s)\Big|_{s\to 1/2} = -\frac{1}{2(s-1/2)} + \dots$$

By shifting the integration contour to P' (which, in fact, lies in the 'forbidden' region) we have to take out the poles. This procedure yields

$$f(x) = -\frac{1}{2} + \frac{1}{2\sqrt{x}} + \frac{1}{2\pi i}\int_{-1/2-i\infty}^{-1/2+i\infty} ds\, x^{-s}\, \frac{1}{\pi^s}\,\Gamma(s)\,\zeta(2s).$$

We now perform the substitution $s \to 1/2 - s$ which returns our integration contour to the original P and obtain

$$= -\frac{1}{2} + \frac{1}{2\sqrt{x}} - \frac{1}{2\pi i}\int_{1-i\infty}^{1+i\infty} ds\, x^{-1/2+s}\, \frac{1}{\pi^{1/2-s}}\,\Gamma(1/2-s)\,\zeta(1-2s).$$

In the next step we use the identity[6]

$$\Gamma(1/2-s)\,\zeta(1-2s) = \frac{1}{\pi^{2s-1/2}}\Gamma(s)\,\zeta(2s) \tag{2.26}$$

and arrive at

$$\begin{aligned} f(x) &= -\frac{1}{2} + \frac{1}{2\sqrt{x}} + \frac{1}{2\pi i\,\sqrt{x}}\int_{1-i\infty}^{1+i\infty} ds\, x^{s}\, \frac{1}{\pi^s}\,\Gamma(s)\,\zeta(2s) \\ &= -\frac{1}{2} + \frac{1}{2\sqrt{x}} + \frac{1}{\sqrt{x}} f(1/x), \end{aligned}$$

which proves the relation (2.24).

2.3 Series Expansion

For numerical evaluation of the Gamma function near some point $z \in \mathbb{C}$, a series expansion around it could be useful. Before doing this we introduce Psi or Digamma function related to the logarithmic derivative of $\Gamma(z)$, that is:

$$\Psi(z) = \frac{\Gamma'(z)}{\Gamma(z)} = \frac{d}{dz}\ln\Gamma(z). \tag{2.27}$$

Below we investigate the main properties of the Psi function which, evidently, are closely related to the properties of the Gamma function.

[6] This identity can be derived using the Hurvitz formula given in e.g. Sect. 13.15 of [3].

Taking the logarithm of both sides of the Weierstrass formula (2.15) and then differentiating, we obtain the series representation of the Psi function:

$$\Psi(z) = -\gamma + \sum_{k=1}^{\infty}\Big(\frac{1}{k} - \frac{1}{z+k-1}\Big). \tag{2.28}$$

We observe that the rhs of the above sum is absolutely convergent for $\text{Re}z > 0$ and hence the function $\Psi(z)$, just like the Gamma function, is an analytic function in this domain.[7] The structure of the series expansion (2.28) leads to the following recurrence relation

$$\Psi(z) = \Psi(z+1) - \sum_{k=1}^{\infty}\Big(\frac{1}{z+k-1} - \frac{1}{z+k}\Big) = \Psi(z+1) - \frac{1}{z}, \tag{2.30}$$

which ensures the analytic continuation of the Psi function to the region $\text{Re}z < 0$ from the region $\text{Re}z > 0$. Indeed, applying the recurrence relation repeatedly, one finds

$$\begin{aligned}\Psi(z) &= \Psi(z+1) - \frac{1}{z} = \Psi(z+2) - \frac{1}{z} - \frac{1}{z+1} = \cdots \\ &= \Psi(z+n) - \frac{1}{z} - \frac{1}{z+1} - \cdots - \frac{1}{z+n-1},\end{aligned}$$

the formula that defines $\Psi(z)$ in the region $-n < \text{Re}z < -n+1$. Continuing this process we obtain $\Psi(z)$ for all $\text{Re}z < 0$, except negative integers $z = -n$ $(n = 0, 1, 2, \cdots)$, which are simple poles of the Psi function.

Obviously, zeros of the Psi function are the extrema of the Gamma function. Since the values

$$\Psi(1) = -\gamma \qquad \text{and} \qquad \Psi(2) = 1 - \gamma$$

immediately follow from the series expansion (2.28) and the recurrence relation (2.30), respectively, $\Psi(z)$ has one zero $\text{Re}z = x_0 \in [1, 2]$ on the positive semiaxis. On the negative semiaxis the Psi function has a single zero between each consecutive negative integers, which are the poles of the Gamma function.

For $\text{Re}z > 0$, the Psi function is represented by the integral

---

[7] If we go on differentiating the relation (2.28) several times, we obtain

$$\Psi^{(n)}(z) = \frac{d^n}{dz^n}\Psi(z) = (-1)^{n+1} n! \sum_{k=1}^{\infty} \frac{1}{(k+z-1)^{n+1}}. \tag{2.29}$$

Clearly, the first derivative $\Psi^{(1)} = \sum_{k=1}^{\infty}(k+z-1)^{-2}$ is convergent in $\text{Re}z > 0$ domain as are all higher derivatives.

$$\Psi(z) = \int_0^\infty \left[e^{-\xi} - (1+\xi)^{-z}\right]\frac{d\xi}{\xi}. \tag{2.31}$$

To establish this result, we start with the double integral

$$I(t) = \int_0^\infty \Bigl(\int_1^t e^{-sz} ds\Bigr) dz = \int_0^\infty \frac{e^{-z} - e^{-tz}}{z}\, dz.$$

On the other hand, calculation of the integrals in the reversed order yields

$$I(t) = \int_1^t \Bigl(\int_0^\infty e^{-sz} dz\Bigr) ds = \int_1^t \frac{dt}{t} = \ln t.$$

Hence we conclude that

$$\int_0^\infty \frac{e^{-z} - e^{-tz}}{z}\, dz = \ln t. \tag{2.32}$$

Next we observe that the first derivative of the Gamma function is

$$\Gamma'(z) = \int_0^\infty e^{-t} t^{z-1} \ln t \, dt.$$

Replacing $\ln t$ in this expression by (2.32) we obtain

$$\begin{aligned}\Gamma'(z) &= \int_0^\infty \Bigl(e^{-\xi} \underbrace{\int_0^\infty t^{z-1} e^{-t} dt}_{\Gamma(z)} - \underbrace{\int_0^\infty t^{z-1} e^{-t(1+\xi)} dt}_{(1+\xi)^{-z}\Gamma(z)}\Bigr)\frac{d\xi}{\xi} \\ &= \Gamma(z) \int_0^\infty \left[e^{-\xi} - (1+\xi)^{-z}\right]\frac{d\xi}{\xi},\end{aligned}$$

which immediately leads to (2.31).

By logarithmic differentiation of the complement (2.6) and duplication (2.12) formulae for the Gamma function we find the corresponding identities for the Psi function:

$$\Psi(1-z) = \Psi(z) + \pi \cot \pi z,$$
$$2\Psi(2z) = \Psi(z) + \Psi\left(z + \frac{1}{2}\right) + 2\ln 2.$$

Last formula being evaluated at $z = 1/2$ yields the value

$$\Psi\left(\frac{1}{2}\right) = -\gamma - 2\ln 2.$$

Next we investigate the Psi and Gamma functions near the point $z = 1$. We begin with derivatives of the Psi function given by the relation (2.29). At $z = 1$ we have

$$\Psi^{(n)}(1) = (-1)^{n+1} n! \zeta(n+1),$$

where the Zeta function $\zeta(k)$ [see the definition (2.25)] is convergent for $k > 1$. Then for the Taylor series expansion we obtain

$$\Psi(1+z) = -\gamma + \sum_{k=2}^{\infty} (-1)^k \zeta(k) z^{k-1} \quad \text{for} \quad |z| < 1.$$

Using term by term integration of the both sides of this relation, one obtains the Taylor expansion of the logarithm of the Gamma function in the form

$$\ln\left[\Gamma(1+z)\right] = -\gamma z + \sum_{k=2}^{\infty} \frac{(-1)^k \zeta(k)}{k} z^k, \qquad |z| < 1.$$

From this series expansion and with a help of the recurrence relation (2.2), the behaviour of $\Gamma(z)$ near a the point $z = 0$ is

$$\Gamma(z) = e^{-\gamma z}\left[\frac{1}{z} + \frac{\zeta(2)}{2} z + O(z^2)\right], \qquad \text{as } |z| \ll 1,$$

which is consistent with $\Gamma(z)$ having a simple pole at $z = 0$.

2.3.1 Hankel Contour Integral Representation

Let us consider the following integral

$$I_\gamma(z) = \int_\gamma d\xi \, (-\xi)^{z-1} \, e^{-\xi},$$

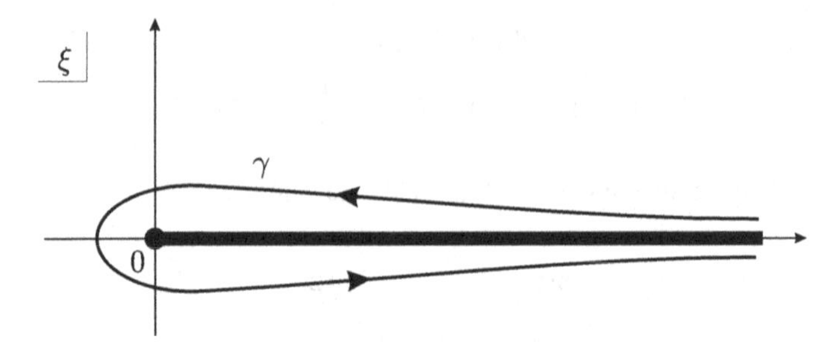

Fig. 2.2 Contour γ used in the Hankel integral representation

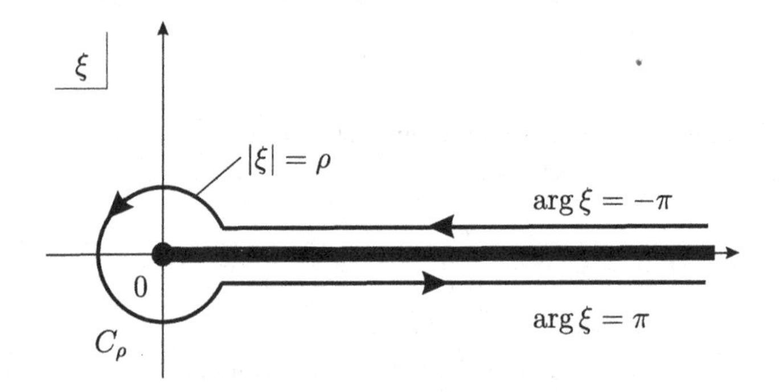

Fig. 2.3 The deformed contour of Fig. 2.2

along the contour γ, which starts at the point $x + i0^+$, $x > 0$ just above the real axis, encircles the coordinate origin counterclockwise and returns to the point $x - i0^+$ just below the real axis, see Fig. 2.2. We fix the branch of the multivalued function $(-\xi)^{z-1}$ by the convention that in $(-\xi)^{z-1} = \exp[(z-1)\ln(-\xi)]$, where the part $\ln(-\xi)$ is purely real on the negative real axis. Thus on γ we have $-\pi \leq \arg(-\xi) \leq \pi$ and the complex plane $\mathbb{C}$ is cut along the positive real semiaxis.

As the next step we deform the contour in such a way that it now consists of two line segments $C_\pm$ along the upper/lower edge of the cut and a circle C_ρ with radius ρ around the coordinate origin as is shown in Fig. 2.3. Then arg $(-\xi) = \mp\pi$, so that $(-t)^{z-1} = e^{\mp i(z-1)}\,\xi^{z-1}$ on the respective contour parts $C_\pm$. We choose the parametrisation on C_ρ to be $(-\xi) = \rho\, e^{i\theta}$, then

$$
\begin{aligned}
I_\gamma(z) &= \int_{C_+} + \int_{C_\rho} + \int_{C_-} = \int_x^\rho d\xi\, e^{-i\pi(z-1)}\,\xi^{z-1}\,e^{-\xi} \\
&\quad + \int_{-\pi}^{\pi} \rho e^{i\theta}\, i d\theta\, (\rho\, e^{i\theta})^{z-1}\, e^{-\rho(\cos\theta + i\sin\theta)} + \int_\rho^x d\xi\, e^{i\pi(z-1)}\,\xi^{z-1}\,e^{-\xi} \\
&= -2i\sin(\pi z)\int_\rho^x d\xi\,\xi^{z-1}\,e^{-\xi} + i\rho^z \int_{-\pi}^{\pi} d\theta\, e^{iz\theta - \rho(\cos\theta + i\sin\theta)}\,.
\end{aligned}
$$

Taking the limit $\rho \to 0$ we infer that for Re$z > 0$

$$I_\gamma(z) = -2i\,\sin(\pi z) \int_0^x d\xi\, \xi^{z-1}\, e^{-\xi}.$$

In the last step we send $x \to \infty$ and obtain

$$\Gamma(z) = -\frac{1}{2i\,\sin(\pi z)} \int_\gamma d\xi\, (-\xi)^{z-1}\, e^{-\xi}, \tag{2.33}$$

which is a well known contour integral representation of the Gamma function due to Hankel. It is important that the contour γ does not pass through the point $\xi = 0$ and therefore the integral in (2.33) is the analytic function of z in the whole complex plane. Hence the Hankel formula (2.33) is valid for all z, except $z = 0, \pm 1, \pm 2, \cdots$, where $\sin(\pi z) = 0$.

2.4 Hypergeometric Series: Basic Properties

2.4.1 Definition and Convergence

The following functional series in the variable z

$$F(a, b; c; z) = \sum_{n=0}^{\infty} \frac{(a)_n (b)_n}{(c)_n} \frac{z^n}{n!} \tag{2.34}$$

are called *hypergeometric series*.[8] Here $(a)_0 = 1$, $(a)_1 = a$, $(a)_2 = a(a+1)$, and generally

$$(a)_n = a(a+1)\cdots(a+n-1) = \frac{\Gamma(a+n)}{\Gamma(a)} \tag{2.35}$$

is the Pochhammer symbol, $(1)_n = n!$.[9] Similar definitions, $(b)_n = \Gamma(b+n)/\Gamma(b)$ and $(c)_n = \Gamma(c+n)/\Gamma(c)$, hold for $(b)_n$ and $(c)_n$ and a, b, and c are, in general, complex parameters. Obviously (2.34) is symmetric with respect to the exchange $a \leftrightarrow b$.

For later convenience we define the general term

$$A_n = \frac{(a)_n (b)_n}{(c)_n n!},$$

[8] In the more recent literature it is sometimes called ${}_2F_1(a, b; c; z)$ in order to indicate the number and arrangement of parameters. We believe, however, that this convention unnecessarily complicates notation.

[9] Remember that $\Gamma(z+1) = z\Gamma(z)$.

so that the hypergeometric series is represented as $F(a, b; c; z) = \sum_{n=0}^{\infty} A_n z^n$. From the standard ratio test for the series convergence

$$\lim_{n\to\infty} \left| \frac{A_{n+1} z^{n+1}}{A_n z^n} \right| = \left[\lim_{n\to\infty} \frac{(n+a)(n+b)}{(n+c)(n+1)} \right] |z| = |z|$$

it follows that the hypergeometric series is absolutely convergent inside the unit circle $|z| < 1$, where it defines an analytic function: the hypergeometric function. As with the Gamma function it is possible to analytically continue the hypergeometric function outside the analyticity domain to which the original definition is limited. We shall do so in later sections.

2.4.2 Special Cases

All the elementary functions and many special functions including all orthogonal polynomials can be obtained as special cases of the hypergeometric function. Here we consider some particularly simple and important examples.
(i) When $b = c$ we recognise the elementary series[10]

$$F(a, b; b; z) = \sum_{n=0}^{\infty} \frac{(a)_n}{n!} z^n = (1-z)^{-a}. \tag{2.36}$$

Interestingly, the above expression does not depend on b any more. This expansion naturally converges in the unit circle $|z| < 1$, but can clearly be extended to the whole complex plane. Then it has one singular point at $z = 1$ and for non-integer values of a a branch cut on $z \in [1, \infty)$. We shall see later that these properties still hold in the general case. In particular, for $a = 1$ one obtains the elementary geometric series:

$$F(1, b; b; z) = \sum_{n=0}^{\infty} z^n = \frac{1}{1-z}. \tag{2.37}$$

(ii) For $b = a + 1/2$ and $c = 3/2$ the following identity

[10] Alternatively one can compute the Taylor coefficients

$$\frac{d^n}{dz^n}(1-z)^{-a} = a\frac{d^{n-1}}{dz^{n-1}}(1-z)^{-a-1} = \ldots = a(a+1)\ldots(a+n-1)(1-z)^{-a-n}$$

and find that

$$\left.\frac{d^n}{dz^n}(1-z)^{-a}\right|_{z=0} = (a)_n.$$

$$F\left(a, \frac{1}{2}+a; \frac{3}{2}; z^2\right) = \frac{1}{2z(1-2a)}\left[(1+z)^{1-2a} - (1-z)^{1-2a}\right] \tag{2.38}$$

holds. In particular, for $a = 1$ and then $b = c$ it reduces to the above case, $F\left(1, 3/2; 3/2; z^2\right) = (1-z^2)^{-1}$. Otherwise using the duplication formula (2.12), the generic coefficient of the series is equal to

$$A_n = \frac{\Gamma(a+n)\Gamma\left(a+\frac{1}{2}+n\right)\Gamma\left(\frac{3}{2}\right)}{\Gamma(a)\Gamma\left(a+\frac{1}{2}\right)\Gamma\left(\frac{3}{2}+n\right)n!} = \frac{\Gamma(2a+2n)}{\Gamma(2a)(2n+1)!},$$

leading to

$$F\left(a, \frac{1}{2}+a; \frac{3}{2}; z^2\right) = \frac{1}{z(2a-1)}\sum_{n=0}^{\infty}\frac{\Gamma(2a-1+2n+1)}{\Gamma(2a-1)(2n+1)!}z^{2n+1}.$$

We observe now that only odd powers of z contribute to the above sum over n. Adding and subtracting sums of even powers of z we obtain

$$\begin{aligned} F(\ldots) = {} & \frac{1}{2z(2a-1)}\left[\sum_{n=0}^{\infty}\frac{\Gamma(2a-1+2n+1)}{\Gamma(2a-1)(2n+1)!}z^{2n+1} + \sum_{n=0}^{\infty}\frac{\Gamma(2a-1+2n)}{\Gamma(2a-1)(2n)!}z^{2n}\right] \\ & + \frac{1}{2z(2a-1)}\left[\sum_{n=0}^{\infty}\frac{\Gamma(2a-1+2n+1)}{\Gamma(2a-1)(2n+1)!}z^{2n+1} - \sum_{n=0}^{\infty}\frac{\Gamma(2a-1+2n)}{\Gamma(2a-1)(2n)!}z^{2n}\right], \end{aligned}$$

which, in turn, can be rewritten as

$$F(\ldots) = \frac{1}{2z(2a-1)}\left[\underbrace{\sum_{k=0}^{\infty}\frac{\Gamma(2a-1+k)}{\Gamma(2a-1)k!}z^k}_{(1-z)^{-2a+1}} - \underbrace{\sum_{k=0}^{\infty}\frac{\Gamma(2a-1+k)}{\Gamma(2a-1)k!}(-z)^k}_{(1+z)^{-2a+1}}\right],$$

where in the last step we have used the above identity (2.36). Interestingly, for $a = (1-n)/2$, $z = \sqrt{5}$ one obtains a relation to the famous *Fibonacci numbers*,[11]

[11] Fibonacci sequence is the series of integer numbers F_n defined by the linear recurrence equation $F_n = F_{n-1} + F_{n-2}$. By definition, the first two numbers are 0 and 1 and each next number is found by adding up the two numbers before it: 0, 1, 1, 2, 3, 5, 8, 13, 21, 34, 55, $\cdots$. The closed form for this sequence is given by

$$F_n = \frac{1}{\sqrt{5}}\left[\left(\frac{1+\sqrt{5}}{2}\right)^n - \left(\frac{1-\sqrt{5}}{2}\right)^n\right].$$

$$F_n = \frac{1}{\sqrt{5}}\left[\left(\frac{1+\sqrt{5}}{2}\right)^n - \left(\frac{1-\sqrt{5}}{2}\right)^n\right] = \frac{n}{2^{n-1}}\, F\left(\frac{1-n}{2}, \frac{2-n}{2}; \frac{3}{2}; 5\right). \tag{2.39}$$

Since one of the numbers $(1-n)/2$, $(2-n)/2$ is always a negative integer (or zero) for $n \geq 1$, (2.39) is in fact a finite sum.

(iii) For $a = -b$ and $c = 1/2$ one can show that

$$F\left(a, -a; \frac{1}{2}; -z^2\right) = \frac{1}{2}\left[(\sqrt{1+z^2}+z)^{2a} + (\sqrt{1+z^2}-z)^{2a}\right]. \tag{2.40}$$

In order to prove it we compare the series expansions in variable z^2 for rhs and lhs separately. For the rhs the Taylor expansion leads to the following series

$$1 + \sum_{k=0}^{\infty} 2^{2k+2} \frac{a^2(a^2-1)(a^2-2^2)\dots(a^2-k^2)}{(2k+2)!} z^{2k+2}. \tag{2.41}$$

The lhs is given by

$$\sum_{n=0}^{\infty} (-1)^n (A)_n z^{2n},$$

where $(A)_0 = 1$ and

$$(A)_n = \frac{(a)_n(-a)_n}{n!(\frac{1}{2})_n} = 2^n \frac{a(a+1)\dots(a+n-1)\overbrace{[-a(1-a)\dots(n-1-a)]}^{(-1)^n a(a-1)\dots(a-n+1)}}{n!\underbrace{(2n-1)!!}_{(2^{-n})(2n)!/n!}},$$

which simplifies to

$$1 + \sum_{n=1}^{\infty} 2^{2n} \frac{a^2(a^2-1)(a^2-2^2)\dots[a^2-(n-1)^2]}{(2n)!} z^{2n}.$$

In this last expression we replace the summation index as $n = k+1$ and obtain the Taylor expansion (2.41) of the rhs.

(iv) The identity

$$F\left(a, -a; \frac{1}{2}; \sin^2 z\right) = \cos(2az) \tag{2.42}$$

follows from (iii) after the substitution $z \to i \sin z$.

(v) In the case $a = b = 1$ and $c = 2$, the Pochhammer symbols are: $(c)_n = (1+n)(a)_n$, $(a)_n = (b)_n = n!$, and $A_n = 1/(1+n)$. One obtains then an interesting relation to the logarithm function,

$$F(1,1;2;z) = \sum_{n=0}^{\infty} \frac{z^n}{1+n} = \sum_{n=1}^{\infty} \frac{z^{n-1}}{n} = -\frac{1}{z} \ln(1-z). \tag{2.43}$$

(vi) When $a = b = 1/2$ and $c = 3/2$

$$F\left(\frac{1}{2}, \frac{1}{2}; \frac{3}{2}; z\right) = \frac{\arcsin\sqrt{z}}{\sqrt{z}}. \tag{2.44}$$

In order to verify that we calculate the generic coefficient in the hypergeometric series:

$$A_n = \frac{\Gamma(n+\frac{1}{2})}{\sqrt{\pi}(2n+1)n!}.$$

Using the duplication formula

$$\Gamma\left(n+\frac{1}{2}\right) \Gamma(n) = 2^{1-2n}\sqrt{\pi}\,\Gamma(2n)$$

one obtains

$$A_n = \frac{(2n)!}{2^{2n}(n!)^2} \frac{1}{2n+1},$$

so that for $0 < z < 1$

$$F\left(\frac{1}{2}, \frac{1}{2}; \frac{3}{2}; z\right) = \sum_{n=0}^{\infty} \frac{(2n)!}{2^{2n}(n!)^2} \frac{z^n}{2n+1} = \frac{1}{\sqrt{z}} \sum_{n=0}^{\infty} \frac{(2n)!}{2^{2n}(n!)^2} \frac{(\sqrt{z})^{2n+1}}{2n+1}$$
$$= \frac{\arcsin\sqrt{z}}{\sqrt{z}}. \tag{2.45}$$

In particular,

$$F\left(\frac{1}{2}, \frac{1}{2}; \frac{3}{2}; 1\right) = \frac{\pi}{2}. \tag{2.46}$$

(vii) When $a = 1/2,\ b = 1$, and $c = 3/2$ one obtains:

$$F\left(\frac{1}{2}, 1; \frac{3}{2}; z^2\right) = \frac{1}{2z} \ln\frac{1+z}{1-z}, \tag{2.47}$$

since because of the coefficient $A_n = \frac{1}{2n+1}$ we have

$$F\left(\frac{1}{2}, 1; \frac{3}{2}; z^2\right) = \sum_{n=0}^{\infty} \frac{z^{2n}}{2n+1} = \frac{1}{2z} \underbrace{2\sum_{n=0}^{\infty} \frac{z^{2n+1}}{2n+1}}_{\ln\frac{1+z}{1-z}}.$$

(viii) By an analytic continuation $F(\ldots, z^2) \to F(\ldots, -z^2)$ of (2.47) one can verify that

$$F\left(\frac{1}{2}, 1; \frac{3}{2}; -z^2\right) = \frac{1}{z} \arctan z. \tag{2.48}$$

For other examples we refer to [8].

2.4.3 Pochhammer Integral and the Gauss Summation Theorem

Let us consider the integral

$$I = \int_0^1 t^{b-1}(1-t)^{c-b-1}(1-zt)^{-a}dt,$$

which is convergent for $|z| < 1$, $\mathrm{Re}b > 0$, and $\mathrm{Re}(c-b) > 0$. Inserting the elementary series, Eq. (2.36),

$$(1-zt)^{-a} = \sum_{n=0}^{\infty} \frac{(a)_n}{n!} t^n z^n,$$

one obtains:

$$I = \sum_{n=0}^{\infty} \frac{(a)_n z^n}{n!} \int_0^1 t^{b+n-1}(1-t)^{c-b-1}dt.$$

We observe that the nth term in this sum is nothing but Euler integral (2.10) with $z = b+n$ and $w = c-b$ and so

$$I = \sum_{n=0}^{\infty} \frac{(a)_n z^n}{n!} \frac{\Gamma(b+n)\Gamma(c-b)}{\Gamma(c+n)} = \frac{\Gamma(c-b)\Gamma(b)}{\Gamma(c)} \sum_{n=0}^{\infty} \frac{(a)_n (b)_n z^n}{(c)_n n!}.$$

In this way we obtain the following integral representation for the hypergeometric function due to Pochhammer

$$F(a, b; c; z) = \frac{\Gamma(c)}{\Gamma(b)\Gamma(c-b)} \int_0^1 t^{b-1}(1-t)^{c-b-1}(1-zt)^{-a}dt. \tag{2.49}$$

Using it we can obtain interesting information about the hypergeometric function near the point $z = 1$. As one can see from Eq. (2.36) this point is a singular point of F at least for some values of the parameters a, b, and c. (We shall see later that it nearly always is a singular point, except when $c - a - b$ is a positive integer.) This does not mean that the limit $\lim_{z\to 1} F(a, b; c; z)$ is necessarily infinite.[12] We defer the investigation of the exact nature of the singularity of F at $z = 1$ to the later section and focus here on the conditions for the existence of the limit $\lim_{z\to 1} F(a, b; c; z)$. These can be studied with the help of the above Pochhammer formula. Indeed, substituting $z = 1$ in Eq. (2.49) leads to the integral of the type (2.10)

$$\int_0^1 t^{b-1}(1-t)^{c-a-b-1}dt = \frac{\Gamma(b)\Gamma(c-a-b)}{\Gamma(c-a)},$$

which is convergent for Re$b > 0$ and Re$(c - a - b) > 0$. It follows that under this condition the limit in question exists and is equal to

$$\lim_{z\to 1} F(a, b; c; z) = \frac{\Gamma(c)\Gamma(c-a-b)}{\Gamma(c-a)\Gamma(c-b)}. \tag{2.50}$$

The condition Re$b > 0$ can be removed, because we can just replace Eq. (2.49) by the version with a and b interchanged, $a \leftrightarrow b$, but the other condition Re$(c-a-b) > 0$ will remain essential. This result is also known as the *Gauss summation theorem*.

We also would like to remark that the general symmetry $a \leftrightarrow b$ of the hypergeometric function is, of course, intact in Eq. (2.50).

2.5 Differential Equations

2.5.1 Hypergeometric Equation: Solutions Around $z=0$

The hypergeometric series (2.34) satisfy the second order differential equation

$$z(1-z)F'' + [c - (a+b+1)z]F' - abF = 0 \tag{2.51}$$

known as the hypergeometric equation. Indeed, substituting $F = \sum_{n=0}^{\infty} A_n z^n$ into this equation, equating powers of z and collecting terms results in a two-term recurrence relation:

$$(n+c)(n+1)A_{n+1} - (n+a)(n+b)A_n = 0.$$

This relation is obviously solved by

---

[12] Just think about the following example: $z = 1$ is a singular, branch point of the function $f(z) = z + \sqrt{1-z}$, yet the limit is well defined and equal to $\lim_{z\to 1} f(z) = 1$.

$$A_n = \frac{(a)_n(b)_n}{(c)_n n!} = \frac{\Gamma(a+n)\Gamma(b+n)\Gamma(c)}{\Gamma(a)\Gamma(b)\Gamma(c+n)n!},$$

which is the general term of the hypergeometric series. Conversely, up to an arbitrary multiplicative constant the hypergeometric function provides a solution to the hypergeometric equation regular at $z \to 0$.

According to the elementary analysis, the hypergeometric equation, being an equation of second order, has two linearly independent solutions. So we must find the second solution, which is linearly independent of $F(a, b; c; z)$. To this end we substitute $F \to z^\alpha F$. Using $F' \to z^\alpha F' + \alpha z^{\alpha-1} F$ and $F'' \to z^\alpha F'' + 2\alpha z^{\alpha-1} F' + \alpha(\alpha-1) z^{\alpha-2} F$, one easily finds

$$z(1-z)F'' + [c + 2\alpha - (a + b + 2\alpha + 1)z]F' + \left[\frac{\alpha(\alpha + c - 1)}{z} - \alpha(a + b + \alpha) - ab\right] F = 0.$$

If we choose $\alpha = 1 - c$ then the $1/z$ term vanishes and the above equation takes the same form as the original Eq. (2.51) but with a new set of parameters which we denote as $(\bar{a}, \bar{b}, \bar{c})$.[13] The parameter $\bar{c}$ is readily seen to be equal to $\bar{c} = 2 - c$. The two remaining parameters satisfy the system of equations

$$\bar{a}+\bar{b} = a+b-2c+2, \qquad \bar{a}\bar{b} = (1-c)(a+b-c+1)+ab = (a-c+1)(b-c+1),$$

which is clearly resolved by

$$\bar{a} = a - c + 1, \qquad \bar{b} = b - c + 1.$$

It follows that the second, linearly independent solution of (2.51) is of the form

$$z^{1-c} F(a - c + 1, b - c + 1; 2 - c; z).$$

We see that the second solution is singular in the limit $z \to 0$.

The general solution of the hypergeometric equation, applicable inside the unit circle $|z| < 1$ is then obtained as a linear combination of the two above solutions

$$A_1 F(a, b; c; z) + B_1 z^{1-c} F(a - c + 1, b - c + 1; 2 - c; z) \qquad (2.52)$$

with arbitrary coefficients A_1 and B_1.

[13] This notation should not be confused with complex conjugate.

2.5.2 Solutions Around $z = 1$ and their Relation to Solutions Around $z = 0$

To investigate the hypergeometric equation around the point $z = 1$ we substitute $z \to 1 - z$, which immediately results in

$$z(1-z)F'' - [a+b+1-c-(a+b+1)z]F' - abF = 0.$$

The equation is again of the same form as Eq. (2.51) with the new set of parameters $\bar{a} = a$, $\bar{b} = b$, and $\bar{c} = a + b + 1 - c$. The general solution, convergent inside the circle $|z - 1| < 1$, can therefore be read off formula (2.52):

$$A_2 F(a, b; a+b+1-c; 1-z) + B_2(1-z)^{c-a-b} F(c-a, c-b; c+1-a-b; 1-z), \tag{2.53}$$

where A_2 and B_2 are again arbitrary coefficients, which are, of course, different from those in Eq. (2.52). We see that, like around $z = 0$ there are two linearly independent solutions around $z = 1$, one regular and one singular.

The circles $|z| < 1$ and $|z - 1| < 1$ obviously overlap. In the overlap region, the two solutions (2.52) and (2.53) represent one and the same function. This means that each of the two particular solutions in Eq. (2.52) should be expressible as a linear combination of the two solutions in Eq. (2.53) and vice versa. In particular, one should have

$$\begin{aligned} F(a, b; c; z) = {} & AF(a, b; a+b+1-c; 1-z) \\ & + B(1-z)^{c-a-b} F(c-a, c-b; c+1-a-b; 1-z), \end{aligned}$$

with this time definite coefficients A and B. To begin with let us assume $\mathrm{Re}(c - a - b) > 0$ and set $z = 1$ in the above. Using the Gauss summation theorem, Eq. (2.50), for the lhs one finds

$$A = \frac{\Gamma(c)\Gamma(c-a-b)}{\Gamma(c-a)\Gamma(c-b)}.$$

Next we set $z = 0$

$$1 = AF(a, b; a+b+1-c; 1) + BF(c-a, c-b; c+1-a-b; 1).$$

Using again the Gauss theorem and identity (2.6), we simplify the first term on the rhs to

$$\begin{aligned} & \frac{\Gamma(c)\Gamma(c-a-b)}{\Gamma(c-a)\Gamma(c-b)} \frac{\Gamma(a+b+1-c)\Gamma(1-c)}{\Gamma(a+1-c)\Gamma(b+1-c)} = \frac{\sin[\pi(c-a)]\sin[\pi(c-b)]}{\sin(\pi c)\sin[\pi(c-a-b)]} \\ & = \frac{\cos[\pi(a-b)] - \cos[\pi(2c-a-b)]}{\cos[\pi(a+b)] - \cos[\pi(2c-a-b)]} = 1 + \frac{\cos[\pi(a-b)] - \cos[\pi(a+b)]}{\cos[\pi(a+b)] - \cos[\pi(2c-a-b)]} \end{aligned}$$

$$= 1 - \frac{\sin(\pi a)\sin(\pi b)}{\sin(\pi c)\sin[\pi(c-a-b)]}.$$

It follows that

$$B = \frac{\Gamma(1-a)\Gamma(1-b)}{\Gamma(1-c)\Gamma(c+1-a-b)}\frac{\sin(\pi a)\sin(\pi b)}{\sin(\pi c)\sin[\pi(c-a-b)]} = \frac{\Gamma(c)\Gamma(a+b-c)}{\Gamma(a)\Gamma(b)}.$$

Collecting the results, we obtain the following relation between the regular solution around $z = 0$ and the two solutions around $z = 1$:

$$\begin{aligned} F(a,b;c;z) = &\frac{\Gamma(c)\Gamma(c-a-b)}{\Gamma(c-a)\Gamma(c-b)}F(a,b;a+b+1-c;1-z) \\ &+ \frac{\Gamma(c)\Gamma(a+b-c)}{\Gamma(a)\Gamma(b)}(1-z)^{c-a-b} \\ &\times F(c-a,c-b;c+1-a-b;1-z). \end{aligned} \tag{2.54}$$

On the grounds of the analytic continuation theorem of Sect. 1.3.6, one can now argue that the restriction $\mathrm{Re}(c-a-b) > 0$ can be removed so that the above formula is valid for all a, b, and c and for $\mathrm{Re}(a+b-c) > 0$ reveals the nature of the divergency of the function $F(a,b;c;z)$ as z approaches 1 from inside the unit circle:

$$F(a,b;c;z) = \frac{\Gamma(c)\Gamma(a+b-c)}{\Gamma(a)\Gamma(b)}\frac{1}{(1-z)^{a+b-c}} + O\left[(1-z)^{1+c-a-b}\right].$$

The remaining three formulas relating solutions in Eqs. (2.52) and (2.53) can be obtained in a similar manner or read off Eq. (2.54) by suitable substitutions of the parameters and the variables z and $1-z$.

2.5.3 Barnes' Integral

We have so far found linearly independent solutions to hypergeometric Eq. (2.51) in terms of hypergeometric series convergent inside the circles $|z| < 1$ and $|z-1| < 1$ and obtained the relations between these solutions by using the Gauss summation theorem. To find solutions outside these circles and in particular for $z \to \infty$, we could, in principle, continue along these lines by studying Eq. (2.51) under the substitution $z \to 1/z$.

It is, however, advantageous to employ a different method here based on the integral representation ideas of Sect. 2.1.2. Clearly, if we set in Eq. (2.5)

$$A(s) = \frac{\Gamma(a+s)\Gamma(b+s)\Gamma(c)}{\Gamma(a)\Gamma(b)\Gamma(c+s)}$$

then we recover the correct Taylor coefficients $A(n) = (a)_n(b)_n/(c)_n$. This gives rise to the integral

$$I = \frac{1}{2\pi i} \int_{-i\infty}^{+i\infty} \frac{\Gamma(a+s)\Gamma(b+s)}{\Gamma(c+s)} \Gamma(-s)(-z)^s ds,$$

where the integration contour is mainly along the imaginary axis but is deformed in such a way as to avoid the three strings of poles $s = n$, $s = -a - n$, and $s = -b - n$ $(n = 0, 1, 2, \ldots)$. This integral is known as *Barnes' integral*. The convergence properties of Barnes' integral are such that it is well defined for all z and the integration contour can be shifted all the way to the right for $|z| < 1$ and all the way to the left for $|z| > 1$ (see [3] for details).

So, for $|z| < 1$ the poles $s = n$ contribute and, in accordance with formula (2.5), we have

$$I = \frac{\Gamma(a)\Gamma(b)}{\Gamma(c)} F(a, b; c; z).$$

For $|z| > 1$, the poles at $s = -a - n$ and $s = -b - n$ contribute, so

$$I = \sum_{n=0}^{\infty} \frac{(-1)^n}{n!} \frac{\Gamma(b-a-n)}{\Gamma(c-a-n)} \Gamma(a+n)(-z)^{-a-n} + (a \leftrightarrow b).$$

Now we use identity (2.6) in order to have n and not $-n$ in the arguments of the Gamma functions:

$$I = \sum_{n=0}^{\infty} \frac{(-1)^n}{n!} \frac{\sin[\pi(c-a-n)]}{\sin[\pi(b-a-n)]} \frac{\Gamma(1-c+a+n)\Gamma(a+n)}{\Gamma(1-a+b+n)} (-z)^{-a-n} + (a \leftrightarrow b).$$

Next we re-write this as

$$I = \frac{\sin[\pi(c-a)]}{\sin[\pi(b-a)]} \frac{\Gamma(a)\Gamma(1-c+a)}{\Gamma(1-a+b)} (-z)^{-a} \sum_{n=0}^{\infty} \frac{(a)_n(1-c+a)_n}{(1-a+b)_n n!} \left(\frac{1}{z}\right)^n + (a \leftrightarrow b)$$

and use identity (2.6) again to obtain

$$I = \frac{\Gamma(a)\Gamma(a-b)}{\Gamma(c-a)} (-z)^{-a} F\left(a, 1-c+a; 1-a+b; \frac{1}{z}\right) + (a \leftrightarrow b).$$

The two ways of calculating Barnes' integral thus result in the analytic continuation formula from $|z| < 1$ to $|z| > 1$ for the hypergeometric function:

$$F(a,b;c;z) = \frac{\Gamma(c)\Gamma(b-a)}{\Gamma(b)\Gamma(c-a)}(-z)^{-a} F\left(a, a+1-c; a+1-b; \frac{1}{z}\right)$$
$$+ \frac{\Gamma(c)\Gamma(a-b)}{\Gamma(a)\Gamma(c-b)}(-z)^{-b} F\left(b, b+1-c; b+1-a; \frac{1}{z}\right). \quad (2.55)$$

2.6 Confluent Hypergeometric Series

Given the very appealing form of the hypergeometric series coefficients it might appear surprising that in Sect. 2.4.2 we did not recover the exponential function as a special case. In fact, e^x can be represented by a distinct class of confluent hypergeometric series, which we would like to discuss next.

2.6.1 *Definition and Differential Equation*

Consider the following functional series in the variable z

$$F(a,c;z) = \sum_{n=0}^{\infty} \frac{(a)_n}{n!(c)_n} z^n, \quad (2.56)$$

where coefficients $(a)_n$ and $(c)_n$ are the Pochhammer symbols introduced in (2.35). According to the standard ratio test we obtain

$$\lim_{n\to\infty}\left|\frac{(a)_{n+1}(c)_n}{(a)_n(c)_{n+1}} \frac{z^{n+1}}{z^n} \frac{n!}{(n+1)!}\right| = \left\{\lim_{n\to\infty}\left|\frac{(a+n)}{(c+n)(n+1)}\right|\right\}|z| \underbrace{\rightarrow}_{\text{finite } z} 0.$$

Therefore the above series defines an analytic function for all finite z. This series, called the *confluent hypergeometric function*,[14] is closely connected to the hypergeometric function and is obtained as a limit of $F(a,b;c;z/b)$ at $b \to \infty$:

$$F(a,c;z) = \lim_{b\to\infty} F\left(a, b; c; \frac{z}{b}\right).$$

Substituting $z \to z/b$ in the hypergeometric equation (2.51) and then taking the limit $b \to \infty$, we find that the confluent function is a solution of the following differential equation

$$zF''(a,c;z) + (c-z)F'(a,c;z) - aF(a,c;z) = 0. \quad (2.57)$$

[14] Sometimes it is also referred to as *Kummer series* or *Kummer function* with the common notation $M(a,c,z)$.

Indeed, substitute the series $F = \sum_{n=0}^{\infty} B_n z^n$ as well as its derivatives

$$F' = \sum_{n=1}^{\infty} n B_n z^{n-1} = \sum_{n=0}^{\infty} (n+1) B_{n+1} z^n \quad \text{and} \quad F'' = \sum_{n=1}^{\infty} (n+1) n B_{n+1} z^{n-1}$$

into (2.57). Then equating powers of z, for coefficients B_n we obtain the recurrence relation

$$(c+n)(n+1)B_{n+1} - (a+n)B_n = 0,$$

which is obviously solved by $B_n = (a)_n/n!(c)_n$ as prescribed by (2.56).

Differential Eq. (2.57) for the confluent hypergeometric function has only two singularities: one simple pole at the point $z = 0$ and an irregular singularity at infinity. It differs from the case of the hypergeometric Eq. (2.51) which has an additional pole at the point $z = 1$. For the Eq. (2.57), this pole therefore merges into one of the above enumerated singularities, namely into the irregular one.[15] The term 'confluent' for the function $F(a, c; z)$ displays precisely this circumstance.

In order to find the second solution of Eq. (2.57) we substitute $F \to z^{1-c} F$. The respective derivatives are $F' \to (1-c)z^{-c} F + z^{1-c} F'$ and $F'' \to z^{1-c} F'' + 2(1-c) z^{-c} F' - c(1-c)z^{-c-1} F$, and one easily obtains the equation

$$zF'' + (2-c-z)F' - (a-c+1)F = 0,$$

[15] Both the hypergeometric Eq. (2.51) and its degenerate form (2.57) is, in fact, a special case of the Riemann's differential equation (see, for instance [3]). The latter allows three simple poles to be situated anywhere in the complex plane z including infinity. In a particular case of two poles fixed at $z = 0$ and $z = \infty$ as well as a third one at z_3, respectively, the Riemann's hypergeometric equation has the following form

$$u'' + \Big(\frac{1-\alpha-\alpha'}{z} + \frac{1-\gamma-\gamma'}{z-z_3}\Big)u' + \Big[-\frac{z_3\,\alpha\alpha'}{z^2(z-z_3)} + \frac{\beta\beta'}{z(z-z_3)} + \frac{z_3\,\gamma\gamma'}{z(z-z_3)^2}\Big]u = 0.$$

The parameters $\alpha, \alpha'; \beta, \beta'; \gamma, \gamma'$ (which are referred to as exponents) satisfy the condition

$$\alpha + \alpha' + \beta + \beta' + \gamma + \gamma' - 1 = 0.$$

Setting $z_3 = 1$ and $\alpha = \gamma = 0$ (observe $\alpha, \beta, \gamma \leftrightarrow \alpha', \beta', \gamma'$ symmetry), the Riemann's equation immediately becomes similar to the hypergeometric Eq. (2.51). Considering another set of exponents, namely setting $\beta = 0$, $\beta' = -z_3$, and $\alpha + \alpha' = 1$ and moving the singular point z_3 to infinity $z_3 \to \infty$ the Riemann's equation transforms to the following one

$$u'' + u' + \Big[\frac{\alpha(1-\alpha)}{z^2} + \frac{\gamma}{z}\Big]u = 0,$$

which is known as the *Whittaker's differential equation*. Further substitution $u(z) = F(z)\, e^{-z} z^{\alpha}$ reduces this equation to the confluent form

$$zF'' + (2\alpha - z)F' - (\alpha - \gamma)F = 0.$$

which has the same form as the original Eq. (2.57), but with a new set of parameters, namely, $a \to a - c + 1$ and $c \to 2 - c$. Hence the second, linearly independent solution takes the form

$$z^{1-c} F(a - c + 1, 2 - c; z).$$

This solution is singular in the limit $z \to 0$.

2.6.2 *Integral Representation*

For Re$c >$ Re$a > 0$, the following integral representation,

$$F(a, c; z) = \frac{\Gamma(c)}{\Gamma(a)\Gamma(c-a)} \int_0^1 e^{zt}\, t^{a-1}(1-t)^{c-a-1} dt \tag{2.58}$$

is valid. In order to prove it we use the integral representation of the Beta function (Euler integral of the first kind), Eq. (2.10). The ratio of the Pochhammer symbols entering the definition (2.56) is expressed as follows

$$\begin{aligned} \frac{(a)_n}{(c)_n} &= \frac{\Gamma(c)}{\Gamma(a)} \frac{\Gamma(a+n)}{\Gamma(c+n)} = \frac{\Gamma(c)}{\Gamma(a)\Gamma(c-a)} B(a+n, c-a) \\ &= \frac{\Gamma(b)}{\Gamma(a)\Gamma(c-a)} \int_0^1 t^{a+n-1}(1-t)^{c-a-1} dt, \end{aligned}$$

where we have used (2.11) to write $B(a+n, c-a) = \Gamma(a+n)\Gamma(c-a)/\Gamma(c+n)$. Inserting this result into the series (2.56) and changing the order of the summation and integration, we finally obtain

$$F(a, c; z) = \frac{\Gamma(c)}{\Gamma(a)\Gamma(c-a)} \int_0^1 t^{a-1}(1-t)^{c-a-1} \underbrace{\sum_{n=0}^{\infty} \frac{(zt)^n}{n!}}_{e^{zt}} dt.$$

With the help of Eq. (2.58), we can derive the following functional relation:

$$F(a, c; z) = e^z F(c - a, c; -z). \tag{2.59}$$

To that end we need only to substitute $t \to 1 - t$ in the integrand of (2.58). Then we have

$$F(a, c; z) = \frac{\Gamma(c)}{\Gamma(a)\Gamma(c-a)} \int_0^1 e^{-z(t-1)} t^{c-a-1}(1-t)^{a-1} dt$$
$$= \frac{\Gamma(c)}{\Gamma(a)\Gamma(c-a)} \frac{\Gamma(c-a)\Gamma(a)}{\Gamma(c)} e^z F(c-a, c; -z) = e^z F(c-a, c; -z).$$

The following recurrence relation

$$aF(a+1, c+1; z) = (a-c)F(a, c+1; z) + cF(a, c; z) \tag{2.60}$$

holds as well. Indeed, using the integral representation Eq. (2.58), we obtain

$$aF(a+1, c+1; z) - cF(a, c; z) = \frac{a\,\Gamma(c+1)}{\Gamma(a+1)\Gamma(c-a)} \int_0^1 e^{zt} t^a (1-t)^{c-a-1} dt$$
$$- \frac{c\,\Gamma(c)}{\Gamma(a)\Gamma(c-a)} \int_0^1 e^{zt} t^{a-1}(1-t)^{c-a-1} dt = -\frac{c\,\Gamma(c)}{\Gamma(a)\Gamma(c-a)} \int_0^1 e^{zt} t^{a-1}(1-t)^{c-a} dt$$
$$= - \underbrace{\frac{c\,\Gamma(c)}{\Gamma(a)\Gamma(c-a)} \frac{\Gamma(a)\Gamma(c-a+1)}{\Gamma(c+1)}}_{=(c-a)} F(a, c+1; z) = (a-c)F(a, c+1; z).$$

2.6.3 Special Cases

Just as for the ordinary hypergeometric function, some of the elementary and special functions can be expressed as special cases of the confluent hypergeometric function. (i) In a particular case of $a = b$ we recognize the exponential function

$$F(a, a; z) = \sum_{n=0}^{\infty} \frac{z^n}{n!} = e^z, \tag{2.61}$$

which is obviously independent on a. The above functional relation (2.59) is, evidently, satisfied: $F(a, a; z) = e^z F(0, a; -z) \equiv e^z$.
(ii) The error function, which is widely used in probability and statistics,

$$\text{erf}(x) = \frac{2}{\sqrt{\pi}} \int_0^x e^{-t^2} dt \tag{2.62}$$

is expressed as

$$\text{erf}(x) = \frac{2x}{\sqrt{\pi}}\, F\Big(\frac{1}{2}, \frac{3}{2}; -x^2\Big). \tag{2.63}$$

It immediately follows from the integral representation (2.58):

$$F\Big(\frac{1}{2}, \frac{3}{2}; -x^2\Big) = \underbrace{\frac{\Gamma(3/2)}{\Gamma(1/2)\Gamma(1)}}_{=1/2} \int_0^1 \frac{e^{-x^2 t}}{\sqrt{t}} dt = \frac{1}{x} \int_0^x e^{-t^2} dt.$$

(iii) When $a = 1$ and $b = 2$, the following identity holds:

$$F(1, 2; 2z) = \underbrace{\frac{\Gamma(2)}{\Gamma(1)\Gamma(1)}}_{=1} \int_0^1 e^{2zt} dt = \frac{1}{2z}(e^{2z} - 1) = e^z \frac{\sinh z}{z}. \tag{2.64}$$

(iv) In Example 1.2 on page 11 we have encountered Bessel function of integer order. It can be generalized to non-integer case, for instance the *modified Bessel function* of the first kind can be represented by the following series[16]

---

[16] The Bessel functions are solutions $y(z)$ of the differential equation [3]

$$\frac{d^2 y}{dz^2} + \frac{1}{z}\frac{dy}{dz} + \Big(1 - \frac{\nu^2}{z^2}\Big) y = 0.$$

The Bessel function of the first kind, which is regular around $z = 0$, is defined by its Taylor series expansion:

$$J_\nu(z) = \Big(\frac{z}{2}\Big)^\nu \sum_{k=0}^{\infty} (-1)^k \frac{z^{2k}}{2^{2k} k! \Gamma(\nu + k + 1)}, \quad |\arg z| < \pi.$$

For non-integer ν, the functions $J_\nu(z)$ and $J_{-\nu}(z)$ are linearly independent, so that $J_{-\nu}(z)$ may serve as a second solution, which has a singularity at $z = 0$. Bessel functions of the second kind, denoted as $N_\nu(z)$ and sometimes called the *Neumann functions*, are related to $J_{-\nu}(z)$ by

$$N_\nu(z) = \frac{1}{\sin \nu\pi}\Big[\cos \nu\pi\, J_\nu(z) - J_{-\nu}(z)\Big], \quad |\arg z| < \pi.$$

For integer $\nu = n$, this solution is defined by taking the $\lim_{\nu \to n} N_\nu(z)$. For a special case of a purely imaginary argument, the Bessel equation reduces to the form

$$\frac{d^2 y}{dz^2} + \frac{1}{z}\frac{dy}{dz} - \Big(\frac{\nu^2}{z^2} + 1\Big) y = 0.$$

The respective solutions, the modified Bessel functions of the first and second kind, are defined by

$$I_\nu(z) = i^{-\nu} J_\nu(iz) \;\text{ and }\; K_\nu(z) = \frac{\pi}{2 \sin \nu\pi}[I_{-\nu}(z) - I_\nu(z)], \tag{2.65}$$

respectively. The $K_\nu(z)$ function is known also as the *Macdonald function*. As functions of a real argument at large values of $|z|$, $I_\nu(z) \sim z^{-1/2}\, e^{|z|}$ and $K_\nu(z) \sim z^{-1/2}\, e^{-|z|}$ are exponentially increasing and decreasing functions, respectively. For more details see for instance [9].

$$I_\nu(z) = \left(\frac{z}{2}\right)^\nu \sum_{k=0}^{\infty} \frac{z^{2k}}{2^{2k} k! \Gamma(\nu + k + 1)}. \tag{2.66}$$

It arises in many problems of applied physics and for $\mathrm{Re}(\nu + 1/2) > 0$ can be related to the confluent hypergeometric function via

$$F\left(\frac{1}{2} + \nu, 1 + 2\nu; 2z\right) = \Gamma(\nu + 1) e^z \left(\frac{2}{z}\right)^\nu I_\nu(z). \tag{2.67}$$

To establish this result, we consider the integral

$$\mathcal{I} = \int_{-1}^{1} dt\, e^{zt} (1 - t^2)^{\nu - \frac{1}{2}}.$$

We observe that the integral representation (2.58), by means of a substitution $t \to (t + 1)/2$, can be rewritten as

$$F(a, c; z) = \frac{2^{1-c} e^{\frac{z}{2}}}{B(a, c - a)} \int_{-1}^{1} e^{\frac{z}{2} t} (1 - t)^{c-a-1} (1 + t)^{a-1} dt,$$

whence at $c = 2a = 2\nu + 1$ and $z \to 2z$

$$\mathcal{I} = \frac{\sqrt{\pi}\, \Gamma(\nu + \frac{1}{2})\, e^{-z}}{\Gamma(\nu + 1)} F\left(\frac{1}{2} + \nu, 1 + 2\nu; 2z\right). \tag{2.68}$$

On the other hand, at $\mathrm{Re}(\nu + \frac{1}{2}) > 0$ the integral $\mathcal{I}$ may be expressed as

$$\mathcal{I} = \left(\frac{2}{z}\right)^\nu \sqrt{\pi}\, \Gamma\left(\nu + \frac{1}{2}\right) I_\nu(z). \tag{2.69}$$

This result is easily proven by an expansion of the exponential in the integrand of $\mathcal{I}$ (notice that the integral $\mathcal{I}$ is an even function of z) in powers of z and a subsequent re-arrangement of the integration and the summation order:

$$\mathcal{I} = 2 \int_{0}^{1} (1 - t^2)^{\nu - \frac{1}{2}} \sum_{k=0}^{\infty} \frac{(zt)^{2k}}{(2k)!} dt = \sum_{k=0}^{\infty} \frac{(z)^{2k}}{(2k)!} B\left(k + \frac{1}{2}, \nu + \frac{1}{2}\right)$$

$$= \sum_{k=0}^{\infty} \frac{(z)^{2k}}{(2k)!} \frac{\overbrace{\Gamma\left(k + \frac{1}{2}\right)}^{\sqrt{\pi}(2k)!/2^{2k}k!} \Gamma\left(\nu + \frac{1}{2}\right)}{\Gamma(\nu + k + 1)} = \sqrt{\pi} \Gamma\left(\nu + \frac{1}{2}\right) \underbrace{\sum_{k=0}^{\infty} \frac{z^{2k}}{2^{2k} k! \Gamma\left(k + \nu + 1\right)}}_{I_\nu(z)\, (2/z)^\nu}.$$

Equating the rhs of (2.68) and (2.69), we obtain the identity (2.67). In particular, for $\nu = 1/2$ from (2.67) and (2.64) we obtain a special case for the modified Bessel function

$$I_{\frac{1}{2}}(z) = \sqrt{\frac{z}{2}}\,\frac{e^{-z}}{\underbrace{\Gamma\left(1+\frac{1}{2}\right)}_{\sqrt{\pi}/2}}\,\underbrace{F(1,2;2z)}_{e^{z}\sinh z/z} = \sqrt{\frac{2}{\pi z}}\sinh z. \tag{2.70}$$

Further important special cases will be given in Sect. 4.10.

2.7 Generalized Hypergeometric Series

The hypergeometric series (2.34) can be generalized to the case of p parameters of a and b-type and q parameters of c-type

$${}_pF_q[(a);(b);z] = \sum_{n=0}^{\infty} \frac{(a_1)_n \cdots (a_p)_n}{(b_1)_n \cdots (b_q)_n} \frac{z^n}{n!}, \tag{2.71}$$

where $(a) = a_1, \ldots, a_p$, $(a_i)_n = a_i(a_i+1)\cdots(a_i+n)$ and $(b_i)_n = b_i(b_i+1)\cdots(b_i+n)$ are the Pochhammer symbols. The series (2.71) converges for all finite z if $p \leq q$, for all $|z| < 1$ at $p = q+1$ and diverges for all $z \neq 0$ if $p > q+1$. It can be easily shown that these series satisfy the following differential equation:

$$\begin{aligned}\Delta(\Delta + b_1 - 1)(\Delta + b_2 - 1)\cdots(\Delta + b_q - 1)F \\ = z(\Delta + a_1)(\Delta + a_2)\cdots(\Delta + a_p)F,\end{aligned} \tag{2.72}$$

where

$$\Delta = z\frac{d}{dz}.$$

The most widespread class of generalized hypergeometric functions is the one with $p = q+1$, a typical representative of which is the 'conventional' series $F(\ldots) = {}_2F_1(\ldots)$ considered above. An analytic continuation beyond the singularity at $z = 1$ can be found with the help of the Barnes' integral (2.5). To that end we define in Eq. (2.5)

$$A(s) = \frac{\prod_{i=1}^{p}\Gamma(a_i + s)}{\prod_{j=1}^{q}\Gamma(b_j + s)}.$$

For $|z| < 1$ we then shift the integration contour to ∞[17] and obtain

[17] Or alternatively close the contour by an semi circle to the right of the integration line of the Barnes' integral.

$$I = \frac{\prod_{i=1}^{q} \Gamma(a_j)}{\prod_{j=1}^{p} \Gamma(b_i)} \, {}_pF_q[(a); (b); z].$$

For $|z| > 1$ we move the integration line to $-\infty$ instead. Then we have to take care of the poles at $s = -a_k - n$, so

$$I = \sum_{k=1}^{p} \sum_{n=0}^{\infty} \frac{\prod_{i=1}^{\prime p} \Gamma(a_i - a_k - n)}{\prod_{j=1}^{q} \Gamma(b_j - a_k - n)} \Gamma(a_k + n) \frac{(-1)^n}{n!} (-z)^{-a_k - n},$$

where the primed product does not contain the term $i = k$. Now we would like to use the identity (2.6) in the form

$$\Gamma(c - n) = \pi(-1)^n / [\sin(\pi c)\, \Gamma(1 - c + n)],$$

in order to rewrite the above expression as

$$I = \sum_{k=1}^{p} (-z)^{-a_k} \sum_{n=0}^{\infty} \prod_{i=1}^{\prime p} \left[\frac{\pi(-1)^n}{\sin[\pi(a_i - a_k)]} \frac{1}{\Gamma(1 - a_i + a_k + n)} \right]$$
$$\times \prod_{j=1}^{q} \left[\frac{(-1)^n}{\pi} \sin[\pi(b_j - a_k)] \Gamma(1 - b_j + a_k - n) \right] \frac{\Gamma(a_k + n)}{n!} (1/z)^n.$$

Further we use the fact that $p = q + 1$ and remove n from the Gamma functions by virtue of $\Gamma(c + n) = \Gamma(c)(c)_n$ to arrive at

$$I = \sum_{k=1}^{p} (-z)^{-a_k} \sum_{n=0}^{\infty} \prod_{i=1}^{\prime p} \frac{\Gamma(a_i - a_k)}{(1 - a_i + a_k)_n} \prod_{j=1}^{q} \frac{(1 - b_j + a_k)_n}{\Gamma(b_j - a_k)} \frac{\Gamma(a_k)(a_k)_n}{n!} (1/z)^n$$
$$= \sum_{k=1}^{p} (-z)^{-a_k} \left[\prod_{i=1}^{\prime p} \Gamma(a_i - a_k) \right] \Gamma(a_k) \left[\prod_{j=1}^{q} \frac{1}{\Gamma(b_j - a_k)} \right]$$
$$\times \sum_{n=0}^{\infty} (a_k)_n \left[\prod_{i=1}^{\prime p} \frac{1}{(1 - a_i + a_k)_n} \right] \left[\prod_{j=1}^{q} (1 - b_j + a_k)_n \right] \frac{1}{n!} (1/z)^n$$
$$= \sum_{k=1}^{p} \frac{\Gamma(a_k) \prod_{i=1}^{\prime p} \Gamma(a_i - a_k)}{\prod_{j=1}^{q} \Gamma(b_j - a_k)} (-z)^{a_k}$$
$$\times \, {}_pF_q \left[a_k, 1 - b_1 + a_k, \ldots, 1 - b_q + a_k; 1 + a_1 + a_k, \ldots, 1 - a_p + a_k; 1/z \right].$$

As required the F-function has $1 + q = p$ parameters of the a and b-type and $p - 1 = q$ parameters of the c-type (because $1 = 1 - a_k + a_k$ parameter is absent). To summarize we obtain the following relation for the analytic continuation

$$ {}_pF_q[(a_i);(b_j);z] = \sum_{k=1}^{p}\prod_{i=1}^{'p}\left[\frac{\Gamma(a_i-a_k)}{\Gamma(a_i)}\right]\prod_{j=1}^{q}\left[\frac{\Gamma(b_j)}{\Gamma(b_j-a_k)}\right](-z)^{-a_k} \tag{2.73} $$
$$ \times\ {}_pF_q\left[a_k,1-b_1+a_k,\ldots,1-b_q+a_k;1+a_1+a_k,\ldots,1-a_p+a_k;1/z\right], $$

where recall that $p = q + 1$. In particular, for $q = 1$ and $p = 2$ we immediately recover the result for the conventional hypergeometric series (2.55).

2.8 Examples from Mathematical Physics

2.8.1 Momentum Distribution Function of Interacting Systems

The calculation of particle distribution functions with respect to their quantum numbers is one of the most important tasks in many-body problems. While pure non-interacting bosonic and fermionic systems are subject to Bose and Fermi distribution functions, the situation changes dramatically as soon as the particles start to interact. The momentum distribution function $n(k)$ is usually calculated as a Fourier transform of the equal time Green's function [10]. In one dimension the problem of interacting fermions is under certain conditions analytically solvable and the corresponding Green's function is given by

$$ G(x) = \frac{1}{2\pi}\frac{1}{(-ix+\delta)^{\alpha}}, \tag{2.74} $$

where δ is a positive infinitesimal and α is a dimensionless interaction parameter. It is $\alpha = 1$ in a non-interacting system and then

$$ n(k) = \int_{-\infty}^{\infty} dx\, e^{ikx}\, G(x) = \int_{-\infty}^{\infty}\frac{dx}{2\pi}\frac{e^{ikx}}{-ix+\delta} $$

can be evaluated by closing in the upper/lower half-plane for positive/negative k. The result, known from the theory of non-interacting Fermi gas, is the Heaviside step-function $n(k) = \Theta(-k)$ defined in (1.26).[18]

For generic α one is confronted with a multivalued integrand. We chose to cut the complex plane along $[-i\delta, -i\infty)$ starting at the branch point $-i\delta$. Next we deform the contour in the way shown in Fig. 2.4. On $C_\pm$ we use $(-z)^\alpha = z^\alpha\, e^{\mp i\pi\alpha}$. Furthermore the integral along C_ρ vanishes. Then $(ix = z)$

[18] We have chosen the Fermi edge to be located at $k = 0$.

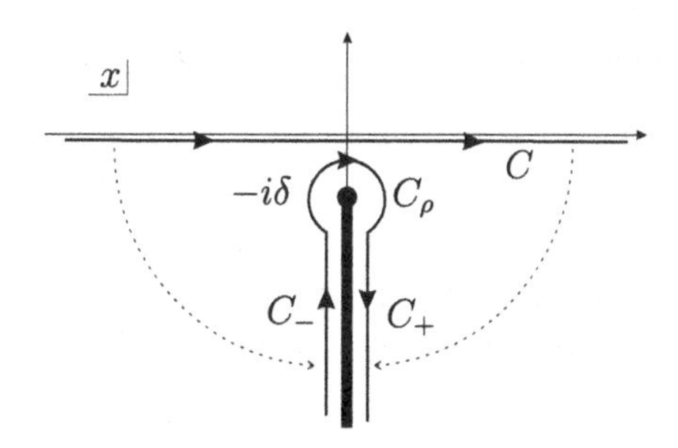

Fig. 2.4 Deformation of the contour for the calculation of $n(k)$

$$
\begin{aligned}
n(k<0) &= \frac{-i}{2\pi}\left[\int_{\infty}^{0} dz\,\frac{e^{-zk}}{(-z)^{\alpha}} + \int_{0}^{\infty} dz\,\frac{e^{-kz}}{(-z)^{\alpha}}\right] \\
&= -\frac{i}{2\pi}\left[-\int_{0}^{\infty} dz\,e^{-kz}\,\frac{e^{-i\pi\alpha}}{z^{\alpha}} + \int_{0}^{\infty} e^{-kz}\,\frac{e^{i\pi\alpha}}{z^{\alpha}}\right] \qquad (2.75)\\
&= \frac{\sin(\pi\alpha)}{\pi}\int_{0}^{\infty} dz\,e^{-kz}\,z^{-\alpha} = \frac{\sin(\pi\alpha)}{\pi}\,|k|^{\alpha-1}\,\Gamma(1-\alpha)\,e^{-|k|\delta}.
\end{aligned}
$$

For positive k the contour can be shifted to infinity along the imaginary positive semiaxis and the integral is identically zero. The limiting case of the non-interacting system is nicely reproduced by the above result as $\sin(\pi\alpha)\Gamma(1-\alpha) = \pi$ for $\alpha \to 1$. The unphysical behaviour of (2.75) for $k \to -\infty$ is due to the fact that the original Green's function (2.74) only captures the long wavelength behaviour of the actual Green's function. Therefore (2.75) is only valid around the Fermi edge.

2.8.2 One–Dimensional Schrödinger Equation

For a genuine one–dimensional quantum mechanical problem, the wave-function must satisfy the equation[19]

$$
-\frac{1}{2}\frac{d^2\psi}{dx^2} + [U(x) - E]\,\psi = 0 \qquad (2.76)
$$

for $x \in (-\infty, \infty)$. The boundary conditions satisfied by $\psi(x)$ are that $\psi(x)$ is bounded for $x \to \pm\infty$. According to the Liouville theorem (see Sect. 1.2.5) such solution cannot be an analytic function everywhere. In fact, for most problems the corresponding Schrödinger equation is an equation with several singularities and thus is solved by some version of hypergeometric function.

There is no explicit solution to this equation for an arbitrary potential $U(x)$. So we shall first discuss general properties of the solutions and then provide several solvable examples illustrating general points. We begin by assuming that $U(x)$ is a

[19] We use units in which $\hbar^2/m = 1$.

continuous function of x and that both limits

$$\lim_{x\to\pm\infty} U(x) = U_{1(2)}$$

exist. Without loss of generality we also set $U_1 \leq U_2$. For large enough x we can approximate $U(x)$ in Eq. (2.76) by its limit U_1 or U_2. The asymptotic form of the solution for large values of the variable therefore is

$$\psi(x) = A_1 e^{ik_1 x} + B_1 e^{-ik_1 x} \quad \text{for} \quad x \to -\infty \tag{2.77}$$

and

$$\psi(x) = A_2 e^{ik_2 x} + B_2 e^{-ik_2 x} \quad \text{for} \quad x \to +\infty,$$

where $k_1 = \sqrt{2(E - U_1)}$, $k_2 = \sqrt{2(E - U_2)}$ and we set $E > U_2$ for now. We note that as the two expressions above are asymptotic forms of one and the same solution $\psi(x)$ to a linear differential equation, the coefficients must obey a linear relation

$$A_2 = \alpha A_1 + \beta B_1,$$

and similarly $B_2 = \alpha' A_1 + \beta' B_1$, where α and β are, in general complex, coefficients (α' and β' will be related to α and β shortly).

Because Eq. (2.76) is real, if $\psi(x)$ is a solution then its complex conjugate $\psi^*(x)$ is also a solution. This fact has two consequences. First, the asymptotic form of $\psi^*(x)$,

$$\psi^*(x) = A_1^* e^{-ik_1 x} + B_1^* e^{ik_1 x} \quad \text{for} \quad x \to -\infty$$

and

$$\psi^*(x) = A_2^* e^{-ik_2 x} + B_2^* e^{ik_2 x} \quad \text{for} \quad x \to +\infty,$$

only differs from that of $\psi(x)$ by re-labeling the constants. So if $A_2 = \alpha A_1 + \beta B_1$ holds for A_2 then $B_2^* = \alpha B_1^* + \beta A_1^*$, or $B_2 = \beta^* A_1 + \alpha^* B_1$, should hold for B_2. We conclude that $\alpha' = \beta^*$ and $\beta' = \alpha^*$. One arrives at the second consequence of $\psi^*(x)$ being a solution by observing that the functions $\psi(x)$ and $\psi^*(x)$ are linearly independent (unless $A_{1(2)} = \pm B_{1(2)}$). We therefore construct the *Wronskian determinant*

$$W[\psi, \psi^*] = \psi \frac{d\psi^*}{dx} - \psi^* \frac{d\psi}{dx}.$$

For the Schrödinger equation it is a constant:

$$\frac{d}{dx} W[\psi, \psi^*] = \psi \frac{d^2\psi^*}{dx^2} - \psi^* \frac{d^2\psi}{dx^2} = 2[E - U(x)]|\psi|^2 - 2[E - U(x)]|\psi|^2 = 0.$$

From the asymptotic form of $\psi(x)$ and $\psi^*(x)$ then follows the identity

$$k_1(|A_1|^2 - |B_1|^2) = k_2(|A_2|^2 - |B_2|^2).$$

The physical interpretation of $W[\psi, \psi^*]$ is that of a probability flow and the above identity reflects the conservation of it. Expressing here A_2 and B_2 via A_1 and B_1, one obtains a constraint:

$$|\alpha|^2 - |\beta|^2 = \frac{k_1}{k_2}.$$

We see that for $E > U_2$ there are two linearly independent bounded solutions to the Schrödinger equation. In quantum mechanics, the solution of the form e^{ikx} $(k > 0)$ is interpreted as a particle wave travelling from left to right and e^{-ikx} — a wave in the opposite direction. Therefore setting $B_2 = 0$ describes a physical situation of an incoming wave (A_1) from the left which is partially reflected (B_1) and partially transmitted (A_2) through the potential $U(x)$. The reflection coefficient R is then defined by the ratio of the reflected to the incoming wave,

$$R = \frac{|B_1|^2}{|A_1|^2} = \frac{|\beta|^2}{|\alpha|^2}, \tag{2.78}$$

while the transmission coefficient T is defined via

$$T = 1 - R = \frac{k_1}{k_2}\frac{1}{|\alpha|^2}.$$

One can show that these coefficients remain the same when the wave is coming from the opposite direction, that accounts for the second linearly independent solution.

As an example we would like to solve Schrödinger equation for a step-like potential of the form

$$U(x) = \frac{U_0}{2}\left[1 + \tanh\left(\frac{x}{2a}\right)\right]. \tag{2.79}$$

With $U_1 = 0$ and $U_2 = U_0$ the above discussion of the asymptotic behaviour is still valid. For the energy $E = \hbar^2 k^2/2m = k^2/2$ we obtain

$$\psi''(x) + \left[k^2 - 2U(x)\right]\psi(x) = 0.$$

In the next step we make the substitution $y = 1/(1+e^{x/a})$ in order to obtain an equation with polynomial coefficients and introduce new parameters $\kappa^2 = (ka)^2 = 2a^2E$, $\lambda^2 = 2a^2U_0$. Then since

$$\frac{d}{dx} = -\frac{y(1-y)}{a}\frac{d}{dy},$$

we obtain

$$y(1-y)\psi'' + (1-2y)\psi' + \left[\frac{\kappa^2}{y(1-y)} - \frac{\lambda^2}{y}\right]\psi = 0.$$

This equation has poles at 0, 1 and ∞ and therefore has hypergeometric solutions. In order to see that explicitly we make the substitution $\psi(y) = y^{\alpha}(1-y)^{\beta} f(y)$, where $\alpha^2 = \lambda^2 - \kappa^2$, and $\beta^2 = -\kappa^2$. This procedure yields

$$\begin{aligned} y(1-y)\, f'' + [(2\alpha+1) - (2\beta+2\alpha+2)y]\ f' \\ - (\alpha+\beta)(\alpha+\beta+1)\, f = 0. \end{aligned}$$

According to (2.51) one particular solution is given by

$$f(y) = C\, F(\alpha+\beta, \alpha+\beta+1; 2\alpha+1; y),$$

where C is a constant. We shall show now, that this solution possesses all required asymptotic properties. For $x \to \infty$ $y \to e^{-x/a} \to 0$ and therefore $f(y) \to C$. For the original wave function we thus obtain $\psi(x) \to C y^{\alpha} \approx C e^{-\alpha x/a}$. Further we are confronted with two different situations:

- $\lambda > \kappa$, then α is a positive real number. In this case the wave function decays exponentially into the step as expected for $E < U_0$.
- $\lambda < \kappa$, then $\alpha = -ik'a$ is purely imaginary. Then $\psi(x) \to C e^{ik'x}$ with $k'^2 = 2(E - U_0)$. So we have a freely propagating plane wave.

On the other hand, for $x \to -\infty$ $y \to 1$ and therefore $1 - y \approx e^{x/a} \to 0$. Here it is more convenient to switch to the analytic continuation prescription (2.54) and rewrite

$$\begin{aligned} &F(\alpha+\beta, \alpha+\beta+1; 2\alpha+1; y) \\ &= \frac{\Gamma(2\alpha+1)\,\Gamma(-2\beta)}{\Gamma(\alpha-\beta)\,\Gamma(\alpha-\beta+1)}\, F(\alpha+\beta, \alpha+\beta+1; 2\alpha+1; 1-y) \\ &\quad + (1-y)^{-2\beta}\, \frac{\Gamma(2\alpha+1)\,\Gamma(2\beta)}{\Gamma(\alpha+\beta)\,\Gamma(\alpha+\beta+1)} F(\alpha-\beta, \alpha-\beta+1; -2\beta+1; 1-y). \end{aligned}$$

Hence, remembering that $1 - y = e^{x/a}$ we get for $x \to -\infty$

$$\psi(x) \to C \left[\frac{\Gamma(2\alpha+1)\,\Gamma(-2\beta)}{\Gamma(\alpha-\beta)\,\Gamma(\alpha-\beta+1)}\, e^{\beta x/a} + \frac{\Gamma(2\alpha+1)\,\Gamma(2\beta)}{\Gamma(\alpha+\beta)\,\Gamma(\alpha+\beta+1)}\, e^{-\beta x/a} \right]$$

Since $\beta = ika$ we indeed obtain for the asymptotic form of the solution a linear superposition of left/right moving plain waves. By comparing it with (2.77) we can then calculate the reflection coefficient as given in (2.78),

$$R = \left| \frac{\Gamma(2\beta)\,\Gamma(\alpha-\beta)\,\Gamma(\alpha-\beta+1)}{\Gamma(-2\beta)\,\Gamma(\alpha+\beta)\,\Gamma(\alpha+\beta+1)} \right|^2 . \tag{2.80}$$

Let us now again consider two different cases:

- $E < U_0$, then $\beta = i\kappa$ is purely imaginary while α is a positive real number. Then the numerator and denominator of (2.80) are complex conjugate numbers and thus $R = 1$ as expected.
- $E > U_0$, then β remains, of course, the same but $\alpha = -i\sigma$ becomes purely imaginary. Then only $|\Gamma(2\beta)/\Gamma(-2\beta)|^2 = 1$ while for the rest using the recurrence relation (2.2) we obtain

$$R = \left|\frac{(\alpha+\beta)\,\Gamma^2(\alpha-\beta+1)}{(\alpha-\beta)\,\Gamma^2(\alpha+\beta+1)}\right|^2 = \left(\frac{\kappa-\sigma}{\kappa+\sigma}\right)^2 \left[\left|\frac{\Gamma(1-i(\kappa+\sigma))}{\Gamma(1+i(\kappa-\sigma))}\right|^2\right]^2.$$

Now we can take advantage of the useful relation

$$|\Gamma(1+ix)|^2 = \frac{\pi x}{\sinh(\pi x)}$$

in order to rewrite the result in the following form

$$R = \left[\frac{\sinh\pi(\kappa-\sigma)}{\sinh\pi(\kappa+\sigma)}\right]^2 = \left[\frac{\sinh\pi(k-k')a}{\sinh\pi(k+k')a}\right]^2,$$

where k and k' are the wave numbers on the left/right side of the barrier.

In the limit $a \to 0$ the scattering potential becomes a sharp step. In this limit no explicit solution in terms of hypergeometric function is required. The corresponding results can of course be recovered from the above equations, see for instance [11].

2.8.3 *Problems which can be Mapped on the Effective 1D Schrödinger Equation*

In some cases the solution of a 1D Schrödinger equation is only possible using a generalisation of the hypergeometric series. For example an equation which arises in the context of a three-body bosonic problem can be written down as [12]

$$\left[\hat{T}(-i\partial_x) - E + e^x\right]\psi(x) = 0, \tag{2.81}$$

where $\hat{T}(-i\partial_x)$ is a differential operator acting as $\hat{T}(-i\partial_x)\psi(x) = \int dx'\,T(x-x')\psi(x')$, where T is the Fourier transform of $\hat{T}$. It can formally be considered as the kinetic energy. Let the Fourier transform of $\hat{L}(x) = \hat{T}(-i\partial_x) - E$ be a rational function,

$$L(k) = P(k)/Q(k),$$

and let $F(k)$ be the image of $\psi(x)$. Then (2.81) is given by

$$P(k)F(k) + Q(k)F(k+i) = 0,$$

and the back transformation yields

$$\left[P(-i\partial_x) + e^x\, Q(-i\partial_x - i)\right]\, \psi(x) = 0.$$

We now want to concentrate on

$$L(k) = \bar{\alpha}_0^2 \frac{k^2 - s_0^2}{k^2 + \alpha_0^2} \frac{k^2 + \beta_1^2}{k^2 + \alpha_1^2}.$$

$L(0) = -E = -\bar{\alpha}_0^2 s_0^2 \beta_1^2/(\alpha_0\alpha_1)^2$ is then the energy eigenvalue. The fact that the above problem is indeed related to the single-particle Schrödingier equation can be demonstrated by the following observation. By expansion of T around $k \to 0$ we see that the leading order term is $\sim k^2$. This is precisely what one obtains for the dispersion relation of a free particle. On the other hand, at large k the dispersion levels off to a constant, which usually happens in the band theory of metals.

With the above assumption in the coordinate representation we obtain

$$\left\{\bar{\alpha}_0^2\,(-\partial_x^2 - s_0^2)(-\partial_x^2 + \beta_1^2) + e^x\,[-(\partial_x + 1)^2 + \alpha_0^2][-(\partial_x + 1)^2 + \alpha_1^2]\right\}\, \psi(x) = 0.$$

In the next step we perform the substitutions $z = e^x/\bar{\alpha}_0^2$ and $\partial_x = z\partial_z \equiv \Delta$ leading to

$$(\Delta + is_0)(\Delta - is_0)(\Delta + \beta_1)(\Delta - \beta_1)\psi(z) = z \prod_{r=0}^{1} (\Delta + 1 + \alpha_r)(\Delta + 1 - \alpha_r)\psi(z).$$

Next we redefine $\psi(z) = z^{i\alpha}\, \phi(z)$, where α is a parameter, then we obtain

$$(\Delta + is_0 + i\alpha)(\Delta - is_0 + i\alpha)(\Delta + i\alpha + \beta_1)(\Delta + i\alpha - \beta_1)\phi(z)$$
$$= z \prod_{r=0}^{1} (\Delta + 1 + \alpha_r + i\alpha)(\Delta + 1 - \alpha_r + i\alpha)\phi(z).$$

Setting α to four different values we obtain four linearly independent solutions of the original equations.

- Take $\alpha = s_0$, then the equation changes to

$$\Delta(\Delta + i2s_0)(\Delta + \beta_1 + is_0)(\Delta - \beta_1 + is_0)\phi(z)$$
$$= z \prod_{r=0}^{1} (\Delta + 1 + \alpha_r + is_0)(\Delta + 1 - \alpha_r + is_0)\phi(z).$$

By comparison with (2.72) we then immediately write down the solution

$$\psi(z) = z^{is_0}\, {}_4F_3\,(1+\alpha_0+is_0, 1-\alpha_0+is_0, 1+\alpha_1+is_0, 1-\alpha_1+is_0; \\ 1+i2s_0, 1+\beta_1+is_0, 1-\beta_1+is_0; -z)$$

- The second independent solution is obtained by the complex conjugation of the above, or by the substitution $s_0 \to -s_0$.
- Next we set $i\alpha = \beta_1$, then the equation is

$$\Delta(\Delta+2\beta_1)(\Delta+\beta_1+is_0)(\Delta+\beta_1-is_0)\phi(z) \\ = z\prod_{r=0}^{1}(\Delta+1+\alpha_r+\beta_1)(\Delta+1-\alpha_r+\beta_1)\phi(z).$$

In this case the solution of the original equation is given by

$$\psi(z) = z^{\beta_1}\, {}_4F_3\,(1+\alpha_0+\beta_1, 1-\alpha_0+\beta_1, 1+\alpha_1+\beta_1, 1-\alpha_1+\beta_1; \\ 1+2\beta_1, 1+\beta_1+is_0, 1+\beta_1-is_0; -z)\,.$$

- The fourth solution is obtained by setting $i\alpha = -\beta_1$. However, the resulting solution is not finite at $z = 0$ and we want to discard it.

The full solution is obtained by a linear combination of the above three possibilities. The coefficients are, as usual, set by the boundary conditions. We would like to remind the reader, that this solution is regular for $|z| < 1$. The analytic continuation for $|z| > 1$ is obtained via the prescription (2.73).

2.9 Problems

Problem 2.1

Upon using identities (2.2) and (2.6) express $|\Gamma(iy)|$ for real y in terms of elementary functions. Compute the asymptotics for $y \to \infty$ and compare it to the Stirling formula. Find the limiting form for $y \to 0$ and explain the result.

Problem 2.2

Euler's formula. Find the value of the product $\prod_{k=1}^{n-1}\Gamma(k/n)$, n being a positive integer. *Hint*: write out the product in the inverse order, multiply and use identity (2.6); to simplify the resulting product of sin's consider the limit $\lim_{z\to 1}(z^n-1)/(z-1)$.

Problem 2.3

Raabe's integral. Calculate the integral

$$I = \int_0^1 dx\, \ln\Gamma(x).$$

Hint: double up the integral, use the transformation $x \leftrightarrow 1-x$ in one of them, reduce the integrand to an elementary function by means of identity (2.6) and integrate by doubling the variable and exploiting the symmetry of the integrand.

Problem 2.4
For $\lambda > 0, 0 < x < 1$, and $-\pi/2 < \alpha < \pi/2$ evaluate the following integrals:

$$(\mathbf{a}): \ I_1(x,\alpha) = \int_0^\infty t^{x-1} e^{-\lambda t \cos\alpha} \sin(\lambda t \sin\alpha)\, dt\,,$$

$$(\mathbf{b}): \ I_2(x,\alpha) = \int_0^\infty t^{x-1} e^{-\lambda t \cos\alpha} \cos(\lambda t \sin\alpha)\, dt\,,$$

$$(\mathbf{c}): \ I_3(x,\alpha) = \int_0^\infty \frac{\sin(\alpha t)}{t^x}\, dt\,, \qquad (\mathbf{d}): \ I_4(x,\alpha) = \int_0^\infty \frac{\cos(\alpha t)}{t^x}\, dt$$

in terms of the Gamma function $\Gamma(x)$.

Problem 2.5
Evaluate the integral

$$I_{\mu\nu} = \int_0^\infty (\sinh x)^\mu\, (\cosh x)^\nu\, dx$$

in terms of the Beta function.

Problem 2.6
Show that the ratio

$$\frac{F(a+1,b;c;z) - F(a,b;c;z)}{F(a+1,b+1;c+1;z)}$$

is an elementary function of z and find this function.

Problem 2.7
Verify the following transformation formulae for the hypergeometric functions:

$$(\mathbf{a}): \quad F(a,b;c;z) = (1-z)^{-a} F\left(a, c-b; c; \frac{z}{z-1}\right)$$
$$= (1-z)^{-b} F\left(b, c-a; c; \frac{z}{z-1}\right),$$
$$(\mathbf{b}): \quad F(a,b;c;z) = (1-z)^{c-a-b} F(c-a, c-b; c; z),$$
$$(\mathbf{c}): \quad F\left(-\nu, \nu+1; 1; \frac{1-z}{2}\right) = \frac{\sqrt{\pi}}{\Gamma(\frac{1-\nu}{2})\Gamma(\frac{2+\nu}{2})} F\left(-\frac{\nu}{2}, \frac{\nu+1}{2}; \frac{1}{2}; z^2\right)$$
$$+ \frac{\sqrt{\pi}\nu z}{\Gamma(\frac{1+\nu}{2})\Gamma(\frac{2-\nu}{2})} F\left(-\frac{\nu-1}{2}, \frac{\nu+2}{2}; \frac{3}{2}; z^2\right).$$

Problem 2.8
Express the following hypergeometric functions

$$(\mathbf{a}): \quad F\left(a, b; \frac{a+b+1}{2}; \frac{1}{2}\right),$$
$$(\mathbf{b}): \quad F(a, b; a-b+1; -1)$$

in terms of the Gamma functions.

Problem 2.9
The six functions $F(a \pm 1, b; c; z)$, $F(a, b \pm 1; c; z)$, $F(a, b; c \pm 1; z)$ are said to be contiguous to the hypergeometric function $F(a, b; c; z)$. There exists a linear relationship between this function and any two contiguous functions with coefficients which are linear polynomials in z or constants. As an example, verify the following two recurrence relations:

$$(\mathbf{a}): \quad cF(a,b;c;z) - (c-a)F(a,b;c+1;z) - aF(a+1,b;c+1;z) = 0,$$
$$(\mathbf{b}): \quad cF(a,b-1;c;z) - cF(a-1,b;c;z) + (a-b)zF(a,b;c+1;z) = 0.$$

Problem 2.10
Prove the quadratic transformations:

$$(\mathbf{a}): \quad F(2a, 2a+1-c; c; z) = (1+z)^{-2a} F\left(a, a+\tfrac{1}{2}; c; \tfrac{4z}{(1+z)^2}\right),$$
$$(\mathbf{b}): \quad F\left(a, a-b+\tfrac{1}{2}; b+\tfrac{1}{2}; z^2\right) = (1+z)^{-2a} F\left(a, b; 2b; \tfrac{4z}{(1+z)^2}\right).$$

Problem 2.11
Determine the Mellin transform and the fundamental strip for the following functions:

$$
\begin{aligned}
(\mathbf{a}):&\quad f(t) = (1+t)^{-a},\\
(\mathbf{b}):&\quad f(t) = \ln(1+t),\\
(\mathbf{c}):&\quad f(t) = (1-t)^{-1}.
\end{aligned}
$$

Problem 2.12
Ramanujan considered a function defined by the following series:

$$
f(x) = \sum_{n=1}^{\infty} \frac{e^{-nx}}{1+e^{-2nx}}.
$$

Using the method presented in Sect. 2.2 derive the following identity:

$$
f(x) = \frac{\pi}{4x} - \frac{1}{4} + \frac{\pi}{x} f(\pi^2/x).
$$

Problem 2.13
Use the definitions of the modified Bessel function and of the Macdonald function given by (2.65) on page 96.

(a): Show that for $\mathrm{Re}(\nu + 1/2) > 0$ the function $I_\nu(z)$ has the following integral representation:

$$
I_\nu(z) = \frac{z^\nu}{2^\nu \Gamma(\nu+\frac{1}{2})\Gamma(\frac{1}{2})} \int_0^\pi \cosh(z\cos\varphi)\sin^{2\nu}\varphi \, d\varphi. \tag{2.82}
$$

(b): Express

$$
I_{\frac{1}{2}}(z), \quad I_{-\frac{1}{2}}(z), \quad K_{\pm\frac{1}{2}}(z)
$$

in terms of elementary functions.

Problem 2.14
Express the following integral

$$
\mathcal{I}_{i\alpha}(\xi) = \int_1^\infty e^{-\xi t} P_{-\frac{1}{2}+i\alpha}(t)\, dt
$$

in terms of the modified Bessel function for $\mathrm{Re}\xi > 0$ and $P_\nu(t) \equiv F(\nu+1, -\nu; 1; \frac{1-t}{2})$ being the special case of the hypergeometric series.[20]

Problem 2.15
The Macdonald function has the following integral representation:

[20] This notation is not casual, see Sect. 4.10 later, and especially Eq. (4.55).

$$K_\nu(z) = \frac{\sqrt{\pi}}{\Gamma(\nu+\frac{1}{2})}\left(\frac{z}{2}\right)^\nu \int\limits_1^\infty e^{-zt}(t^2-1)^{\nu-1/2}dt.$$

Using it calculate the integral

$$g(\nu) = \int\limits_0^\infty e^{-at} K_\nu(\beta t)\, t^{\mu-1} dt$$

under the assumptions $\mathrm{Re}(\mu \pm \nu) > 0$ and $\mathrm{Re}\, a > 0$.

Answers:
Problem 2.1: $|\Gamma(iy)|^2 = \pi/y \sinh \pi y$; $|\Gamma(iy)| = \sqrt{2\pi/y} e^{-\pi y/2}[1 + O(y)]$ as $y \to \pm\infty$; $|\Gamma(iy)| = 1/y + O(y^0)$ as $y \to 0^+$.
Problem 2.2: $(2\pi)^{(n-1)/2}/\sqrt{n}$.
Problem 2.3: $\ln\sqrt{2\pi}$.
Problem 2.4 (a): $\Gamma(x)\sin(\alpha x)/\lambda^x$; **(b):** $\Gamma(x)\cos(\alpha x)/\lambda^x$; **(c):** $\mathrm{sgn}(\alpha)\Gamma(1-x)|\alpha|^{x-1} \cos(\pi x/2)$; **(d):** $\Gamma(1-x)|\alpha|^{x-1}\sin(\pi x/2)$.
Problem 2.5: $\frac{1}{2}B\left(\frac{\mu+1}{2}, \frac{-\nu-\mu}{2}\right)$.
Problem 2.6: bz/c.
Problem 2.8:

$$(\mathbf{a}): \quad \frac{\sqrt{\pi}\Gamma(\frac{a+b+1}{2})}{\Gamma(\frac{a+1}{2})\Gamma(\frac{b+1}{2})}; \quad (\mathbf{b}): \quad \frac{2^{-a}\sqrt{\pi}\Gamma(1+a-b)}{\Gamma(1+\frac{a}{2}-b)\Gamma(\frac{a+1}{2})}.$$

Problem 2.11 (a): $F_M(s) = \Gamma(s)\Gamma(a-s)/\Gamma(a), \quad 0 < \mathrm{Re}(s) < a$; **(b)**: $F_M(s) = \pi/s\sin(\pi s), \quad -1 < \mathrm{Re}(s) < 0$; **(c)**: $F_M(s) = \pi\cot(\pi s), \quad 0 < \mathrm{Re}(s) < 1$.
Problem 2.13 (b): $\sqrt{2/\pi z}\ \sinh z$, $\sqrt{2/\pi z}\ \cosh z$, $\sqrt{\pi/2z}\ e^{-z}$.
Problem 2.14: $\sqrt{2/\pi\xi}\ K_{i\alpha}(\xi)$.
Problem 2.15:

$$\frac{\sqrt{\pi}(2\beta)^\nu}{\Gamma(\mu+\frac{1}{2})(a+\beta)^{\nu+\mu}}\Gamma(\mu+\nu)\Gamma(\mu-\nu)F\left(\nu+\mu, \nu+\frac{1}{2}; \mu+\frac{1}{2}; \frac{a-\beta}{a+\beta}\right).$$

Chapter 3
Integral Equations

Integral equations arise in many problems of mathematical physics. It is not our intention to give an exhaustive discussion of all available techniques. We would rather like to concentrate on those in which complex integration plays an essential role. Special attention is paid to Wiener–Hopf decomposition and its application to singular integral equations.

3.1 Introduction

3.1.1 Classification

We keep to the one-dimensional case throughout and consider the equation

$$\varphi(x)f(x) = g(x) + \lambda \int_a^b k(x, y)f(y)dy, \tag{3.1}$$

where the functions $\varphi(x)$, the *kernel* $k(x, y)$ and the driving or inhomogeneous term $g(x)$ are given while the function $f(x)$ is to be determined. It is almost the most general form of a linear integral equation.[1] This is generally known as *Fredholm equation*. However, when the kernel has a special property that $k(x, y) = 0$ for all $y > x$ so that the upper limit b can be replaced by x, the equation is called a *Volterra equation*.

When the function $\varphi(x)$ is identically zero, it is convenient to re-define $g(x) \to -g(x)$ and write

[1] The path of integration from a to b on the real axis could be extended to an arc, or a collection of arcs, or a closed contour in the complex plane.

A. O. Gogolin (edited by E. G. Tsitsishvili and A. Komnik), *Lectures on Complex Integration*, Undergraduate Lecture Notes in Physics, DOI: 10.1007/978-3-319-00212-5_3,

$$\int_a^b k(x,y)f(y)dy = g(x),$$

(we suppressed λ here, see below for explanation) which is known as the integral equation of the first kind. Clearly this can also be interpreted as an inversion problem. On the other hand if the function $\varphi(x)$ is positive for all $x \in [a,b]$, then it can be eliminated from the general second kind equation by a re-scaling $f \to f/\sqrt{\varphi}$

$$f(x) = \tilde{g}(x) + \int_a^b \tilde{k}(x,y)f(y)dy,$$

where the new kernel and driving term are

$$\tilde{k}(x,y) = \frac{k(x,y)}{\sqrt{\varphi(x)\varphi(y)}}, \quad \tilde{g}(x) = \frac{g(x)}{\sqrt{\varphi(x)}},$$

where λ is omitted again. If the kernel is symmetric, $k(x,y) = k(y,x)$, then this property is preserved by the above scaling by the above re-scaling. The hermiticity $k(x,y) = \bar{k}(y,x)$ is preserved as well provided that $\varphi(x)$ is real. Therefore in the general theory of symmetric (hermitian) kernels we can safely suppress the auxiliary function $\varphi(x)$. Note, however, that if the kernel belongs to a narrower class and is only a function of the difference $x - y$ or of the product, then the re-scaling will generally violate such properties making the presence of $\varphi(x)$ non-trivial.

Finally, when $g(x)$ is identically zero, then

$$\varphi(x)f(x) = \lambda \int_a^b k(x,y)f(y)dy,$$

and we end up with a variant of the eigenvalue problem. The meaning of the parameter λ is therefore that of the spectral parameter. Note that λ can always be formally absorbed in the definition of the kernel, that is why we shall mostly suppress it except when it is useful to keep it. Mostly this is convenient in two cases—for the eigenvalue problem and for the perturbative treatment of the next section.

3.1.2 Resolvent

Although the general Fredholm theory is outside the scope of these lectures, certain basic elements of it, which are useful for the further analysis, are covered in this section.

For $\lambda = 0$, the solution $f(x) = g(x)/\varphi(x)$ is immediate. It is therefore natural to begin the study of the Fredholm equation perturbatively, or iteratively, in λ. The reasoning below is valid for a wide class of non-singular kernels, hence we suppress $\varphi(x)$ in this section (but keep λ).

The systematic iterative investigation of

$$f(x) = g(x) + \lambda \int_a^b k(x, y) f(y) dy$$

consists in seeking the solution in the form of the perturbative expansion in λ

$$f(x) = f_0(x) + \lambda f_1(x) + \lambda^2 f_2(x) + \ldots$$

Substituting this expansion and equating the powers of λ, we obtain iterative formulas for $f_n(x)$,

$$f_0(x) = g(x), \quad f_1(x) = \int_a^b k(x, y) f_0(y) dy, \quad f_2(x) = \int_a^b k(x, y) f_1(y) dy,$$

and generally

$$f_n(x) = \int_a^b k(x, y) f_{n-1}(y) dy.$$

Because of the structure of these equations it makes sense to define the iterated kernels via

$$k_1(x, y) = k(x, y), \quad k_n(x, y) = \int_a^b k_{n-1}(x, y_1) k(y_1, y) dy_1,$$

leading to

$$k_2(x, y) = \int_a^b k(x, y_1) k(y_1, y) dy_1,$$

$$k_3(x, y) = \int_a^b k_2(x, y_1) k(y_1, y) dy_1 = \int_a^b \left[\int_a^b k(x, y_2) k(y_2, y_1) dy_2 \right] k(y_1, y) dy_1$$

$$= \int_a^b \int_a^b k(x, y_2)k(y_2, y_1)k(y_1, y)dy_1 dy_2,$$

and generally

$$k_n(x, y) = \int_a^b \int_a^b ... \int_a^b k(x, y_{n-1})k(y_{n-1}, y_{n-2})...k(y_2, y_1)k(y_1, y)dy_1 dy_2...dy_{n-1}.$$

For a non-singular kernel, the order of integrations is not important.

We see that the following identity

$$k_{n+m}(x, y) = \int_a^b k_n(x, t)k_m(t, y)dt$$

holds. Indeed, k_n contains $n - 1$ integrations and k_m contains $m - 1$ integrations, one integration over t needed to form k_{n+m}, remains.

By construction,

$$f_n(x) = \int_a^b k_n(x, y)g(y)dy,$$

so if we define the function

$$R(x, y; \lambda) = k_1(x, y) + k_2(x, y)\lambda + k_3(x, y)\lambda^2 + ... = \sum_{n=0}^{\infty} k_{n+1}(x, y)\lambda^n,$$

called *resolvent* or *Neumann series*, then the solution is:

$$f(x) = g(x) + \lambda \int_a^b R(x, y; \lambda)g(y)dy. \qquad (3.2)$$

This solution is unique as long as the resolvent is well defined. In fact, one can show that the resolvent is well defined unless λ approaches the eigenvalue of

$$f(x) = \lambda \int_a^b k(x, y)f(y)dy,$$

when it is singular (and has a pole or a branch cut depending on whether the spectrum is discrete or continuous). Therefore, a further discussion of solutions in the case when

λ is an eigenvalue would require a discussion of the spectrum of integral equations, which is not appropriate at this point. Instead we continue with examples and refer the reader interested in Fredholm theory of spectra for non-singular kernels to [13, 14].

3.2 Product Kernels $k(x, y) = k(xy)$

The very fact that the kernel of the Fourier transformation is e^{ixy}, makes kernels of this type worthy of investigation. The most general linear equation with such a kernel is

$$\varphi(x)f(x) = g(x) + \lambda \int_a^b k(xy)f(y)dy.$$

This is not solvable generally, so we limit our discussion to solvable cases for the most of which $\varphi = 1$ and the integration is either over the whole real axis or the semiaxis.

Degenerate kernel. The simplest solvable example is the case of a degenerate kernel $k(xy) = h(x)w(y)$. Then the solution of the above equation ($\varphi = 1$) readily obtains in the form

$$f(x) = g(x) + \frac{\lambda\lambda_1}{\lambda_1 - \lambda} g_1 h(x), \quad \lambda_1 = \Big[\int_a^b w(t)h(t)dt\Big]^{-1}, \quad g_1 = \int_a^b g(t)w(t)dt.$$

Here one supposes that both integrals exist and the characteristic value $\lambda_1 \neq \lambda$. If $\lambda_1 = \lambda$ the solution reduces to $f(x) = g(x) + Ch(x)$ (C is an arbitrary constant) at $g_1 = 0$ and there are no solutions at $g_1 \neq 0$.

3.2.1 The Fourier Equation

We refer to the following

$$f(x) = g(x) + \lambda \int_{-\infty}^{\infty} e^{ixy} f(y)dy \tag{3.3}$$

as the Fourier integral equation. There are different ways to approach this problem. We begin with the most direct method of building up the resolvent. We have

$$k_1(x, y) = e^{ixy}$$

and so calculate

$$k_2(x, y) = \int_{-\infty}^{\infty} e^{ixy_1} e^{iy_1 y} dy_1 = 2\pi\delta(x + y),$$

$$k_3(x, y) = \int_{-\infty}^{\infty} [2\pi\delta(x + y_1)] e^{iy_1 y} dy_1 = (2\pi)e^{-ixy},$$

$$k_4(x, y) = \int_{-\infty}^{\infty} \left[(2\pi)e^{-ixy_1}\right] e^{iy_1 y} dy_1 = (2\pi)^2\delta(x - y),$$

$$k_5(x, y) = \int_{-\infty}^{\infty} \left[(2\pi)^2\delta(x - y_1)\right] e^{iy_1 y} dy_1 = (2\pi)^2 e^{ixy},$$

$$k_6(x, y) = \int_{-\infty}^{\infty} \left[(2\pi)^2 e^{ixy_1}\right] e^{iy_1 y} dy_1 = (2\pi)^3\delta(x + y),$$

$$k_7(x, y) = \int_{-\infty}^{\infty} \left[(2\pi)^3\delta(x + y_1)\right] e^{iy_1 y} dy_1 = (2\pi)^3 e^{-ixy},$$

$$k_8(x, y) = \int_{-\infty}^{\infty} \left[(2\pi)^3 e^{-ixy_1}\right] e^{iy_1 y} dy_1 = (2\pi)^4\delta(x - y),$$

and so on. Clearly, we have a repeating pattern here leading to

$$\begin{aligned}
k_{4m+1}(x, y) &= (2\pi)^{2m} e^{ixy}, \\
k_{4m+2}(x, y) &= (2\pi)^{2m+1}\delta(x + y), \\
k_{4m+3}(x, y) &= (2\pi)^{2m+1} e^{-ixy}, \\
k_{4m+4}(x, y) &= (2\pi)^{2m+2}\delta(x - y),
\end{aligned}$$

for integer $m = 0, 1, 2\ldots$ Therefore, computing the geometric series, we obtain the resolvent of the Fourier equation in the form:

$$\begin{aligned}
R(x, y; \lambda) &= \sum_{m=0}^{\infty} \Big[(2\pi)^{2m} e^{ixy}\lambda^{4m} + (2\pi)^{2m+1}\delta(x + y)\lambda^{4m+1} \\
&\quad + (2\pi)^{2m+1} e^{-ixy}\lambda^{4m+2} + (2\pi)^{2m+2}\delta(x - y)\lambda^{4m+3}\Big] \\
&= \frac{1}{1 - (2\pi\lambda^2)^2}\left[e^{ixy} + 2\pi\lambda\delta(x + y) + 2\pi\lambda^2 e^{-ixy} + (2\pi)^2\lambda^3\delta(x - y)\right].
\end{aligned}$$

Consequently, for $(2\pi\lambda^2)^2 \neq 1$, we have the solution to the Fourier equation

$$f(x) = \frac{1}{1-(2\pi\lambda^2)^2}\left\{g(x) + 2\pi\lambda^2 g(-x) + \lambda[G(x) + 2\pi\lambda^2 G(-x)]\right\},$$

which can be verified by substituting it back into the Eq. (3.3) [$G(x)$ is the Fourier transform of $g(x)$]. Leaving aside the natural question about what happens if $(2\pi\lambda^2)^2 = 1$ for the moment, we observe that the form of the above solution (and indeed the resolvent) is strongly suggestive of an exploitation of the parity ($x \leftrightarrow -x$) properties, see next section.

3.2.2 Reduction to Integration over the Semiaxis

We consider a more general equation here

$$f(x) = g(x) + \lambda \int_{-\infty}^{\infty} k(xy) f(y) dy \tag{3.4}$$

and use unique properties of the product kernel to reduce the integration to $x \in [0, \infty)$.

First of all note that any function, real or complex, can be written as a sum of its symmetric and anti-symmetric parts:

$$f(x) = f_c(x) + f_s(x),$$

where

$$f_c(x) = \frac{1}{2}\left[f(x) + f(-x)\right],$$

and

$$f_s(x) = \frac{1}{2}\left[f(x) - f(-x)\right].$$

Substituting this back into the integral Eq. (3.4) and replacing $y \to -y$ where it is negative, one immediately obtains:

$$f_c(x) + f_s(x) = g_c(x) + g_s(x) + 2\lambda \int_0^{\infty} k_c(xy) f_c(y) dy + 2\lambda \int_0^{\infty} k_s(xy) f_s(y) dy.$$

Now we 'reflect' by sending $x \to -x$ and get

$$f_c(x) - f_s(x) = g_c(x) - g_s(x) + 2\lambda \int_0^\infty k_c(xy) f_c(y) dy - 2\lambda \int_0^\infty k_s(xy) f_s(y) dy.$$

By adding and subtracting the above two relations one finds the pair of equations

$$f_c(x) = g_c(x) + 2\lambda \int_0^\infty k_c(xy) f_c(y) dy,$$

$$f_s(x) = g_s(x) + 2\lambda \int_0^\infty k_s(xy) f_s(y) dy. \tag{3.5}$$

Note that Eq. (3.4) is equivalent to (3.5): Eqs. (3.5) follow from Eq. (3.4) and vice versa. The remaining problem can be solved by Mellin transform. But let us first return to the Fourier Eq. (3.3) in the previous section. The above pair of equations now reads

$$f_c(x) = g_c(x) + 2\lambda \int_0^\infty \cos(xy) f_c(y) dy,$$

$$f_s(x) = g_s(x) + 2i\lambda \int_0^\infty \sin(xy) f_s(y) dy,$$

and the solution decomposes as

$$f(x) = \frac{1}{1 - 2\pi\lambda^2} [g_c(x) + \lambda G_c(x)] + \frac{1}{1 + 2\pi\lambda^2} [g_s(x) + \lambda G_s(x)].$$

Taking into account that

$$G(x) = \int_{-\infty}^\infty e^{ixy} [g_c(y) + g_s(y)] dy = 2 \int_0^\infty \cos(xy) g_c(y) dy + 2i \int_0^\infty \sin(xy) g_s(y) dy$$

we have finally

$$f_c(x) = \frac{1}{1 - 2\pi\lambda^2} \left[g_c(x) + 2\lambda \int_0^\infty \cos(xy) g_c(y) dy \right],$$

$$f_s(x) = \frac{1}{1 + 2\pi\lambda^2} \left[g_s(x) + 2i\lambda \int_0^\infty \sin(xy) g_s(y) dy \right].$$

Note that the cosine solution is precisely the same as given in [15], with the redefinition $\lambda \to \lambda/\sqrt{2\pi}$.

3.2.3 Fox's Equation

This is an equation with a product kernel and an integration over the semiaxis,

$$f(x) = g(x) + \int_0^\infty k(xy) f(y) dy, \tag{3.6}$$

where, as usual for a general kernel, we absorb λ into the definition of $k(xy)$.

Applying Mellin transform (2.20) results in

$$F_M(s) - G_M(s) = \int_0^\infty dy f(y) \int_0^\infty dx x^{s-1} k(xy) = K_M(s) \int_0^\infty \frac{dy}{y^s} f(y),$$

or

$$F_M(s) = G_M(s) + K_M(s) F_M(1-s).$$

By sending $s \to 1-s$ we obtain another equation,

$$F_M(1-s) = G_M(1-s) + K_M(1-s) F_M(s),$$

which is equivalent to the previous one. Eliminating $F_M(1-s)$ we obtain the image of the solution,

$$F_M(s) = \frac{G_M(s) + K_M(s) G_M(1-s)}{1 - K_M(s) K_M(1-s)}.$$

Applying the inversion formula (2.26) we obtain the formal solution:

$$f(x) = \frac{1}{2\pi i} \int_{c-i\infty}^{c+i\infty} \frac{G_M(s) + K_M(s) G_M(1-s)}{1 - K_M(s) K_M(1-s)} x^{-s} ds. \tag{3.7}$$

3.3 Difference Kernels: The Non-singular Case

This section we would like to devote to integral equations with a kernel which depends on the difference between the arguments, $k(x, y) = k(x-y)$.

3.3.1 Infinite Interval

Probably the simplest case is the equation of the kind

$$f(x) = g(x) + \int_{-\infty}^{\infty} k(x-y)\, f(y) dy, \tag{3.8}$$

where the integration is performed along the whole real axis [λ is again absorbed into the definition of $k(x-y)$]. The rhs is a convolution of two functions, which factorises after a Fourier transformation (1.20). Then we immediately obtain

$$F(q) = G(q) + K(q)\, F(q), \tag{3.9}$$

and finally

$$f(x) = \int_{-\infty}^{\infty} \frac{dq}{2\pi}\, e^{-iqx}\, \frac{G(q)}{1-K(q)}. \tag{3.10}$$

In general it requires some work to find out whether a given integral equation is of a such very convenient form and thus can be easily solved. Sometimes by a simple substitution it is possible to deform the integration interval to the one over the whole real axis. In the example below such a substitution even results in a difference kernel.

Example 3.1
Solve the integral equation

$$f(x) = g(x) + \int_{-\infty}^{0} dy\, k(x,y)\, f(y),$$

where

$$g(x) = \frac{x}{1+x^2} \quad \text{and} \quad k(x,y) = \frac{4U}{\pi} \frac{x}{(x+y)^2 + U^2(x-y)^2}.$$

Equations of such structure often arise in the context of Bethe Ansatz solutions of integrable many-particle models [16]. We use the substitution $x = -e^{-\lambda}$ (while x is usually the conventional momentum, λ is referred to as *rapidity*), then we obtain

$$f(\lambda) = g(\lambda) + \int_{-\infty}^{\infty} d\lambda'\, k(\lambda - \lambda')\, f(\lambda'),$$

with

$$g(\lambda) = -\frac{e^{-\lambda}}{1+e^{-2\lambda}} \quad \text{and} \quad k(\lambda) = -\frac{4U}{\pi} \frac{e^{-\lambda}}{(1+e^{-\lambda})^2 + U^2(1-e^{-\lambda})^2}.$$

The Fourier transform of the source term is computed in Problem 1.7, see (5.1) on p. 230,

$$G(k) = -\int_{-\infty}^{\infty} d\lambda \, \frac{e^{i\lambda k} \, e^{-\lambda}}{1 + e^{-2\lambda}} = -\frac{\pi}{2\cosh(\pi k/2)},$$

with the fundamental strip being $-1 < \operatorname{Im} k < 1$. The Fourier transform of the kernel can also be calculated for arbitrary U, see also Problem 1.7, but it is very simple for $U = 1$ since in this case it is equal to the source term up to a prefactor,

$$K(k) = -\frac{1}{\cosh(\pi k/2)}.$$

Thus, according to (3.10), the solution of the equation is

$$f(\lambda) = -\frac{1}{4} \int_{-\infty}^{\infty} dk \, \frac{e^{-ik\lambda}}{\cosh(\pi k/2) + 1}.$$

Formally (neglecting the above condition $-1 < \operatorname{Im} k < 1$) we can now close the contour in the lower half plane and calculate the integral by residua.[2] Alternatively one can map the integral onto the one of the form treated in Sect. 1.3.2 by a substitution $z = e^{\pi k/2}$. The result is then

$$f(\lambda) = -\frac{\lambda}{\pi \sinh(2\lambda)}.$$

In some cases the solution of the Eq. (3.8) may prove to be more convenient in terms of a resolvent $R(x, y, \lambda)$, see Eq. (3.2):

$$f(x) = g(x) + \lambda \int_{-\infty}^{\infty} R(x - y, \lambda) g(y) dy.$$

For the considered here case of the infinite interval and the difference kernel, the function $R(t, \lambda)$ is easily calculated. It can be shown that the relation (3.9) between $F(q)$, $G(q)$ and $K(q)$ can be rewritten in the form:

$$F(q) = G(q) + \lambda G(q) \, \frac{K(q)}{1 - \lambda K(q)}.$$

Now we perform the Fourier transform of the both parts of the resolvent equation

$$F(q) = G(q) + \lambda R(q, \lambda) G(q) \, .$$

Thus the Fourier transform of the resolvent is determined by

[2] The integral itself is convergent for $-\pi/2 < \operatorname{Im} \lambda < \pi/2$.

$$R(q,\lambda) = \frac{K(q)}{1-\lambda K(q)}.$$

The inversion formula then yields

$$R(t,\lambda) = \frac{1}{2\pi}\int_{-\infty}^{\infty} R(q,\lambda)e^{-iqt}dq = \frac{1}{2\pi}\int_{-\infty}^{\infty} e^{-iqt}\frac{K(q)}{1-\lambda K(q)}dq.$$

Example 3.2
Solve the integral Eq. (3.8) with the kernel

$$k(x-y) = e^{-\alpha|x-y|}, \qquad \alpha > 0.$$

The respective Fourier transform is

$$K(q) = \int_{-\infty}^{\infty} e^{-\alpha|t|}\, e^{iqt}\, dt = \frac{2\alpha}{\alpha^2+q^2}.$$

According to the above definition, the Fourier transform of the resolvent is given by

$$R(q,\lambda) = \frac{2\alpha}{q^2+\alpha^2-2\alpha\lambda},$$

whence

$$R(t,\lambda) = \frac{1}{2\pi}\int_{-\infty}^{\infty} R(q,\lambda)e^{-iqt}dq = \frac{\alpha}{\pi}\int_{-\infty}^{\infty} \frac{e^{-iqt}}{q^2+\alpha^2-2\alpha\lambda}\,dq.$$

The above integral is convergent at $\lambda < \alpha/2$ and can then easily be calculated by the residue technique. The integrand has two simple poles on the imaginary axis of the complex plane q at points $q = \pm i\sqrt{\alpha^2-2\alpha\lambda}$ and the integration contour must be closed in the lower/upper half plane at $t > 0$ and $t < 0$, respectively. As a result we find

$$R(t,\lambda) = \alpha\frac{e^{-|t|\sqrt{\alpha^2-2\alpha\lambda}}}{\sqrt{\alpha^2-2\alpha\lambda}}$$

and finally

$$f(x) = g(x) + \frac{\alpha\lambda}{\sqrt{\alpha^2-2\alpha\lambda}}\int_{-\infty}^{\infty} e^{-|x-y|\sqrt{\alpha^2-2\alpha\lambda}}g(y)dy.$$

When in the Eq. (3.8) $g(x)$ is identically zero (homogeneous equation)

$$f(x) = \lambda \int_{-\infty}^{\infty} k(x-y) f(y) dy, \tag{3.11}$$

we deal with some variant of the eigenvalue problem which requires a different approach. Now the solution may be found in the form:

$$f(x) = e^{ax},$$

where the parameter a is generally a complex number. Inserting the exponential function into both parts of the Eq. (3.11) and substituting $s = x - y$, we obtain the following equation for the parameter a:

$$\lambda \int_{-\infty}^{\infty} k(s) e^{-as} ds = 1. \tag{3.12}$$

If this equation has a root with the multiplicity r, differentiating it with respect to a we obtain:

$$\int_{-\infty}^{\infty} k(s) e^{-as} s^m ds = 0 \qquad (m = 1, 2, ..., r-1). \tag{3.13}$$

This means that in this case not only the function e^{ax}, but also functions xe^{ax}, ..., $x^{r-1}e^{ax}$ are solutions of a homogeneous Eq. (3.11). In order to prove it we insert the function $x^m e^{ax}$ in the Eq. (3.11). Performing the same change of the variable as before and using the identity (3.13), we get the original Eq. (3.12):

$$1 = \lambda \int_{-\infty}^{\infty} k(s) e^{-as} \left(1 - \frac{s}{x}\right)^m ds = \lambda \sum_{p=0}^{m} \frac{m!}{p!(m-p)!} \frac{(-1)^p}{x^p} \int_{-\infty}^{\infty} k(s) e^{-as} s^p ds$$

$$= \lambda \int_{-\infty}^{\infty} k(s) e^{-as} ds + \lambda \sum_{p=1}^{m} \frac{m!}{p!(m-p)!} \frac{(-1)^p}{x^p} \underbrace{\int_{-\infty}^{\infty} k(s) e^{-as} s^p ds}_{=0} = \lambda \int_{-\infty}^{\infty} k(s) e^{-as} ds.$$

Example 3.3
Solve the integral equation

$$f(x) = \lambda \int_{-\infty}^{\infty} e^{-|x-y|} f(y)\, dy.$$

For the solution in the form $f(x) \sim e^{ax}$, the constant a can be computed from the equation

$$\lambda \int_{-\infty}^{\infty} e^{-|s|} e^{-as}\, ds = 1 \quad \rightarrow \quad \frac{2\lambda}{(1-a^2)} = 1.$$

Hence we have two simple roots $a = \pm\sqrt{1-2\lambda}$ merging to a double root $a = 0$ at $\lambda = 1/2$. The solution in question is:

$$f(x) = C_1 e^{\sqrt{1-2\lambda}\,x} + C_2 e^{-\sqrt{1-2\lambda}\,x} \quad \text{for } \lambda \neq \frac{1}{2}$$

and

$$f(x) = C_1 + C_2 x \quad \text{for } \lambda = \frac{1}{2},$$

where C_1, C_2 are arbitrary constants. Note that for eigenvalues the restriction $\mathrm{Re}(\sqrt{1-2\lambda}) < 1$ must be imposed. If this requirement is not satisfied the integral in the rhs of the original integral equation is ill-defined.

The asymptotic behavior of $f(x)$ at large $|x|$ depends on the magnitude of the eigenvalue λ. At $\lambda > 1/2$ the eigenfunctions are bounded and may be written in the form: $\sin(\sqrt{2\lambda-1}\,x)$ and $\cos(\sqrt{2\lambda-1}\,x)$. But at $\lambda < 1/2$ they are the exponentially growing functions of $x \to \pm\infty$ at any choice of the constants C_1 and C_2.

Example 3.4
Solve the integral equation

$$\varphi(x) = \frac{\lambda}{2} \int_{-\infty}^{\infty} \frac{\varphi(y)}{\cosh\left[\frac{1}{2}(x-y)\right]}\, dy.$$

Taking $\varphi(y) \sim e^{-ay}$, the parameter a is found from the following equation

$$\frac{\lambda}{2} \int_{-\infty}^{\infty} \frac{e^{-az}}{\cosh(z/2)}\, dz = 1.$$

By setting $e^z = (1-t)/t$ we get[3]

$$\frac{1}{2} \int_{-\infty}^{\infty} \frac{e^{-az}}{\cosh(z/2)}\, dz = \int_0^1 t^{a-\frac{1}{2}}(1-t)^{-a-\frac{1}{2}}\, dt = B\Big(\frac{1}{2}+a, \frac{1}{2}-a\Big),$$

[3] Obviously, this integral can also be calculated directly by means of the contour integration technique.

where B is the Beta function [see the definition (2.10)]. Using the representation of B in terms of Gamma functions Eq. (2.11) we obtain

$$\frac{1}{2}\int_{-\infty}^{\infty} \frac{e^{-az}}{\cosh(z/2)}\, dz = \Gamma\left(\frac{1}{2}+a\right)\Gamma\left(\frac{1}{2}-a\right) = \frac{\pi}{\cos(\pi a)}.$$

The parameter a is thus determined as a solution of the equation

$$\frac{\pi\lambda}{\cos(\pi a)} = 1, \quad |\mathrm{Re}\, a| < 1/2.$$

For a special case of $\lambda = 1/\pi$, there is a double solution at $a = 0$ and the solution of the integral equation is

$$\varphi(x) = C_1 + C_2 x.$$

On the other hand, when $\lambda = 1/2\pi$ the solution of the integral equation is

$$\varphi(x) = C_1 e^{-x/3} + C_2 e^{x/3}$$

3.3.2 *Wiener-Hopf Technique*

Surprisingly, the situation is much more complicated if the integration on the rhs of the Eq. (3.8) extends over the semiaxis only. The corresponding solution technique—the *Wiener–Hopf method*—is deeply rooted in the theory of analytic functions. The original equation with the kernel

$$k(x) = \frac{1}{2}\int_{|x|}^{\infty} \frac{e^{-\xi}}{\xi}\, d\xi \tag{3.14}$$

arose in the theory of radiation transport and especially radiative equilibrium and describes temperature distribution within a star atmosphere [17].[4] For a detailed and highly insightful derivation see for instance [18].

Let us consider a homogeneous Wiener-Hopf integral equation

[4] In a slightly different form, namely as a problem

$$f(x) = (1/2)\int_0^{\infty} dy \int_0^1 e^{-|x-y|/\mu}\, \frac{d\mu}{\mu} f(y) \tag{3.15}$$

it is referred to as *Milne problem* [4].

$$f(x) = \int_0^\infty dy\, k(x-y) f(y). \tag{3.16}$$

Here the unknown function $f(x)$ is defined on the semiaxis $x > 0$ while the kernel $k(x)$ is defined on the entire real axis $-\infty < x < \infty$.

We want the above equation to become algebraic with respect to the one-sided Fourier transforms F_+, F_- defined in (1.29)–(1.30), just as it is the case for the equations from the previous subsection. Since our $f(x)$ is only defined for positive x the integral

$$\int_0^\infty dy\, k(x-y) f(y) = g_-(x)$$

for $x < 0$ defines an additional unknown function $g_-(x)$. Thus we have

$$\int_0^\infty dy\, k(x-y) f(y) = \begin{cases} f(x), & x > 0 \\ g_-(x), & x < 0. \end{cases}$$

Using our notation $f_+(x)$ as in (1.31) we can also write this as

$$\int_0^\infty dy\, k(x-y) f_+(y) = f_+(x) + g_-(x), \tag{3.17}$$

because $f_+(x < 0) = 0$ and $g_-(x > 0) = 0$ by construction.

Now take the Fourier transform as defined in (1.20) of both sides of (3.17), which gives:

$$K'(s)\, F_+(s) = F_+(s) + G_-(s),$$

where we slightly change notation and call the Fourier transform of $k(x)$ $K'(s)$. F_+ and G_- is the Fourier transform of f_+ and g_-, respectively. By redefinition $K(s) = 1 - K'(s)$ the equation simplifies,

$$K(s)\, F_+(s) + G_-(s) = 0.$$

If the integral equation has an inhomogeneous term, then also its Fourier transform $P(s)$ will enter the equation,

$$K(s) F_+(s) + G_-(s) = P(s). \tag{3.18}$$

This is the general algebraic form of the Wiener-Hopf problem. The functions involved are characterized as follows (see Fig. 3.1):

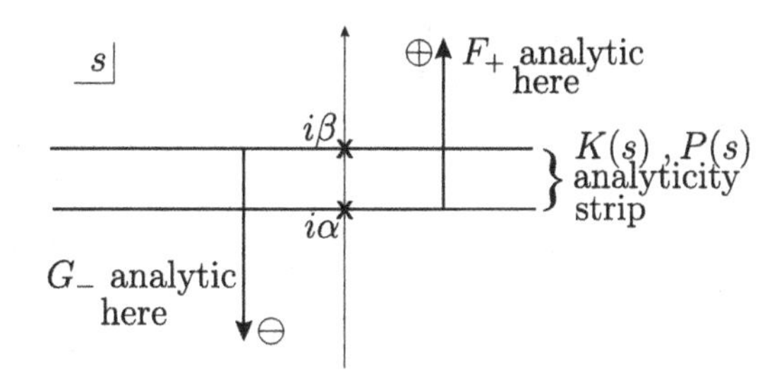

Fig. 3.1 Analyticity domains of functions in Eq. (3.18)

(i) $F_+(s)$ is supposed to be analytic in $s_2 > \alpha$ (we again use the notation we have introduced in Sect. 1.4.2: $s = s_1 + is_2$ with purely real $s_{1,2}$, α is the exponential bound), thus in the upper half-plane, or, more precisely in the area which we denote by $\oplus$, see Fig. 3.1, but is otherwise an unknown function.
(ii) $G_-(s)$ is supposed to be analytic in $s_2 < \beta$, which is basically the lower half-plane, or the area which we shall denote by $\ominus$, see Fig. 3.1, but is otherwise an unknown function.
(iii) $K'(s)$ and $P(s)$ are given and are assumed to be analytic in the strip $\alpha < s_2 < \beta$. When $s \to \infty$ in the strip, we require that K', P grow not faster than algebraically, that means $K', P \leq A|s|^N$ holds.

Our goal is to find $f_+(x)$, which is accomplished by computing $F_+(s)$ first and then act by inversion. However, we have two unknowns, F_+ and G_-, and seemingly only one Eq. (3.18). We shall see that the WH equation encodes in fact two equations, so both F_+ and G_- can be found. The Wiener–Hopf technique is based on the two ideas: the factorisation or product decomposition and the sum decomposition of an analytic function defined in a strip in the complex plane.

The solution proceeds in several stages. In the first step we perform the Wiener-Hopf product decomposition of the kernel. Given a function $K(s)$, analytic in the strip $\alpha < s_2 < \beta$, we can decompose it as follows:

$$K(s) = K_+(s)K_-(s), \tag{3.19}$$

where $K_+(s)$ is analytic in the $\oplus$ region, $s_2 > \alpha$, and $K_-(s)$ is analytic in the $\ominus$ region, $s_2 < \beta$, see Fig. 3.2. We shall formalize this and identify required conditions

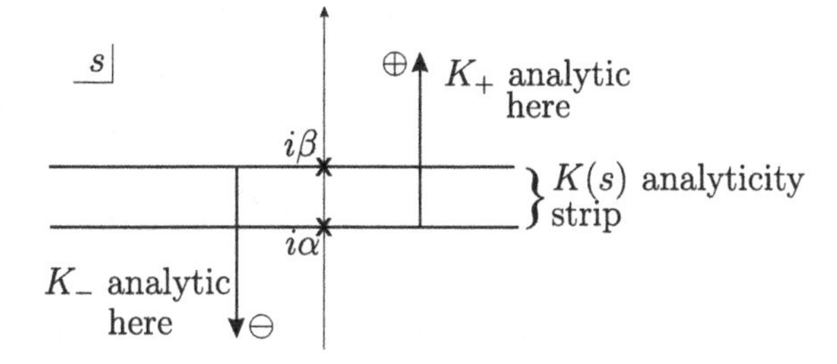

Fig. 3.2 Analyticity strip of the kernel (3.19)

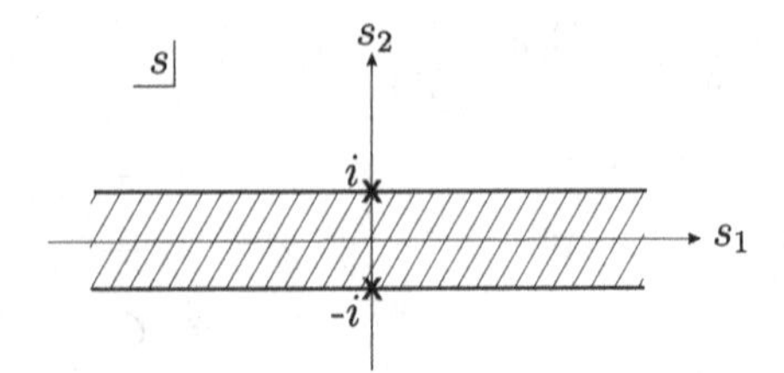

Fig. 3.3 Analyticity strip for the function of Example 3.5

on zeros of K as well as on arg K later on and start with several comprehensible examples.

Example 3.5
The function

$$K(s) = \frac{1}{s^2+1}$$

is analytic in the strip $-1 < s_2 < 1$, see Fig. 3.3. But we can write it as

$$K(s) = \frac{1}{s^2+1} = \frac{1}{s+i}\frac{1}{s-i},$$

so that

$$K_+(s) = \frac{1}{s+i},$$

which is analytic for $s_2 > -1$ in the $\oplus$–region, see Fig. 3.2 with $\alpha = -1$ and

$$K_-(s) = \frac{1}{s-i},$$

which is analytic for $s_2 < 1$ in the $\ominus$–region, see Fig. 3.2 with $\beta = 1$.

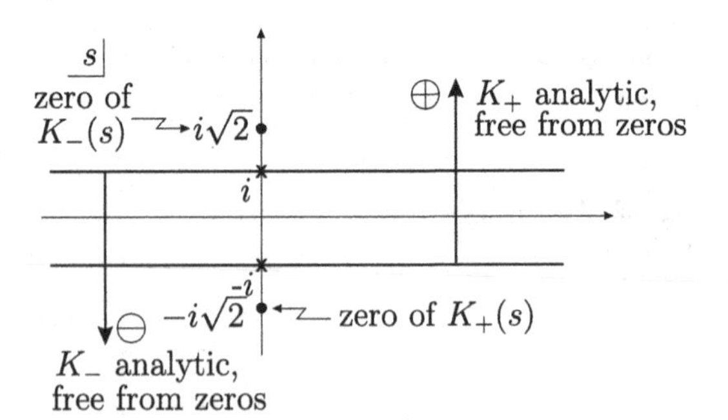

Fig. 3.4 Analyticity domains of the function from Example 3.6

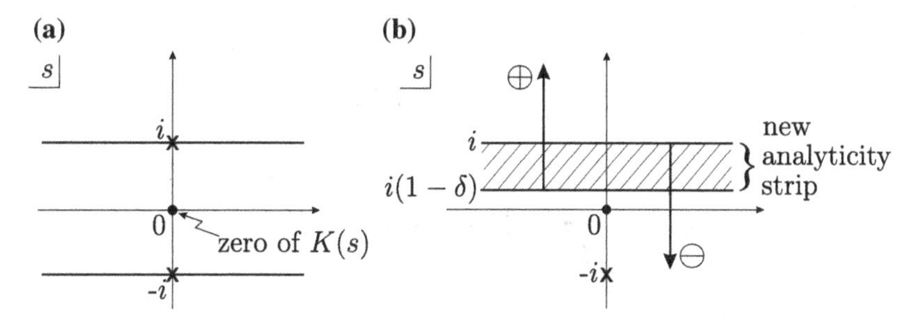

Fig. 3.5 Analyticity domains for the function from Example 3.7

Example 3.6

$$K(s) = \frac{s^2+2}{s^2+1}.$$

In this example $K(s)$ has zeros at $s = \pm i\sqrt{2}$, as is shown in Fig. 3.4. In general we require that (we shall see later why): **(i)** all zeros of $K(s)$ lie outside the analyticity strip; **(ii)** $K_\pm(s)$ be not only analytic in $\oplus$, $\ominus$ regions, that is free from singularities, but also free from zeros in the respective analyticity regions, see Fig. 3.4. So we have to decompose as

$$K(s) = \frac{s^2+2}{s^2+1} = \frac{s+i\sqrt{2}}{s+i}\,\frac{s-i\sqrt{2}}{s-i} = K_+(s)\;K_-(s).$$

Example 3.7

$$K(s) = \frac{s^2}{s^2+1}.$$

The original analyticity strip is $-1 < s_2 < 1$, but $K(s)$ has a double zero there, at $s = 0$, see Fig. 3.5a. Therefore we redefine the analyticity strip to get rid of zeros inside the strip (in practice the way to redefine the strip is usually prescribed by the properties of other functions involved), see Fig. 3.5b, and decompose as

$$K(s) = \frac{s^2}{s^2+1} = \underbrace{\frac{s^2}{s+i}}_{K_+(s)}\;\underbrace{\frac{1}{s-i}}_{K_-(s)}.$$

In fact, any rational function $K(s)$ can by decomposed by using these prescriptions. This can be shown by inspection.

We follow similar rules for construction of $K_\pm(s)$ in the case of functions with branch cuts, as one can see in the following example.

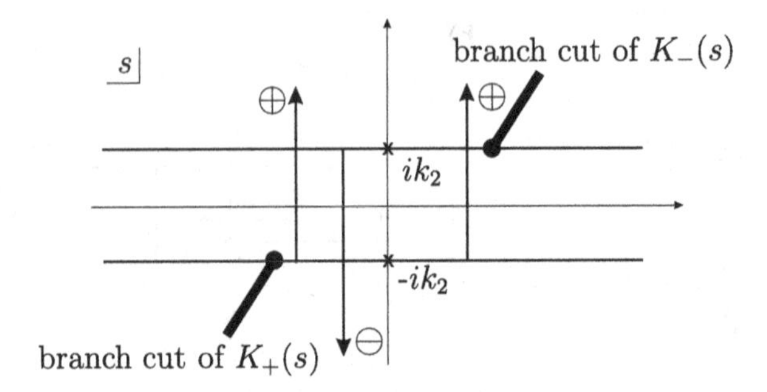

Fig. 3.6 Analyticity domains for the functions from Example 3.8

Example 3.8

$$K(s) = \sqrt{s^2 - k^2},$$

where $k = k_1 + ik_2$, $k_{1,2} > 0$. Let

$$K(s) = (s + k)^{1/2} (s - k)^{1/2},$$

so that the natural choice would be

$$K_+(s) = (s + k)^{1/2} \quad \text{and} \quad K_-(s) = (s - k)^{1/2},$$

see Fig. 3.6.

Now we would like to return to our original Eq. (3.18). Using the above product decomposition (3.19) we obtain

$$K_+(s)K_-(s)F_+(s) + G_-(s) = P(s),$$

which we further modify by dividing it through $K_-(s)$,

$$K_+(s)F_+(s) + \frac{G_-(s)}{K_-(s)} = \frac{P(s)}{K_-(s)} \equiv R(s).$$

Here we see why the absence of zeros is essential for the functions $K_\pm$. Very conveniently, the lhs is a sum of a 'plus'- and a 'minus'-function. However, the analytic properties of the rhs $R(s)$ are not clear yet.

Obviously, it would be very helpful to have a sum decomposition of the function $R(s)$:

$$R(s) = R_+(s) + R_-(s),$$

analytic in the strip $\alpha < s_2 < \beta$. It turns out that in many cases to produce for a given function a sum decomposition of this kind is as easy as to compute a product decomposition. We shall formalize this statement and relate to the product decomposition using the logarithm function later. To get the feeling for how to perform the procedure we take a look on the following example.

Example 3.9
The function

$$R(s) = \frac{1}{s^2 + 1}$$

is analytic in $-1 < s_2 < 1$. We have

$$R(s) = \frac{1}{(s+i)(s-i)} = \frac{-1/2i}{s+i} + \frac{1/2i}{s-i}.$$

Then, obviously,

$$R_+(s) = \frac{i}{2(s+i)} \quad \text{and} \quad R_-(s) = -\frac{i}{2(s-i)}.$$

The analyticity strip has the geometry shown in Fig. 3.3. Please observe that we have dropped the requirement on $R_\pm$ being free from zeros as we do not divide by them.

With the above decomposition we can now go over to the last stage of solving the Wiener-Hopf equation,

$$K_+(s)F_+(s) + \frac{G_-(s)}{K_-(s)} = R_+(s) + R_-(s).$$

Rewrite this to bring all 'plus'– and 'minus'–functions to one side:

$$K_+(s)F_+(s) - R_+(s) = R_-(s) - \frac{G_-(s)}{K_-(s)} \equiv E(s). \tag{3.20}$$

Each side of the equation analytically continues the other to define a function $E(s)$ analytic in the entire s–plane, see Fig. 3.7. This is a remarkable result.

The analyticity of $E(s)$ helps us to make some useful statements concerning its properties. First of all, according to the Liouville theorem from Sect. 1.2.5, if $E(s)$ is analytic for all s and bounded then $E(s) = \text{const}$. In particular, if under the above conditions $E(s) \to 0$ for $s \to \infty$ then $E(s) = 0$ for all s. On the other hand, if $E(s)$ is analytic from all s apart from $s = \infty$, where it exhibits algebraic growth, that is if $E(s) = O(s^N)$ as $s \to \infty$, then $E(s)$ is a polynomial of degree N. The proof of this result can be found in Section V of [3].

In practice the polynomial $E(s)$ is obtained from the asymptotic behaviour of F_+/G_- as $s \to \infty$. If, for example, the lhs as well as rhs of (3.20) is $O(s)$, then also $E(s) = O(s)$. The large s-behavior of Fourier transform is, in turn, determined by the behavior of $f_+(x \to 0^+)$ and $g_-(x \to 0^-)$ at small x.

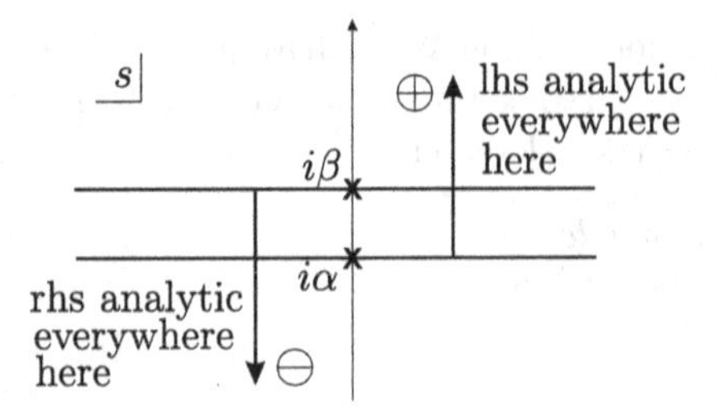

Fig. 3.7 Analyticity domains of lhs and rhs of Eq. (3.20)

In order to better understand the workings of the method we turn to yet another example.

Example 3.10
Solve the integral equation

$$f(x) = \frac{1}{2}\int_{0}^{\infty} e^{-|x-t|} f(t)dt, \quad x \geq 0 \quad \text{with} \quad k(x) = \frac{1}{2}e^{-|x|}. \tag{3.21}$$

This is a homogeneous problem, so f is defined up to a multiplicative constant, $f \to Af$. We want to know whether a non-trivial solution ($f \neq 0$) exists and whether it is unique up to the factor A. We use our notation

$$f_+(x) = \begin{cases} f(x), & x > 0 \\ 0, & x < 0 \end{cases}$$

and

$$g_-(x) = \begin{cases} 0, & x > 0 \\ \frac{1}{2}\int_0^{\infty} dte^{-|x-t|} f(t), & x < 0. \end{cases}$$

Then we have

$$\frac{1}{2}\int_{0}^{\infty} dte^{-|x-t|} f_+(t) = f_+(x) + g_-(x) = \begin{cases} f_+(x), & x > 0 \\ g_-(x), & x < 0. \end{cases}$$

We apply the ordinary Fourier transform to both sides of this equation, which leads to

$$\int_{0}^{\infty} dxe^{isx} f_+(x) + \int_{-\infty}^{0} dxe^{isx} g_-(x) = F_+(s) + G_-(s)$$

and

$$\frac{1}{2}\int_{-\infty}^{\infty} dx\, e^{isx} \int_{0}^{\infty} dt e^{-|x-t|} f_+(t) = \int_{0}^{\infty} dt f_+(t)\, \frac{1}{2}\int_{-\infty}^{\infty} dx\, e^{isx} e^{-|x-t|}.$$

In the last step we have changed the order of integrations. Next we perform the substitution $x \to x + t$ and arrive at

$$= \int_{0}^{\infty} dt f_+(t) e^{ist}\, \frac{1}{2}\int_{-\infty}^{\infty} dx\, e^{isx} e^{-|x|} = F_+(s)\, K'(s).$$

The Fourier transform of the kernel was already calculated in Example 1.14, see (1.37)

$$K'(s) = \frac{1}{s^2+1}. \tag{3.22}$$

$K'(s)$ is analytic in the strip $-1 < s_2 < 1$, see Fig. 3.3. In the Fourier space the Wiener-Hopf problem thus reads

$$\frac{1}{s^2+1} F_+(s) = F_+(s) + G_-(s),$$

or

$$\frac{s^2}{s^2+1}\, F_+(s) + G_-(s) = 0. \tag{3.23}$$

We know that $K'(s)$ and $K(s) = 1 - K'(s)$ are analytic in the strip $-1 < s_2 < 1$, $K(s)$ has a zero there though. To proceed we need to know the analyticity domains for the functions $G_-(s)$ and $F_+(s)$.

- We recall that

$$g_-(x) = \frac{1}{2}\int_{0}^{\infty} dt e^{-|x-t|} f_+(t) \text{ for } x < 0.$$

But for $x < 0, t > 0$ we have $x - t < 0$, so we can write $|x - t| = t - x$ and

$$g_-(x) = \frac{1}{2}\int_{0}^{\infty} dt e^{-(t-x)} f_+(t) = e^x\, \frac{1}{2}\int_{0}^{\infty} dt e^{-t} f_+(t) = e^x\, C,$$

where C is as yet unknown constant. Hence the integral

$$G_-(s) = \int_{-\infty}^{0} dx e^{isx} C e^x = \frac{-i\, C}{s-i}$$

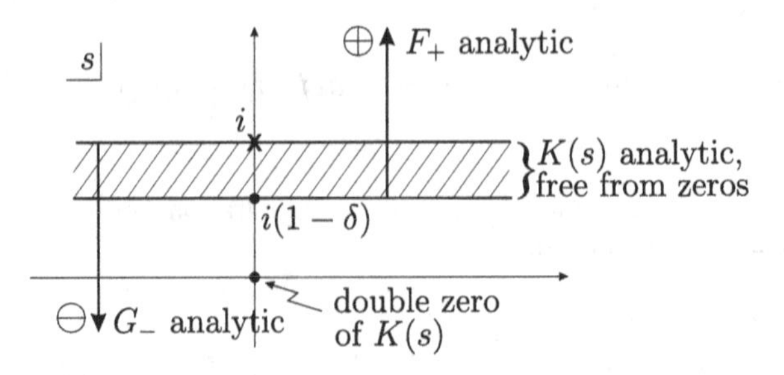

Fig. 3.8 Analyticity domains for the functions from Example 3.10

converges for $s_2 < 1$ and $G_-(s)$ is analytic for $\mathrm{Im}s < 1$, that means that in our notation the exponential bound index is $\beta = 1$.

- As far as $F_+(s)$ is concerned we obviously need the integral

$$\frac{1}{2}\int_0^\infty dt e^{-|x-t|} f_+(t)$$

to converge. Assuming $f_+(x)$ to be integrable (for example continuous) for finite x this amounts to limiting the growth of $f_+(x)$ as $x \to \infty$ and the divergence of $f_+(x)$ as $x \to 0^+$.

■ $x \to \infty$: $f_+(x)$ should grow not faster than $\sim e^x$, therefore we require

$$|f_+(x)| < Ae^{+(1-\delta)x}$$

for $x \to \infty$ and $0 < \delta < 1$. Hence $F_+(s)$ is analytic for $\mathrm{Im}s > 1 - \delta$, or in our notation the exponential bound index is $\alpha = 1 - \delta$.

■ $x \to 0^+$: Here $f_+(x)$ should diverge slower then $1/x$ and we just assume $f_+(x \to 0^+) =$ constant which translates into $F_+(s) \sim if_+(0)/s$ as $s \to \infty$.

Let us now return to the Wiener-Hopf problem (3.23) with the analyticity strip $\alpha < \mathrm{Im}s < \beta$; $\alpha = 1 - \delta$, $\beta = 1$, see Fig. 3.8. We factorize in the following way

$$K(s) = \frac{s^2}{s^2+1} = \frac{s^2}{s+i}\frac{1}{s-i}.$$

The involved multipliers $K_+(s)$ and $K_-(s)$ are analytic in the same domains as $F_+(s)$ and $G_-(s)$, respectively, and are free from zeros in their analyticity domains:

$$K_+(s) = \frac{s^2}{s+i} \quad \text{and} \quad K_-(s) = \frac{1}{s-i}.$$

Double zero is a part of K_+ as it occurs in the ⊖ region at $s = 0$. Next we rewrite

$$\frac{s^2}{s+i}\frac{1}{s-i}\,F_+(s)+G_-(s)=0,$$

and divide by $K_-(s)=1/(s-i)$:

$$\frac{s^2}{s+i}\,F_+(s)+(s-i)G_-(s)=0.$$

We take all the 'plus'–functions to the lhs and all the 'minus'–functions to the rhs of the equation:

$$\frac{s^2}{s+i}\,F_+(s)=-(s-i)G_-(s)\equiv E(s), \tag{3.24}$$

which defines the function $E(s)$. It is analytic in the entire s-plane apart from, perhaps, $s=\infty$. The Liouville theorem tells us then that the function $E(s)$ is, at most, a polynomial $E(s)=O(|s|^N)$ of degree N. To determine the degree N we need to investigate $s\to\infty$ behavior of the lhs and the rhs of (3.24). In fact we know that $G_-(s)\sim 1/s$ as $s\to\infty$ and therefore

$$-(s-i)G_-(s)=O(|s|^0)\ \text{ as }\ s\to\infty,$$

which is essentially a constant. We also know that $F_+\sim 1/s$ and since $s^2/(s+i)=O(|s|)$ we have

$$\frac{s^2}{s+i}\,F_+(s)=O(|s|^0)\ \text{ as }\ s\to\infty,$$

which is a constant as well. It follows that $E(s)$ is a constant $E(s)=A$ for all s. Therefore the solution of (3.24) is

$$F_+(s)=A\,\frac{s+i}{s^2}.$$

To finally determine $f_+(x)$ we use the inversion formula

$$f_+(x)=\frac{1}{2\pi}\int_P dse^{-isx}F_+(s)$$

with the contour P lying above all singularities of $F_+(s)$, as is shown in Fig. 3.9.

- For $x<0$, $|e^{-isx}|=e^{s_2x}\to 0$ in the upper half-plane Im$s>0$. Therefore we close up in the upper half-plane, see Fig. 3.10a, then

$$f_+(x)=0,\quad x<0$$

as required.

- For $x > 0$ closing the contour in the lower half-plane, see Fig. 3.10b, pics up the double pole residue

$$f_+(x) = \frac{A}{2\pi} \int_P dse^{-isx} \frac{s+i}{s^2} = \frac{A}{2\pi}(-2\pi i)\mathrm{Res}\Big(\frac{s+i}{s^2}\, e^{-isx}\Big)\Big|_{s=0}.$$

The residue is, as always, the coefficient of $1/s$ in the Laurent expansion:

$$\frac{s+i}{s^2}\, e^{-isx} = \frac{s+i}{s^2}\Big(1 - isx + ...\Big) = \frac{i+s+sx+...}{s^2} = \frac{i}{s^2} + \frac{1+x}{s} + ...$$

and thus is equal to $1 + x$.

Therefore the solution of Eq. (3.21) is

$$f(x) = A_0(1+x), \tag{3.25}$$

where A_0 $(= -iA)$ is an arbitrary constant which can not be determined as the equation is homogeneous.

Example 3.11
Let us now try to solve an inhomogeneous problem

$$y(x) - 2\int_0^\infty dte^{-|x-t|}y(t) = -2xe^{-x}, \tag{3.26}$$

for $x > 0$. Following the general procedure we introduce for the unknown function $y(x)$:

$$y_+(x) = \begin{cases} y(x), & x > 0 \\ 0, & x < 0. \end{cases}$$

Then the equation can be rewritten in the following form:

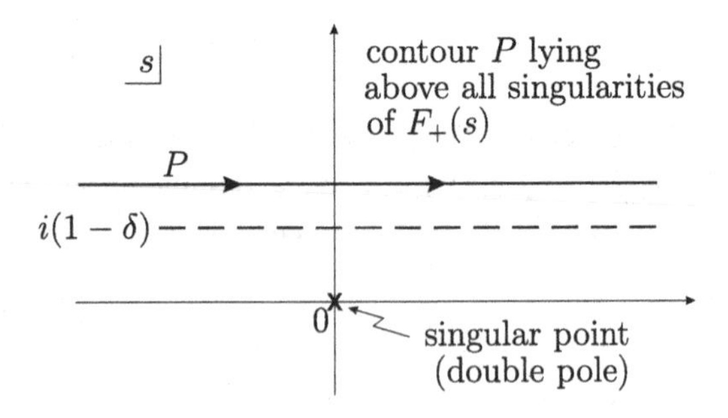

Fig. 3.9 Illustration to the inversion formula of the function $F_+(s)$, page 137

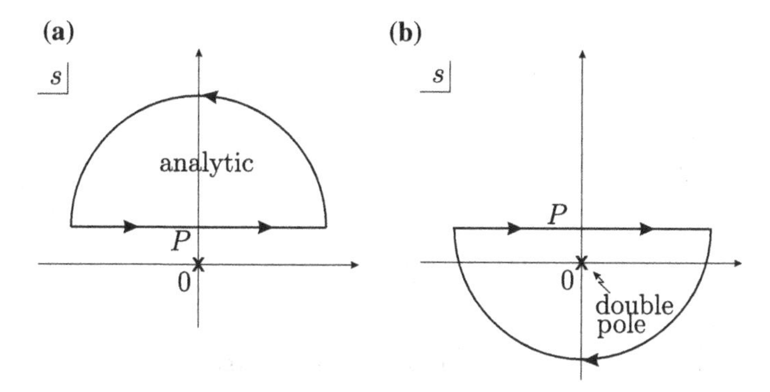

Fig. 3.10 Different ways to close the contour in the integral calculation in Example 3.10

$$2\int_0^\infty dte^{-|x-t|}y_+(t) = \begin{cases} y_+(x) + 2xe^{-x}, & x > 0 \\ g_-(x), & x < 0, \end{cases} \tag{3.27}$$

where $g_-(x)$ is again another unknown function defined by this relation. The Fourier transform of $y_+(x)$ is

$$Y_+(s) = \int_0^\infty dx\, e^{isx}\, y_+(x)$$

and the function $y_+(x)$ satisfies the same restrictions as $f_+(x)$ of the previous example, in particular the integral

$$\int_0^\infty dte^{-|x-t|}y_+(t)$$

must converge. That again leads to the following asymptotic properties:

- for $x \to 0^+$: taking this limit in (3.26) we realise that $y_+(x) = \text{constant}$, therefore $Y_+(s) \to 0$ as $s \to \infty$, which is seen by performing a partial integration;
- for $x \to \infty : |y_+(x)| < Ae^{(1-\delta)x},\ 0 < \delta < 1$, hence $Y_+(s)$ is analytic for $\text{Im} s > 1 - \delta = \alpha$.

As in the previous example we can write

$$g_-(x) = 2\int_0^\infty dte^{-|x-t|}\, y_+(t) = 2e^x \int_0^\infty dte^{-t}\, y_+(t),$$

since here $x < 0$ and $t > 0$. Hence $g_-(x) = C_1\, e^x$ and $G_-(s) = C_1/(s-i)$, where C_1 is some constant. It follows that:

- $G_-(s) = O(1/s)$ for $s \to \infty$,
- $G_-(s)$ is analytic where the integral $\int_{-\infty}^{0} dx e^{isx} e^{x}$ converges absolutely; that is for $\text{Im} s < 1 = \beta$.

The new element is the inhomogeneous term $p(x) = -2xe^{-x}$, which we need to Fourier transform:

$$P(s) = -2 \int_{0}^{\infty} dx e^{isx} x e^{-x}.$$

We have $|e^{isx}e^{-x}| = e^{-(\text{Im}\, s+1)x}$, so this integral converges for $\text{Im} s > -1$ and is a 'plus'–function. We calculate it by parts and obtain

$$P(s) = -\frac{2}{(s+i)^2}.$$

Also we know that the Fourier transform of the kernel up to the prefactor is related to (3.22). Therefore the Fourier transform of (3.27) is

$$\frac{4}{s^2+1} Y_+(s) = Y_+(s) - \frac{2}{(s+i)^2} + G_-(s).$$

Hence the Wiener-Hopf problem reads

$$\frac{s^2-3}{s^2+1} Y_+(s) + G_-(s) = \frac{2}{(s+i)^2}.$$

The function $Y_+(s)$ is analytic in the upper half-plane, the function $G_-(s)$ is analytic in the lower half-plane, and the kernel and the inhomogeneous term are analytic in the strip $(1-\delta) < \text{Im} s < 1$, see Fig. 3.11.

In the next step we decompose the kernel K as a product of a 'plus'– and a 'minus'–functions

$$K(s) = 1 - K'(s) = \frac{s^2-3}{s^2+1} = K_+(s)\, K_-(s),$$

so that

$$K_+(s) = \frac{s^2-3}{s+i} \quad \text{and} \quad K_-(s) = \frac{1}{s-i}.$$

Please observe that the real axis zeros go with $K_+(s)$. By substitution of this decomposition into the original equation and subsequent division by $K_-(s)$ we then arrive at

$$\frac{s^2-3}{s+i} Y_+(s) + (s-i)\, G_-(s) = \frac{2(s-i)}{(s+i)^2} \equiv R(s).$$

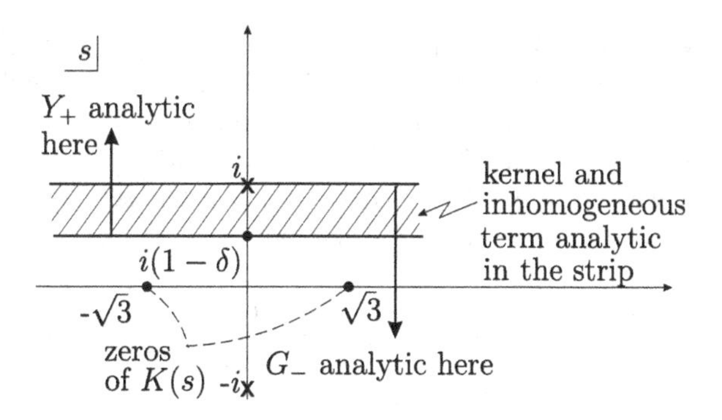

Fig. 3.11 Analyticity domain of the function from Example 3.11

The next natural step is to do the sum decomposition of the inhomogeneous term. However, $R(s)$ is already a 'plus'–function. Indeed, we have $R(s) = P(s)/K_-(s)$. $P(s) = 2/(s+i)^2 = P_+(s)$ is a 'plus'–function by construction and analytic in $\mathrm{Im}s > -1$. The function $K_-(s)$ is a 'minus'–function, not a 'plus'–function. But $1/K_-(s) = s - i$ happens to be a 'plus'–function. So we conclude that[5]

$$R(s) = \frac{2(s-i)}{(s+i)^2} = R_+(s), \quad R_-(s) = 0.$$

Separation of all the $\pm$-functions to the one side then yields

$$\frac{s^2-3}{s+i}\,Y_+(s) - \frac{2(s-i)}{(s+i)^2} \quad = \quad -(s-i)G_-(s),$$

which defines the function $E(s)$. At $s \to \infty$ the rhs approaches a constant because $G_-(s)|_{s\to\infty} \sim 1/s$. Likewise, the lhs levels off to a constant because $Y_+(s)|_{s\to\infty} \sim 1/s$ as well. Thus $E(s) \equiv A$ and is a constant for all s.

Therefore our solution is

$$Y_+(s) = A\,\frac{s+i}{s^2-3} + \frac{2(s-i)}{(s+i)(s^2-3)} = \frac{(A-i)(s+i)^2 + i(s^2-3)}{(s+i)(s^2-3)}. \tag{3.28}$$

In order to obtain $y_+(x)$ we resort to the inversion formula:

$$y_+(x) = \frac{1}{2\pi}\int_P dse^{-isx}\,Y_+(s),$$

with the contour P lying above all singularities of $Y_+(s)$.

---

[5] Please note that the zero at $s = i$ is unimportant because we never have to divide anything by $R_+(s)$. See also discussion above.

- For $x < 0$, $|e^{-isx}| = e^{\mathrm{Im}\, s\, x} \to 0$ for $\mathrm{Im} s > 0$, therefore we close the contour in the upper half-plane, see Fig. 3.12a. The integrand has no singularities there, hence $y_+(x) = 0$ for $x < 0$ as required.
- For $x > 0$, $|e^{-isx}| = e^{\mathrm{Im}\, s\, x} \to 0$ for $\mathrm{Im} s < 0$, hence we close the contour, that we denote by $P_\cup$, in the lower half-plane, see Fig. 3.12b. There are three single poles at $s = -i, \pm\sqrt{3}$. Therefore

$$\begin{aligned} y_+(x) &= \frac{1}{2\pi} \int_{P_\cup} ds \left[\frac{(A-i)(s+i)^2 + i(s^2-3)}{(s+i)(s+\sqrt{3})(s-\sqrt{3})}\, e^{-isx} \right] \\ &= \frac{1}{2\pi}(-2\pi i) \sum \mathrm{Res}\Big[....\Big]\Big|_{s=-i,\, \pm\sqrt{3}}, \end{aligned}$$

where $\Big[....\Big]$ represents the integrand. We have

$$\begin{aligned} \mathrm{Res}\Big[....\Big]\Big|_{s=-i} &= i e^{-x}, \\ \mathrm{Res}\Big[....\Big]\Big|_{s=-\sqrt{3}} &= (A-i)\frac{-\sqrt{3}+i}{-2\sqrt{3}}\, e^{i\sqrt{3}x} = (A-i)\frac{1-i/\sqrt{3}}{2}\, e^{i\sqrt{3}x}, \end{aligned}$$

and also

$$\mathrm{Res}\Big[....\Big]\Big|_{s=\sqrt{3}} = (A-i)\frac{\sqrt{3}+i}{2\sqrt{3}}\, e^{-i\sqrt{3}x} = (A-i)\frac{1+i/\sqrt{3}}{2} e^{-i\sqrt{3}x}.$$

We observe that for the sum of the two last terms we get

$$\mathrm{Res}\Big[....\Big]\Big|_{s=-\sqrt{3}} + \mathrm{Res}\Big[....\Big]\Big|_{s=\sqrt{3}} = (A-i)\Big[\cos\sqrt{3}x + \frac{1}{\sqrt{3}}\sin\sqrt{3}x\Big].$$

Therefore our solution is

$$y_+(x) = e^{-x} - i(A-i)\Big[\cos\sqrt{3}x + \frac{1}{\sqrt{3}}\sin\sqrt{3}x\Big]. \tag{3.29}$$

This is a general solution and it contains an arbitrary constant A. To fix A extra conditions are required. For example one often has to impose (for instance for physical reasons) that

$$\lim_{x\to+\infty} y_+(x) = 0.$$

Then the constant A is fixed at $A = i$ and we simply have

$$y_+(x) = e^{-x}. \tag{3.30}$$

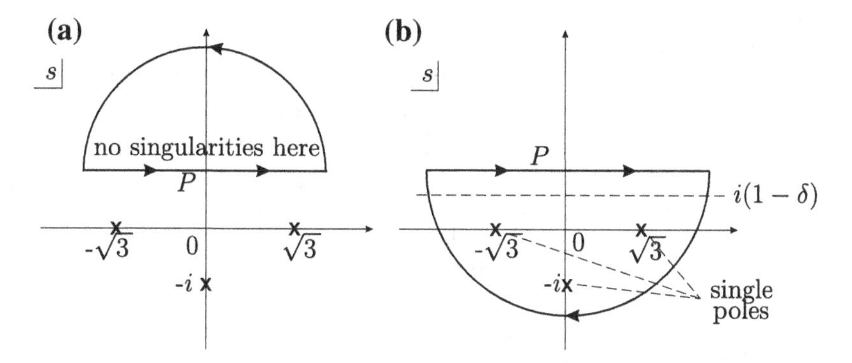

Fig. 3.12 Two ways to close the contour in the calculation of the inverse in Example 3.11

Alternatively we might recall the explicit form of $Y_+(s)$, given in (3.28). The oscillating terms in (3.29) come from the real-axis poles $s = \pm\sqrt{3}$. To get rid of those poles we can set $A = i$ and obtain

$$Y_+(s) = \frac{i}{s+i},$$

which immediately leads to (3.30).

Now we would like to generalize our exposition. There is a number of excellent books discussing the formal side of the method, see for instance [19, 20]. However, we would rather adopt a physicists' way of doing things and first ask the important question about a generic way of constructing the sum and product decompositions, that we have encountered. By answering this question we also supply some very useful information about the existence of such decompositions, an issue more to the taste of mathematicians.

We start with a function $R(s)$ which is analytic in the strip $\alpha < \mathrm{Im}s < \beta$ and $R(s) \to 0$ when $s \to \infty$ in the strip.[6] Let us consider the contour $C = L_1 + L_2 + \Sigma_+ + \Sigma_-$ shown in Fig. 3.13. Then in accordance with the Cauchy integral formula for $\alpha < \alpha_1 < \mathrm{Im}s < \beta_1 < \beta$ and $\sigma_- < \mathrm{Re}s < \sigma_+$ we have

$$R(s) = \frac{1}{2\pi i}\int\limits_C \frac{d\xi}{\xi - s}\, R(\xi).$$

Let us take the two end segments Σ_+ and Σ_- to ∞ and $-\infty$ respectively, $\sigma_\pm \to \pm\infty$. Since $R(s) \to 0$ for $s \to \infty$ in the strip, we have[7]

$$\int_{\Sigma_\pm} \to 0 \quad \text{as} \quad \sigma_\pm \to \pm\,\infty.$$

[6] The condition of vanishing at ∞ can be relaxed, we would like to keep it for simplicity though.

[7] In fact, it is sufficient to require that $R(s)$ be bounded on infinity.

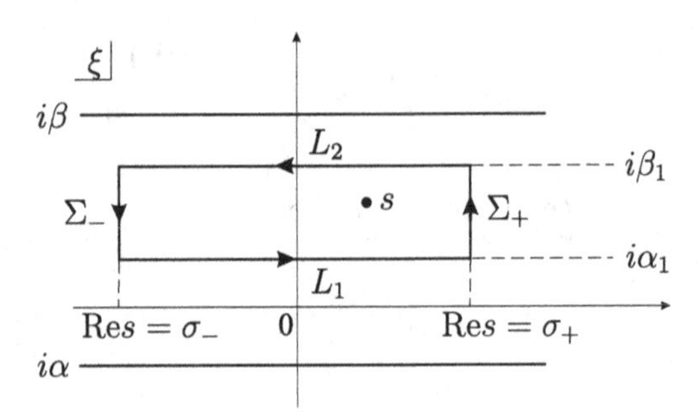

Fig. 3.13 Contour used for the decomposion of function $R(s)$

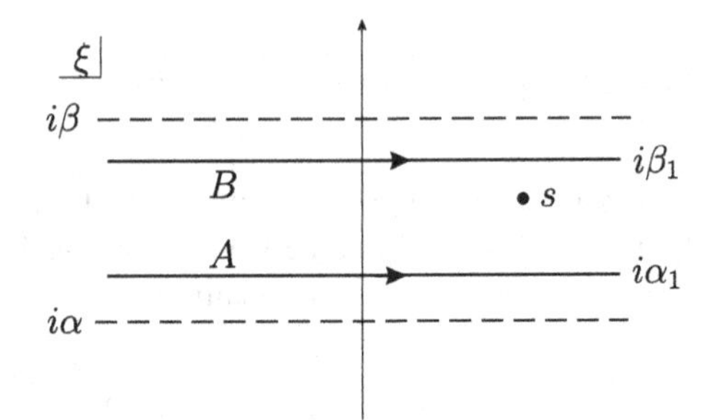

Fig. 3.14 The contour of Fig. 3.13 after sending $\Sigma_{\pm}$ to infinities

Hence the contour integral simplifies to

$$R(s) = \frac{1}{2\pi i}\int\limits_{A}\frac{d\xi}{\xi - s}R(\xi) - \frac{1}{2\pi i}\int\limits_{B}\frac{d\xi}{\xi - s}R(\xi),$$

where A and B are the remnants of $L_{1,2}$ contours as shown in Fig. 3.14. Now we make a crucial observation that the function $R(s)$ splits as $R_+ + R_-$, where

$$R_+(s) = \frac{1}{2\pi i}\int\limits_{A}\frac{d\xi}{\xi - s}R(\xi) \tag{3.31}$$

is analytic above the contour A [see Fig. 3.15a] while

$$R_-(s) = -\frac{1}{2\pi i}\int\limits_{B}\frac{d\xi}{\xi - s}R(\xi) \tag{3.32}$$

is analytic below the contour B [see Fig. 3.15b]. By construction for both integrals the point s lies to one side of the respective contour. A crossing of contours is not allowed. The above procedure yields the sum decomposition we were looking for.

The product decomposition can be deduced from the sum decomposition. Let us suppose that $K(s)$ is analytic for $\alpha < \text{Im}s < \beta$ and we want to split it into a product

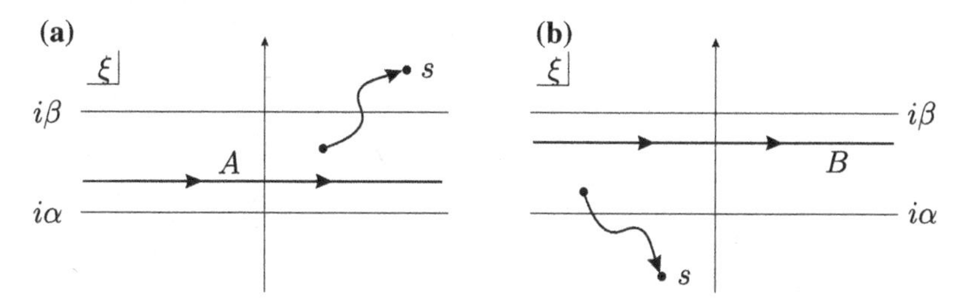

Fig. 3.15 Analyticity domain of function R_+, s. Eq. (3.31) (**a**), analyticity domain of function R_-, s. Eq. (3.32) (**b**)

$$K(s) = K_+(s)K_-(s)$$

of $K_\pm$, which are analytic for $\text{Im}s > \alpha$ and $\text{Im}s < \beta$, respectively. This problem is equivalent to the sum decomposition problem. Indeed,

$$\ln K(s) = \ln K_+(s) + \ln K_-(s).$$

Therefore the product decomposition problem for the function $K(s)$ is the sum decomposition problem for the function $\ln K(s)$, and we immediately obtain

$$K_+(s) = \exp\left[\frac{1}{2\pi i}\int\limits_A \frac{d\xi}{\xi - s}\ln K(\xi)\right] \tag{3.33}$$

and

$$K_-(s) = \exp\left[-\frac{1}{2\pi i}\int\limits_B \frac{d\xi}{\xi - s}\ln K(\xi)\right]. \tag{3.34}$$

Obviously, it follows that $\ln K(s)$ should satisfy the conditions of the sum decomposition. That is why $K(s)$ should not have zeros in the fundamental strip. Moreover, the use of logarithm also entails that $\arg K(s)$ should vary less than 2π along paths A and B. To illustrate the found decomposition procedures we now consider several examples.

Example 3.12
Find the sum decomposition for

$$R(s) = \frac{1}{s^2 + 1}$$

in the strip $-1 < \text{Im}s < 1$.

The function vanishes in the limit $s \to \infty$ in all directions. According to the above prescriptions for the function analytic for $s_2 > -1$ we have

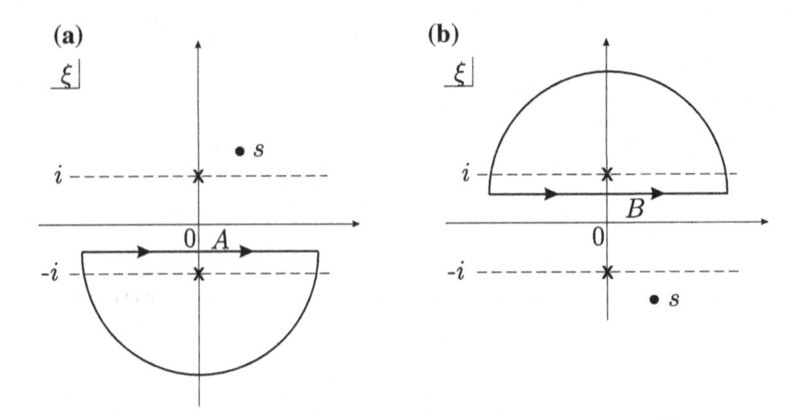

Fig. 3.16 Two ways to close the contour in Examples 3.12 and 3.13

$$R_+(s) = \frac{1}{2\pi i}\int\limits_A \frac{d\xi}{\xi - s}\frac{1}{\xi^2+1}.$$

Close the contour in the lower half-plane [see Fig. 3.16a], then

$$R_+(s) = \frac{1}{2\pi i}\,(-2\pi i)\;\text{Res}[....]\Big|_{\xi=-i} = -\frac{1}{(-i-s)(-2i)} = \frac{i/2}{s+i}.$$

Note that alternatively one can close the contour in the upper half-plane. Then

$$\begin{aligned} R_+(s) &= \frac{1}{2\pi i}(2\pi i)\Big[\text{Res}[....]\Big|_{\xi=i} + \text{Res}[....]|_{\xi=s}\Big] = \frac{1}{s^2+1} + \frac{1}{(i-s)(2i)} \\ &= \frac{1}{(s+i)(s-i)} - \frac{1}{(2i)(s-i)} = \frac{2i-(s+i)}{(2i)(s^2+1)} = \frac{-1/2i}{s+i}, \end{aligned}$$

which is essentially the same result as far as the analytic properties are concerned. The integral for the 'minus'–function is

$$R_-(s) = -\frac{1}{2\pi i}\int\limits_B \frac{d\xi}{\xi - s}\;\frac{1}{\xi^2+1}.$$

Closing the contour in the upper half-plane, see Fig. 3.16b, we obtain[8]

$$R_-(s) = -\frac{1}{2\pi i}\;(2\pi i)\text{Res}[....]|_{\xi=i} = -\frac{1}{(i-s)(2i)} = \frac{-i/2}{s-i}.$$

---

[8] Closing the contour in the lower half-plane gives, of course, the same result.

Evidently, the relation $R_+(s) + R_-(s) = R(s)$ holds. With $R_+(s)$ at hand there is no necessity to explicitly calculate $R_-(s)$ as $R_-(s) = R(s) - R_+(s)$.

The next example is slightly more complicated.

Example 3.13
We want to decompose the following function:

$$R(s) = \frac{e^{-is}}{s^2+1}, \tag{3.35}$$

again in the strip $-1 < \text{Im}s < 1$.

Let us calculate

$$R_+(s) = \frac{1}{2\pi i}\int_A \frac{d\xi}{\xi - s}\frac{e^{-i\xi}}{\xi^2+1}.$$

We have $|e^{-i\xi}| = e^{\text{Im}\,\xi} \to 0$ for $\text{Im}\xi \to -\infty$. So here we can not close the contour in the upper half-plane and must close it in the lower one, as shown in Fig. 3.16a:

$$R_+(s) = \frac{1}{2\pi i}(-2\pi i)\text{Res}[....]|_{\xi=-i} = \frac{\frac{i}{2}e^{-1}}{s+i}.$$

(cf. previous example). The 'minus'–part is given by

$$R_-(s) = R(s) - R_+(s) = \frac{e^{-is}}{s^2+1} - \frac{\frac{i}{2}e^{-1}}{s+i}.$$

We can understand this result without calculating the integral. Observe that in (3.35) $e^{-is} = e^{-i\text{Re}\,s}e^{\text{Im}\,s}$. It decays exponentially for $\text{Im}s \to -\infty$ but diverges at $\text{Im}s \to +\infty$. Hence e^{-is} is a 'minus'–function, analytic everywhere in the lower half-plane including $\text{Im}s \to -\infty$ (remember that a condition for the 'minus'–function is not only analyticity but also boundedness at infinity). Next write

$$R(s) = \frac{\frac{i}{2}e^{-is}}{s+i} + \frac{\frac{-i}{2}e^{-is}}{s-i}.$$

The second term is a proper 'minus'–function. The first term is neither a 'plus'– nor a 'minus'–function. However, it would be a 'minus'–function but for the pole at $s = -i$. To get a proper 'minus'–function subtract this pole off:

$$\frac{\frac{i}{2}e^{-is}}{s+i} = \frac{\frac{i}{2}e^{-is}}{s+i} - \frac{\frac{i}{2}e^{-1}}{s+i} + \frac{\frac{i}{2}e^{-1}}{s+i}.$$

Thus

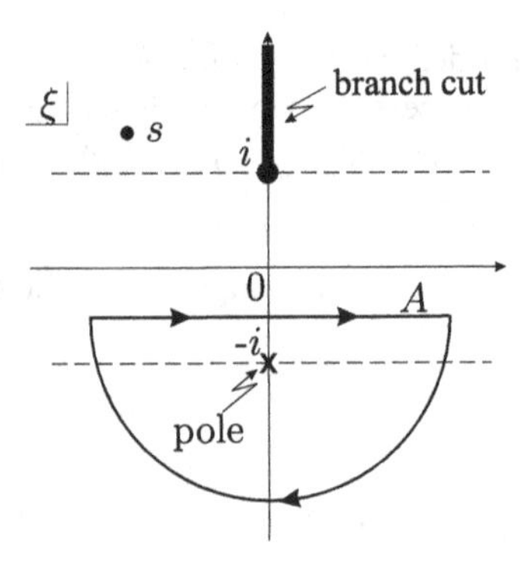

Fig. 3.17 Illustration of how to close the contour in Example 3.14

$$R(s) = \underbrace{\frac{\frac{i}{2}e^{-1}}{s+i}}_{R_+(s)} + \underbrace{\frac{\frac{i}{2}(e^{-is} - e^{-1})}{s+i} - \frac{\frac{i}{2}e^{-is}}{s-i}}_{R_-(s)}.$$

The decomposition can be accomplished in essentially the same way also for functions with branch cut singularities. We do that in the next example.

Example 3.14
Compute the decomposition of the function

$$R(s) = \frac{(s-i)^{1/2}}{s+i}$$

in the strip $-1 < \text{Im}s < 1$. For the 'plus'–function we have to compute the following integral

$$R_+(s) = \frac{1}{2\pi i}\int_A \frac{d\xi}{\xi - s}\,\frac{(\xi - i)^{1/2}}{\xi + i}.$$

To avoid the branch cut close the contour in the lower half-plane, as is shown in Fig. 3.17:

$$R_+(s) = \frac{1}{2\pi i}(-2\pi i)\text{Res}[....]|_{\xi=-i} = \frac{(-2i)^{1/2}}{s+i}.$$

Then $R_-(s) = R(s) - R_+(s)$ and still has a branch cut in the upper half-plane. On the other hand, we can apply ideas from the previous example. We observe that $R(s)$ is almost a 'minus'–function but for the pole at $s = -i$, so subtract the pole off which yields

$$R(s) = \frac{(s-i)^{1/2}}{s+i} = \underbrace{\frac{(s-i)^{1/2}}{s+i} - \frac{(-2i)^{1/2}}{s+i}}_{\text{'minus'}-\text{function}} + z\underbrace{\frac{(-2i)^{1/2}}{s+i}}_{\text{'plus'}-\text{function}},$$

making the decomposition complete.

If the function $K(s)$ contains a finite number of zeros s_i with multiplicities α_i in the analyticity strip, the factorization procedure in general case is slightly more complicated. In this case we introduce an auxiliary function

$$\Phi(s) = \ln\left[\frac{(s^2+b^2)^{N/2}}{\prod_i (s-s_i)^{\alpha_i}}\, K(s)\right], \qquad b \geq |\alpha|, |\beta|. \tag{3.36}$$

Here the denominator $\prod_i$ compensates the zeros of $K(s)$ in the strip. The numerator guarantees the asymptotic behavior at infinity. For that reason N is a total number of zeros s_i including their multiplicity. The positive constant b is introduced in order to avoid any additional zeros inside the strip. Hence the auxiliary function $\Phi(s)$ satisfies all required conditions: $\Phi(s)$ is analytic and has no zeros inside the strip, and tends to the true asymptotics of $K(s)$ as $|s| \to \infty$ in the strip.

To see how this factorisation procedure works in practice, we consider the original Milne problem (3.15),

$$K(s) = 1 - \frac{1}{2}\int_{-\infty}^{+\infty} e^{isx}\left(\int_0^1 e^{-\frac{|x|}{\mu}}\,\frac{d\mu}{\mu}\right)dx.$$

Since $e^{-|x|} = e^{-x}$ at $x > 0$ and $e^{-|x|} = e^{x}$ at $x < 0$, for the double integral in the rhs we have

$$\frac{1}{2}\int_{-\infty}^{+\infty} \ldots.\, dx = \frac{1}{2}\left[\int_{-\infty}^{0} e^{isx}\left(\int_0^1 e^{\frac{x}{\mu}}\,\frac{d\mu}{\mu}\right)dx + \int_0^{+\infty} e^{isx}\left(\int_0^1 e^{-\frac{x}{\mu}}\,\frac{d\mu}{\mu}\right)dx\right]$$

and changing in both integrals the order of the integration and integrating then over x we obtain

$$= \frac{1}{2}\int_0^1 \frac{d\mu}{\mu}\left(\frac{1}{is+\frac{1}{\mu}} - \frac{1}{is-\frac{1}{\mu}}\right) = \int_0^1 \frac{d\mu}{s^2\mu^2+1} = \frac{\arctan(s)}{s} = \frac{1}{2is}\ln\frac{1+is}{1-is}.$$

So the function which we wish to factorise is

$$K(s) = \frac{s - \arctan(s)}{s}.$$

$K(s)$ is analytic in the strip $-1 < \operatorname{Im} s < 1$ (that is $|\alpha| = |\beta| = 1$) where it has double zero at $s = 0$ (so that $s_i = 0$, $\alpha_i = 2$, $N = 2$). Unlike in the simpler case of Example 3.7, the correct factorization cannot be guessed. Therefore we must perform the above prescribed procedure explicitly. So we introduce the auxiliary function

$$\Phi(s) = \ln\Big[\frac{s^2+1}{s^2} K(s)\Big]$$

and decompose it as $\Phi(s) = \Phi_+(s) - \Phi_-(s)$, in accordance with the Cauchy integral formula, where

$$\Phi_+(s) = \frac{1}{2\pi i} \int\limits_{-\infty - i\gamma}^{+\infty - i\gamma} \ln\Big[\frac{\xi^2+1}{\xi^2}\Big(1 - \frac{\arctan \xi}{\xi}\Big)\Big] \frac{d\xi}{\xi - s} \tag{3.37}$$

and $\Phi_-(s)$ contains the same integrand but the integration limits are $-\infty + i\gamma$ and $+\infty + i\gamma$. Then the kernel $K(s)$ is decomposed as $K(s) = K_+(s)K_-(s)$ with

$$K_+(s) = \frac{s^2}{s+i}\, e^{\Phi_+(s)}, \quad K_-(s) = \frac{1}{s-i}\, e^{-\Phi_-(s)}. \tag{3.38}$$

Note that $\Phi(\xi)$ is an even function of ξ. Therefore for $s \neq 0$, taking $\gamma = 0$ the integral (3.37) reduces to

$$\Phi_+(s) = \frac{s}{\pi i} \int\limits_0^{\infty} \ln\Big[\frac{\xi^2+1}{\xi^2}\Big(1 - \frac{\arctan \xi}{\xi}\Big)\Big] \frac{d\xi}{\xi^2 - s^2}$$

and can be evaluated numerically. For a special case of $s = 0$, the singular point $\xi = 0$ can be avoided by a semicircle in the lower half-plane and letting its radius to zero we obtain

$$\Phi_+(0) = \frac{1}{2} \lim_{\rho \to 0} \ln\Big[\frac{\rho^2+1}{\rho^2}\Big(1 - \frac{\arctan \rho}{\rho}\Big)\Big] = -\ln \sqrt{3}.$$

One very interesting example of a useful kernel regularization is discussed in Refs. [21, 22]:

Example 3.15
(*Spitzer's equation*) Solve the homogeneous equation

$$f(x) = \frac{\lambda}{\pi} \int_0^{\infty} dy \frac{f(y)}{1 + (x-y)^2}, \quad x \geq 0 \tag{3.39}$$

for the special case $\lambda = 1$.

As usual we proceed with the Fourier transformation of both sides,

$$K'(s)\, F_+(s) = F_+(s) + G_-(s),$$

where the transform of the kernel were calculated in Example 1.14, page 51,

$$K'(s) = \lambda e^{-|s|}.$$

We observe that since the kernel is of algebraic nature the analyticity strip of the transform collapses to a line—the real axis. Thus

$$K(s)\,F_+(s) + G_-(s) = 0 \quad \text{with} \quad K(s) = 1 - \lambda e^{-|s|}.$$

In the next step we compute an appropriate factorization of $K(s)$. Here we are confronted with a difficulty at $\lambda = 1$. Namely $K(s)$ has a cusp at $s = 0$, see Fig. 3.18, so that taking a logarithm in accordance with the prescriptions (3.33) and (3.34) creates an inconvenient singularity. We can circumvent this difficulty by the following trick. We first observe that similarly to the transformation (3.36) we can perform a 'sub factorization' of the kind:

$$K(s) = \psi_a(s)\,\psi_b(s)$$

with appropriately chosen $\psi_{a,b}(s)$. For instance we can use $\psi_a(s) = |s|$ and

$$\psi_b(s) = \frac{1 - e^{-|s|}}{|s|}.$$

This particular choice of $\psi_a(s)$ has a cusp at $s \to 0$ as well but this is inessential since this kind of function can easily be factorized by inspection without using (3.33) and (3.34). The advantage of the new $\psi_b(s)$ is that it is well behaved at $s \to 0$. It is useless at $s \to \infty$ though as it would lead to logarithmically divergent integrand in the decomposition prescription. This can be helped by taking

$$\psi_b(s) = \frac{\sqrt{1+s^2}}{|s|}\left(1 - e^{-|s|}\right), \quad \psi_a(s) = \frac{|s|}{\sqrt{1+s^2}}.$$

The $\pm$-factorization of $\psi_a(s)$ is achieved by taking

$$\psi_a(s) = \psi_{a+}(s)/\psi_{a-}(s), \quad \psi_{a-}(s) = \frac{\sqrt{s-i}}{\sqrt{s}}, \quad \psi_{a+}(s) = \frac{\sqrt{s}}{\sqrt{s+i}}.$$

The multivaluedness of the latter two functions is fixed by the standard choice $\pm i = e^{\pm i\pi/2}$. In order to factorize $\psi_b(s)$ we use (3.33), then[9]

[9] The computation of the first part is simple after making sure that

$$\int_0^\infty dz\,\frac{\ln z^2}{z^2+\zeta^2} = \frac{\pi}{\zeta}\ln\zeta, \quad \int_0^\infty dz\,\frac{\ln(1+z^2)}{z^2+\zeta^2} = \frac{\pi}{\zeta}\ln(1+\zeta).$$

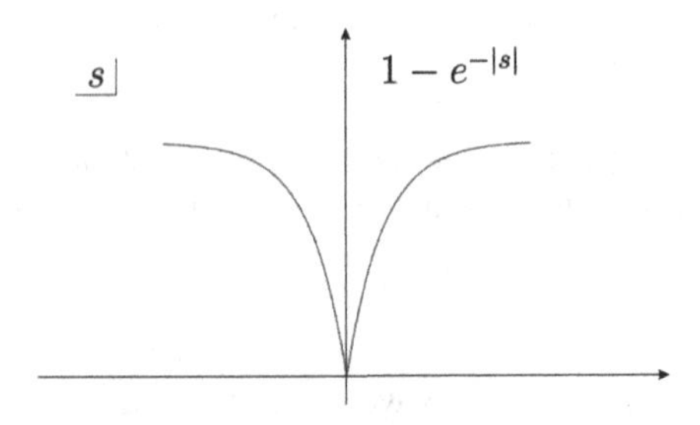

Fig. 3.18 Behaviour of the kernel of Spitzer's equation around $s = 0$

$$\ln \psi_{b+}(s) = \frac{1}{2\pi i}\int_{-\infty}^{\infty} dz\, \frac{1}{z-s} \ln \left[(1 - e^{-|z|}) \frac{\sqrt{1+z^2}}{|z|} \right]$$
$$= \frac{1}{2}\ln(1 + i/s) + \frac{s}{i\pi}\int_0^{\infty} dz \frac{\ln(1-e^{-z})}{z^2 - s^2}. \tag{3.40}$$

The last integral can be computed with the help of a parameter-dependent integral (we use $s = i\zeta$)

$$I(\alpha) = \frac{\zeta}{\pi}\int_0^{\infty} dz \frac{\ln(1 - e^{-\alpha z})}{z^2 + \zeta^2}.$$

Upon differentiation with respect to α we obtain an integral related to the Psi function,[10]

$$I'(\alpha) = \frac{\zeta}{\pi}\int_0^{\infty} dz \frac{z}{e^{\alpha z} - 1}\frac{1}{\zeta^2 + z^2} = \frac{\zeta}{2\pi}\ln\left(\frac{\alpha\zeta}{2\pi}\right) - \frac{1}{2\alpha} - \frac{\zeta}{2\pi}\Psi\left(\frac{\alpha\zeta}{2\pi}\right).$$

Therefore after integrating up we obtain

$$I(\alpha) = \left(\frac{\alpha\zeta}{2\pi} - \frac{1}{2}\right)\ln\left(\frac{\alpha\zeta}{2\pi}\right) - \ln\Gamma\left(\frac{\alpha\zeta}{2\pi}\right) + C,$$

where C is an α-independent constant. Since obviously $I(\alpha \to \infty) \to 0$ this constant can be computed using the asymptotics of the Gamma function, given by the Stirling's formula (2.9). Then $C = (1/2)\ln(2\pi)$ and the last integral of (3.40) is given by

$$I(1) = \left(\frac{\zeta}{2\pi} - \frac{1}{2}\right)\ln\left(\frac{\zeta}{2\pi}\right) - \frac{\zeta}{2\pi} + \frac{1}{2}\ln(2\pi) - \ln\Gamma\left(\frac{\zeta}{2\pi}\right).$$

From this we find

$$K_+(\zeta) = \psi_{a+}(\zeta)\,\psi_{b+}(\zeta) = \sqrt{\zeta}\left(\frac{\zeta}{2\pi}\right)^{\zeta/2\pi - 1} e^{-\zeta/2\pi}\,\Gamma^{-1}(\zeta/2\pi).$$

[10] For full details see for instance Sect. 12.32 of [3].

Therefore we have now all ingredients of our fundamental form

$$K_+(s)\,F_+(s) = -K_-(s)\,G_-(s),$$

which defines some analytic function $E(s)$, but not necessarily at $s = 0$ (which is our 'special' point by construction of the factorization), where it can have a pole. Let us assume that $F_+(s)$ is constant in the limit $s \to \infty$, then, since $K_+(\zeta \to \infty) = 1/\sqrt{2\pi}\ E(s)$ is a constant as well (remember that $\zeta = -is$).

But let us now suppose that we are looking for a function $F_+(s)$, which approaches zero as $s \to \infty$ and is bounded at $s \to 0$, so that $|s|F_+(s) \sim O(|s|^0)$. Since $K_+(\zeta \to 0) \sim \sqrt{\zeta}$, for the function $E(s)$ we have at least the limiting behaviour $\sim s^{-1/2}$. The only possibility to satisfy this constraint is to endow $E(s)$ with a pole and to plainly set it to be $1/s$. Then the solution of the original Eq. (3.39) is given by the following integral

$$f_+(x) \sim \int_{\gamma-i\infty}^{\gamma+i\infty} d\zeta\, \left(\frac{e}{\zeta}\right)^{\zeta+1/2} e^{2\pi x\zeta}\,\Gamma(\zeta).$$

This integral cannot be calculated analytically. But the leading asymptotics for large x is accessible and is given by $\sim \sqrt{x}$. This makes the integral in (3.39) absolutely convergent.

3.3.3 Exponential Kernel

Here we first discuss at some length a simple but educational example provided by the generalized version of the Eq. (3.21) from the Example 3.10 of the previous section, which we now formulate as an eigenvalue problem,

$$f(x) = \lambda \int_0^\infty e^{-|x-y|} f(y)dy. \tag{3.41}$$

We begin by repeating the first steps of the Wiener–Hopf technique. The ordinary Fourier transform of the above equation is

$$K(s)\,F_+(s) + G_-(s) = 0,$$

where the notation $K(s)$ is, according to the last section, reserved for

$$K(s) = 1 - \lambda \int_{-\infty}^{\infty} e^{isx-|x|}dx = \frac{s^2 + (1-2\lambda)}{s^2+1},$$

which is absolutely convergent and analytic in the strip $-1 < \mathrm{Im}s < 1$, $F_+(s)$ is the right-hand Fourier transform of $f(x)$ and $G_-(s)$ is the left-hand transform of

$$g_-(x) = \lambda \int_0^\infty e^{-|x-y|} f(y)dy = \left[\lambda \int_0^\infty e^{-y} f(y) \right] e^x = f_+(0)e^x.$$

Here we made use of the fact that, in the definition of $g_-(x)$, $x < 0$ and $y > 0$ so that $|x - y| = y - x$ and the exponential factorises. As a result the function $g_-(x)$ is known explicitly, up to a multiplicative factor.[11] Consequently the Fourier transform

$$G_-(s) = f_+(0) \int_{-\infty}^0 e^{(is+1)x} dx = -\frac{if_+(0)}{s - i}$$

is absolutely convergent and analytic in $\mathrm{Im}s < 1$. As to the function $f_+(x)$ the y-integral in the integral equation must converge, so it has to be bounded

$$|f_+(x)| < Ae^{(1-\delta)x}$$

when $x \to +\infty$, where A is a positive constant and $0 < \delta < 1$. The natural analyticity strip for this problem, where all three functions entering the algebraic form of the Wiener–Hopf equation are analytic, is therefore $1 - \delta < \mathrm{Im}s < 1$. The next question is how the zeros of $K(s)$ are situated with respect to the analyticity strip. There are three distinct regions of λ:

(i) $2\lambda - 1 > 0$ ($\lambda > 1/2$), where the zeros lie on the real axis and so below the analyticity strip (clearly $\lambda = 1/2$ is a special point when the two zeros merge in a double zero at $s = 0$);
(ii) $-1 < 2\lambda - 1 < 0$ ($0 < \lambda < 1/2$), where the zeros are now on the imaginary axis but still below the analyticity strip;
(iii) $2\lambda - 1 < -1$ ($\lambda < 0$), where the zeros are again on the imaginary axis but now one of them is situated above the strip.

These three regions need be considered separately.

(i) Put $2\lambda - 1 = p^2$ with p real. The appropriate Wiener–Hopf decomposition is:

$$K(s) = \frac{(s-p)(s+p)}{s+i} \frac{1}{s-i} = K_+(s)K_-(s).$$

The entire function $E(s)$ (the Liouville polynomial) is, as always, defined via

$$\frac{s^2 - p^2}{s+i} F_+(s) = -(s-i)G_-(s) \equiv E(s).$$

[11] This is on no account a generic feature and is absent for more complicated kernels.

From the asymptotic form of both the rhs and the lhs of this equation we immediately conclude that the Liouville polynomial is a constant: $E(s) = if_+(0)$ for all s. Therefore

$$F_+(s) = if_+(0)\frac{s+i}{s^2 - p^2}.$$

Using the Fourier inversion formula after a simple residue calculation we arrive at

$$f_+(x) = \frac{if_+(0)}{2\pi} \int\limits_P \frac{(u+i)e^{-iux}}{u^2 - p^2} du = f_+(0)\Big[\cos(px) + \frac{\sin(px)}{p}\Big]. \quad (3.42)$$

Note that the limit $p \to 0$ is perfectly well defined leading to the solution

$$f_+(x) = f_+(0)(1+x)$$

for the special case $\lambda = 1/2$, which we already have encountered in (3.25).

(ii) In this case we put $2\lambda - 1 = -\kappa^2$ with κ real such that $0 < \kappa < 1$. The rest of the calculation is very similar to the above because the zeros are still below the analyticity strip as in case **(i)** so that the kernel decomposition proceeds in the same way with the substitution of p^2 by $-\kappa^2$. Therefore we can immediately write down:

$$F_+(s) = if_+(0)\frac{s+i}{s^2 + \kappa^2}.$$

Fourier inversion then yields

$$f_+(x) = \frac{if_+(0)}{2\pi} \int\limits_P \frac{(s+i)e^{-isx}}{s^2 + \kappa^2} ds = f_+(0)\left[\cosh(\kappa x) + \frac{1}{\kappa}\sinh(\kappa x)\right]. \quad (3.43)$$

We observe that the limit $\kappa \to 0$ is still well defined.

(iii) Again $2\lambda - 1 = -\kappa^2$ but $\kappa > 1$. The zero at $s = i\kappa$ is now above the analyticity strip, so the kernel decomposition here is:

$$K(s) = \frac{s+i\kappa}{s+i}\frac{s-i\kappa}{s-i}$$

and the Liouville polynomial is determined by

$$\frac{s+i\kappa}{s+i}F_+(s) = -\frac{s-i}{s-i\kappa}G_-(s) \equiv E(s).$$

We see that for $s \to \infty$ both the rhs and the lhs tend to zero. Thus $E(s)$ is identically zero for all s and there is no non-trivial solution to the original equation for these values of λ.

To summarise, the spectrum consists of the interval $\lambda \in (0, 1/2)$, where the solutions are growing exponentially but not fast enough as to make the integral equation divergent, the point $\lambda = 1/2$, where the solution grows algebraically, and the interval $\lambda \in (1/2, \infty)$ with oscillating solutions.

An alternative approach which we are going to present next, would allow to see how these results are connected with the spectral theory and quantum mechanics. Indeed, consider a more general equation on a finite interval:

$$f(x) = \lambda \int_a^b e^{-|x-y|} f(y) dy. \tag{3.44}$$

The original problem can then be recovered in the limit $a = 0$, $b \to \infty$. Differentiating once with respect to x we obtain

$$f'(x) = -\lambda \int_a^b \operatorname{sgn}(x-y) e^{-|x-y|} f(y) dy.$$

For the second derivative we obviously get

$$f''(x) = \lambda \int_a^b \left[e^{-|x-y|} - 2\delta(x-y) \right] f(y) = (1 - 2\lambda) f(x).$$

We see that our integral equation reduces to a second order ODE

$$f''(x) + (2\lambda - 1) f(x) = 0, \tag{3.45}$$

which is a consequence of the fact that the exponential kernel itself satisfies the ODE

$$\left(\frac{d^2}{dx^2} - 1 \right) e^{-|x|} = -2\delta(x).$$

The general solution of the ODE (3.45) is

$$f(x) = \begin{cases} Ae^{ipx} + Be^{-ipx}, & \lambda > 1/2, \\ A + Bx, & \lambda = 1/2, \\ Ae^{\kappa x} + Be^{-\kappa x}, & \lambda < 1/2. \end{cases}$$

Any function $f(x)$ satisfying the integral equation will also satisfy the ODE but not vice versa. For a general kernel there are no obvious boundary conditions for the ODE and the only way to proceed is to substitute the candidate solution back into the integral equation. However, for a particular case of the exponential kernel there is a simplification. Indeed, from the first derivative formula above we see that at the ends of the interval

$$f'(a) = -\lambda \int_a^b \mathrm{sgn}(a-y)e^{-|a-y|} f(y)dy = \lambda \int_a^b e^{-|a-y|} f(y)dy = f(a)$$

and similarly

$$f'(b) = -\lambda \int_a^b \mathrm{sgn}(b-y)e^{-|b-y|} f(y)dy = -f(b).$$

Thus, the two twisted boundary conditions

$$f'(a)/f(a) = 1, \quad f'(b)/f(b) = -1$$

at the ends of the interval single out those solutions of the ODE which also solve the integral equation. Let us impose slightly more general boundary conditions

$$f'(a)/f(a) = -f'(b)/f(b) = \eta,$$

where $\eta > 0$.[12] Imposing these boundary conditions is straightforward. Starting with $\lambda > 1/2$, we have

$$f'(a) = ip\left(Ae^{ipa} - Be^{-ipa}\right) = \eta\left(Ae^{ipa} + Be^{-ipa}\right),$$

so re-arranging and adding a similar condition at $x = b$ we arrive at

$$\begin{aligned} (\eta - ip)e^{ipa} A + (\eta + ip)e^{-ipa} B &= 0, \\ (\eta + ip)e^{ipb} A + (\eta - ip)e^{-ipb} B &= 0. \end{aligned} \tag{3.46}$$

The respective determinant must vanish resulting in (we define $l = b - a$)

$$\tan(pl) = \frac{2\eta p}{p^2 - \eta^2}, \tag{3.47}$$

[12] These boundary conditions are generated by the equation

$$f(x) = \lambda \int_a^b e^{-\eta|x-y|} f(y)dy.$$

For the ODE to look the same we need to modify the relation between λ and p as $\eta^2 - 2\eta\lambda = p^2$ and similarly for κ. The factor η, as long as it is positive, can be re-absorbed by the re-scaling $x \to x/\eta$, $p \to \eta p$, and $\lambda \to \eta\lambda$.

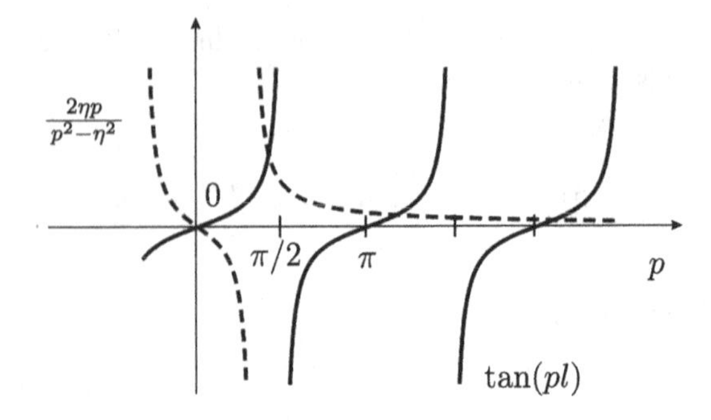

Fig. 3.19 Graphical solution of the equation (3.47) for $\lambda > 1/2$ (real p)

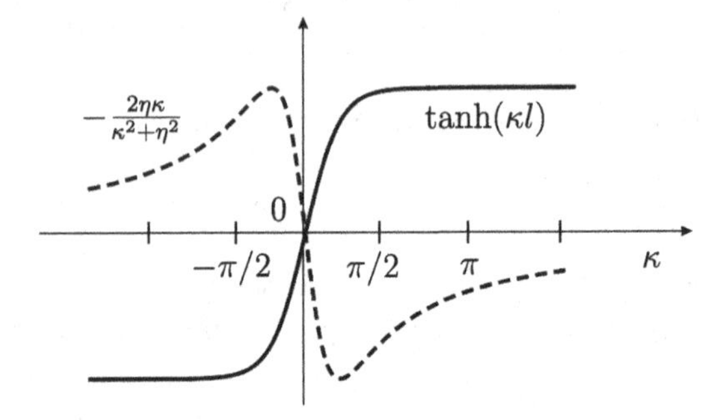

Fig. 3.20 Graphical solution for the situation $\lambda < 1/2$ (imaginary $p = i\kappa$)

which is a transcendental equation for p and so for λ. Presented graphically it can be seen that this equation has infinite number of solutions constituting a (countable) increasing sequence $\{\lambda\} = \{\lambda_0 < \lambda_1 < \lambda_2 ...\}$, see Fig. 3.19. This is as expected since the exponential kernel over a finite interval is completely regular and its spectrum satisfies all the standard propositions of Fredholm theory. For $\lambda < 1/2$, a similar calculation leads to the equation

$$\tanh(\kappa l) = -\frac{2\eta\kappa}{\kappa^2 + \eta^2},$$

which has no solutions (remember that $\eta > 0$) apart from $\kappa = 0$ which is not really a solution as it is also required that $A = -B$, see Fig. 3.20. In the special case $\lambda = 1/2$ we have $f(x) = A + Bx$ and $f'(x) = B$, so the corresponding determinant requires $l = -2/\eta$, which is a contradiction since $l > 0$.

We now want to recover the solutions to the original Eq. (3.41) by putting $a = 0$, which is easy, and taking the limit $b\ (= l)$ to infinity, which is more delicate. For $l \to \infty$, discrete p_n solutions to Eq. (3.47) clearly become dense in the interval $p \in (0, \infty)$ corresponding to $\lambda \in (1/2, \infty)$. The boundary condition $f'(0)/f(0) = \eta$ remains intact. In order to see what happens to the second boundary condition, we calculate the rhs of (3.44) assuming an exponential solution e^{ipy} (by standard integration, putting $\eta = 1$ for simplicity):

$$\lambda \int_a^b e^{-|x-y|} e^{ipy} dy = e^{ipx} - \frac{1}{2}(1 - ip)e^{ipa} e^{a-x} - \frac{1}{2}(1 + ip)e^{ipb} e^{x-b},$$

so, requiring that there are no real exponentials in $\int_a^b e^{-|x-y|}(Ae^{ipy} + Be^{-ipy})dy$, we would have obtained

$$\begin{aligned} e^{a}(1 - ip)e^{ipa} A + e^{a}(1 + ip)e^{-ipa} B &= 0, \\ e^{-b}(1 + ip)e^{ipb} A + e^{-b}(1 - ip)e^{-ipb} B &= 0, \end{aligned} \tag{3.48}$$

instead of Eq. (3.46). For finite a and b the two pairs of equations are equivalent. In the limit $b \to \infty$, however, Eq. (3.48) takes precedence and the second boundary condition disappears. So, in this limit we do not have the determinant condition any more but only $f'(0)/f(0) = \eta$ [the first line in Eq. (3.48)], that can be satisfied for all p provided that the relation

$$\frac{B}{A} = \frac{p + i\eta}{p - i\eta}$$

holds. Writing the solution out in the real form, we see that

$$f(x) = f_p(x) = \cos(px) + \frac{\eta}{p} \sin(px),$$

up to a multiplicative constant (which is itself not important but its dependence on p is, as we shall see shortly). We have no second boundary condition any more, therefore we can satisfy $f'(0)/f(0) = \eta$ with $f(x) = 1 + \eta x$ for the special case $\lambda = 1/2$, and with

$$f_\kappa(x) = \cosh(\kappa x) + \frac{\eta}{\kappa} \sinh(\kappa x) \tag{3.49}$$

for $0 < \lambda < 1/2$.

These are precisely the solutions (3.43) found previously by the Wiener-Hopf technique. What we learn from the differential equation analysis, however, is that solutions for $\lambda \in (0, 1/2]$ do not have any analog (do no exist) on a finite interval and are specific to the problem with strictly semi-infinite integration range. On the other hand the solutions for $\lambda \in (1/2, \infty)$ evolve more or less continuously from the finite interval to the half-infinite case. One therefore expects that the latter solutions (but not the former) describe quantum mechanics of the operator

$$\psi''(x) + p^2 \psi(x) = 0, \tag{3.50}$$

which is simply that of a free particle on a half-line with the boundary condition $f'(0)/f(0) = \eta$. Such solutions would have to form a complete and ortho-normal set, the latter condition being

$$\int_0^\infty \psi_p^*(x)\psi_{p'}(x)dx = \delta(p-p'). \tag{3.51}$$

We now write

$$\psi_p(x) = N_p f_p(x) = N_p \left[\cos(px) + \frac{\eta}{p}\sin(px)\right],$$

which satisfies the boundary condition, where N_p is a normalisation constant we determine next. Recalling that the ortho-normality for the cosine Fourier transform implies that ($x > 0$)

$$\int_0^\infty \cos(px)\cos(p'x)dx = \frac{\pi}{2}\delta(p-p')$$

and similarly for sine and that the two are mutually orthogonal, one easily finds

$$\int_0^\infty \psi_p^*(x)\psi_{p'}(x)dx = \frac{\pi}{2}\left(1+\frac{\eta^2}{p^2}\right)N_p^2\delta(p-p').$$

Therefore, with

$$N_p = \sqrt{\frac{2}{\pi}}\frac{p}{\sqrt{p^2+\eta^2}},$$

the standard completeness condition holds:

$$\int_0^\infty \psi_p^*(x)\psi_p(y)dp = \delta(x-y). \tag{3.52}$$

If we rewrite this resolution of the identity operator as an identity for an arbitrary function $f(x)$ then we arrive at the equation, which can be understood as an expansion formula

$$f(x) = \frac{2}{\pi}\int_0^\infty \frac{p^2 f_p(x)}{p^2+\eta^2}dp \int_0^\infty f_p(y)f(y)dy, \tag{3.53}$$

which is the same as the version of (3.1.1) given in Sect. 4.1 of [23] with the identification $p^2 = \lambda$ and $\eta = \cot\alpha$. Let us qualify the argument involved here. We can verify by direct integration that the set $\{f_p\}$ is orthogonal. Therefore the set $\{\psi_p\}$ is also ortho-normal with respect to the standard normalisation, Eq. (3.51), which fixes the normalisation function N_p. Now assuming the set $\{\psi_p\}$ being, in addition, complete, then also Eqs. (3.52) and (3.53) will hold with N_p pre-determined. However, we do not know whether the set $\{\psi_p\}$ is complete. That does not follow from the ortho-normality alone, and so we should check this by a direct integration or prove it in some other independent way. We shall do so shortly, in conjunction with the following question.

The integral equation only allows for $\eta > 0$, otherwise the kernel would grow exponentially. The respective quantum mechanical problem Eq. (3.50), on the other hand, allows for any real η in the boundary condition. The next question is therefore: are the quantum mechanical solutions for $\eta < 0$ of the same nature as those for $\eta > 0$? We see that the set $\{\psi_p\}$ continues to satisfy the ODE, the boundary condition, and remains ortho-normal for $\eta < 0$. The extra feature, compared with the situation for $\eta > 0$, is that one of the functions (3.49) at the special value of $\kappa = \eta$,

$$f_\eta(x) = e^{\eta x} = e^{-|\eta| x},$$

becomes a decaying exponential and so is an admissible quantum mechanical solution, which we recognise as a bound state. The normalised version is

$$\psi_\eta(x) = \sqrt{2|\eta|}e^{-|\eta|x}, \quad \int_0^\infty |\psi_\eta(x)|^2 dx = 1.$$

If $\psi_\eta(x)$ is a legitimate quantum mechanical solution, then it should be orthogonal to all continuum spectrum solutions $\psi_p(x)$. Indeed we see immediately that

$$\int_0^\infty e^{-|\eta|x}\left[\cos(px) - \frac{|\eta|}{p}\sin(px)\right]dx = 0$$

for all p. The set $\{\psi_p(x)\}$ therefore can not be a complete set, as the function $\psi_\eta(x)$ can not be expanded in terms of $\{\psi_p(x)\}$. Instead $\{\psi_\eta(x), \psi_p(x)\}$ will be the complete set provided we have not missed any more solutions. The direct check is accomplished by calculating the integral [cf. with (3.53)]

$$I(x, y) = \frac{2}{\pi}\int_0^\infty \frac{p^2 dp}{p^2 + \eta^2}\left[\cos(px) + \frac{\eta}{p}\sin(px)\right]\left[\cos(py) + \frac{\eta}{p}\sin(py)\right].$$

Opening up and simplifying we obtain

$$I(x, y) = \frac{1}{\pi} \int_0^\infty \left[\cos p(x-y) + \cos p(x+y) - \frac{2\eta^2}{p^2+\eta^2} \cos p(x+y) \right.$$
$$\left. + \frac{2\eta p}{p^2+\eta^2} \sin p(x+y) \right] dp.$$

The first term produces the delta function $\delta(x-y)$, the second term is equal to zero (remember that we are solving a problem on a semi-axis, so $x, y > 0$), the third term is a standard integral, after writing t for $x+y$ we have

$$\frac{2}{\pi} \int_0^\infty \frac{\cos(pt)dp}{p^2+\eta^2} = \frac{1}{|\eta|} e^{-|\eta|t},$$

then differentiating this result with respect to t accounts for the last term ($t > 0$)

$$\frac{2}{\pi} \int_0^\infty \frac{p \sin(pt)dp}{p^2+\eta^2} = e^{-|\eta|t}.$$

Collecting the results, we obtain

$$I(x, y) = \delta(x-y) + \eta[1 - \mathrm{sgn}(\eta)]\, e^{-|\eta|(x+y)}.$$

Put another way, the completeness condition for all η is

$$\int_0^\infty \psi_p(x)\psi_p(y)dp + \left\{ \begin{array}{cc} 0, & \eta > 0 \\ \psi_\eta(x)\psi_\eta(y), & \eta < 0 \end{array} \right\} = \delta(x-y).$$

Accordingly, the expansion formula (3.53) does not apply for $\eta < 0$ but should be modified as

$$f(x) = 2|\eta| f_\eta(x) \int_0^\infty f_\eta(y) f(y)dy + \frac{2}{\pi} \int_0^\infty \frac{p^2 f_p(x)}{p^2+\eta^2} dp \int_0^\infty f_p(y) f(y)dy.$$

To complete the discussion of Eq. (3.41) we consider the corresponding full fledged equation of the second kind with an inhomogeneity,

$$f(x) = p(x) + \lambda \int_0^\infty e^{-|x-y|} f(y)dy, \tag{3.54}$$

for the most interesting case of $\lambda < 0$, which is outside the spectrum of the homogeneous equation where a unique solution to (3.54) is expected. The kernel decomposition is the same as above, case (iii) on the page 155, naturally the only difference with the analysis of the eigenvalue equation is the appearance of the driving term on the rhs of the algebraic form. We assume $p(x)$ to be defined only for $x > 0$ [to be consistent with the lhs of the original Eq. (3.54)], thus its Fourier transform is a 'plus'–function $P_+(s)$. The equation is then

$$K_+(s)\,F_+(s) + \frac{G_-(s)}{K_-(s)} = R(s) = \frac{P_+(s)}{K_-(s)} = \frac{s-i}{s-i\kappa}P_+(s).$$

It is $R(s)$ which we need to sum-decompose. The function $P_+(s)$ is analytic in the strip and above it. We have $K_-(s) \to 1$ as $s \to \infty$, so the difference between $R(s)$ and $P_+(s)$ is significant at finite s only and this difference is the presence of the pole at $s = i\kappa$. Indeed, without it the function $R(s)$ would be a 'plus'–function. Therefore this is a good example of decomposition by inspection—in order to make a 'plus'–function out of $R(s)$ one has to subtract this pole and add it back on which will be the part of the 'minus'–counterpart:

$$R(s) = \underbrace{\frac{s-i}{s-i\kappa}P_+(s) - \frac{i(\kappa-1)}{s-i\kappa}P_+(i\kappa)}_{\text{'plus'}-\text{function}} + \underbrace{\frac{i(\kappa-1)}{s-i\kappa}P_+(i\kappa)}_{\text{'minus'}-\text{function}}\,.$$

The Liouville polynomial is found from

$$K_+(s)\,F_+(s) - R_+(s) = R_-(s) - G_-(s)/K_-(s) = E(s)$$

and is identically zero but, unlike the eigenvalue problem, the solution is finite and is given by

$$F_+(s) = \frac{R_+(s)}{K_+(s)} = \frac{s^2+1}{s^2+\kappa^2}P_+(s) - \frac{i(\kappa-1)(s+i)}{s^2+\kappa^2}P_+(i\kappa).$$

Rearranging terms and using the inversion formula we obtain:

$$f(x) = \frac{1}{2\pi}\int_P e^{-isx}\left[P_+(s) - \frac{i(\kappa-1)(s+i)}{s^2+\kappa^2}P_+(i\kappa) + \frac{2\lambda}{s^2+\kappa^2}\int_0^\infty e^{isy}\,p(y)dy\right]ds.$$

The contour P of the inversion formula lies below all singularities of $F_+(s)$ and can be taken to be in the analyticity strip. Completing the contour in the separate integrals above according to the sign of exponentials yields

$$f(x) = p(x) + \frac{\lambda}{\kappa}\frac{\kappa - 1}{\kappa + 1}P(i\kappa)e^{-\kappa x} + \frac{\lambda}{\kappa}\int_0^\infty e^{-\kappa|x-y|}\,p(y)dy.$$

This solution implies the resolvent

$$R(x, y; \lambda) = \frac{1}{\kappa}\frac{\kappa - 1}{\kappa + 1}e^{-\kappa(x+y)} + \frac{1}{\kappa}e^{-\kappa|x-y|}.$$

Unsurprisingly, this is exactly the *Green's function* of the operator

$$\left(\frac{d^2}{dx^2} + \kappa^2\right) R(x, y; \lambda) = -\delta(x - y),$$

with the boundary condition

$$[dR(x, y; \lambda)/dx]/R(x, y; \lambda)|_{x=0} = 1$$

and can alternatively be expressed as

$$R(x, y; \lambda) = \int_0^\infty \frac{\psi_p(x)\psi_p(y)}{p^2 + \kappa^2}dp,$$

using the previously obtained complete set $\{\psi_p(x)\}$.

Many other interesting applications of the Wiener-Hopf methods can be found in Ref. [19].

3.4 Singular Integral Equations

As we have seen in the previous section integral equations of the form

$$f(x) = p(x) + \int_0^\infty dt\, k(x - t)\, f(t)$$

are solvable using the Wiener-Hopf technique (λ is now assumed to be the part of the kernel). This technique required that $k(x) = O(e^{-\alpha|x|})$, $\alpha > 0$ as $x \to \pm\infty$, i.e. the kernel must be exponentially bounded for the analyticity strip to exist. If $k(x)$ decays on ∞, but not exponentially fast, for instance like $k(x) = O(1/|x|^\mu)$, $\mu > 0$, then the analyticity strip collapses to a line. This is one reason why it is so important to study this situation.

Another reason emerges if one looks for solutions of the following equation:

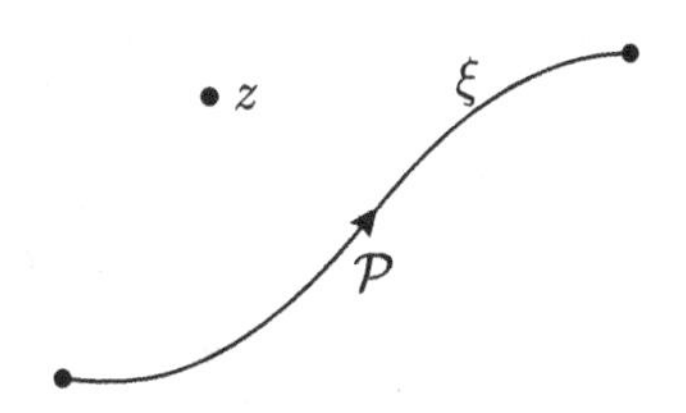

Fig. 3.21 Smooth path of integration between two different points in the complex plane. The respective Cauchy-type integral determines a function $F(z)$ analytic everywhere apart from the points on this path

$$f(x) = p(x) + \int_a^b dt\, k(x-t)\, f(t), \tag{3.55}$$

with both a and b finite. It turns out that such integral equations are not solvable in general. A notable exception and, indeed an exactly solvable one, is when the kernel is of the special form

$$k(x-t) = \frac{1}{x-t}. \tag{3.56}$$

For obvious reasons such equations are called *singular integral equations (SIE)*. To solve them we need to extend our Wiener-Hopf technique. This extension turns out to be the same as the one which is necessary to solve integral equations with an analyticity line instead of a strip.

We start by generalising the principal value integral concept which we have introduced in Sect. 1.3.3. We consider a smooth path $\mathcal{P}$, see Fig. 3.21, and a continuous function $f = f(\xi)$ on this path $\xi \in \mathcal{P}$. Let us define an associated function $F(z)$ via the Cauchy-type integral

$$F(z) = \frac{1}{2\pi i} \int_{\mathcal{P}} \frac{d\xi}{\xi - z}\, f(\xi).$$

According to Cauchy's theorem $F(z)$ is analytic everywhere apart from the points on $\mathcal{P}$.

Next we investigate the behavior of $F(z)$ as z approaches a point z_0 on the path $\mathcal{P}$, which is not an endpoint. In general we will get different results depending on whether $z \to z_0$ on one side of the path or the other. Therefore we define two different functions. Looking along $\mathcal{P}$ in the direction of integration, the limiting value of $F(z)$ as $z \to z_0$ from the left is denoted $F_+(z_0)$ while the limiting value as $z \to z_0$ from the right is denoted by $F_-(z_0)$.

To examine $F_+(z_0)$ we deform the path of integration as shown in Fig. 3.22a. This function F_+ breaks up into two contributions

$$F_+(z_0) = \frac{1}{2\pi i} \lim_{\varepsilon \to 0^+} \int_{\mathcal{P}_\varepsilon} \frac{d\xi}{\xi - z_0} f(\xi) + \frac{1}{2\pi i} \lim_{\varepsilon \to 0^+} \int_{C_\varepsilon^+} \frac{d\xi}{\xi - z_0} f(\xi),$$

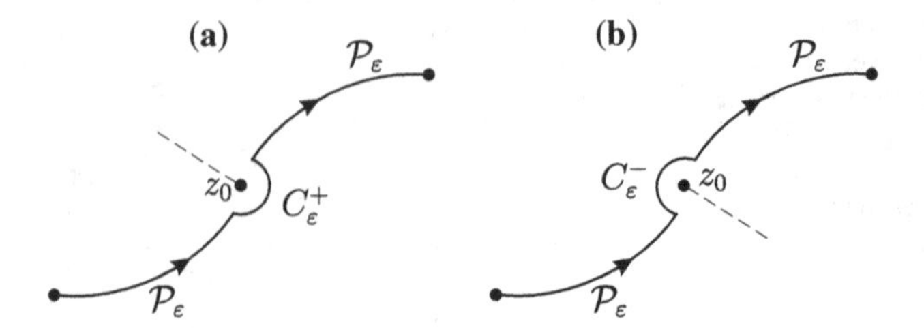

Fig. 3.22 The integration path is deformed in a vicinity of the point z_0, once the point z_0 is bypassed from the right (**a**), and another time from the left (**b**)

where C_ε^+ is a semi-circle of radius ε, $\mathcal{P}_\varepsilon$ is path $\mathcal{P}$ with a section of 2ε around z_0 missed out. The first integral is obviously the principal part as defined in (1.17),

$$\frac{1}{2\pi i}\lim_{\varepsilon\to 0^+}\int\limits_{\mathcal{P}_\varepsilon}\frac{d\xi}{\xi - z_0}f(\xi) = \frac{1}{2\pi i}\mathrm{P}\int\limits_{\mathcal{P}}\frac{d\xi}{\xi - z_0}f(\xi) \equiv F_p(z_0),$$

while the second integral becomes half a residue

$$\lim_{\varepsilon\to 0^+}\int\limits_{C_\varepsilon^+}\frac{d\xi}{\xi - z_0}f(\xi)\bigg|_{\xi=z_0+\varepsilon e^{i\varphi}} = \lim_{\varepsilon\to 0}\int\limits_{\varphi_0}^{\pi+\varphi_0}\frac{i\varepsilon d\varphi e^{i\varphi}}{\varepsilon e^{i\varphi}}f(z_0 + \varepsilon e^{i\varphi}) = i\pi f(z_0).$$

Consequently, for the left-handed limit we get

$$F_+(z_0) = F_p(z_0) + \frac{1}{2}f(z_0).$$

To determine $F_-(z_0)$ we deform the contour as shown in Fig. 3.22b so that

$$F_-(z_0) = \frac{1}{2\pi i}\lim_{\varepsilon\to 0^+}\int\limits_{\mathcal{P}_\varepsilon}\frac{d\xi}{\xi - z_0}f(\xi) + \frac{1}{2\pi i}\lim_{\varepsilon\to 0^+}\int\limits_{C_\varepsilon^-}\frac{d\xi}{\xi - z_0}f(\xi).$$

The first integral ($\mathcal{P}_\varepsilon \to \mathcal{P}$) is the Cauchy principal value $F_p(z_0)$ and the second integral is again half a residue $i\pi f(z_0)$ but now the integration is clockwise so we have a minus sign:

$$F_-(z_0) = F_p(z_0) - \frac{1}{2}f(z_0).$$

Adding and subtracting the formulae for $F_\pm$ we can write

$$\begin{cases} F_+(z_0) + F_-(z_0) = 2F_p(z_0), \\ F_+(z_0) - F_-(z_0) = f(z_0). \end{cases} \tag{3.57}$$

These results are referred to as *Plemelj formulae*.

By the above procedure we have solved the following problem: given a continuous function $f(\xi)$ on a smooth path $\mathcal{P}$, find a function $G(z)$ analytic for all $z \notin \mathcal{P}$, vanishing at infinity, $G(z \to \infty) \to 0$, and with a given jump discontinuity

$$f(z_0) = G_+(z_0) - G_-(z_0), \tag{3.58}$$

on $\mathcal{P}$. The solution to this problem is unique and is given by the integral

$$G(z) = \frac{1}{2\pi i} \int_{\mathcal{P}} \frac{d\xi}{\xi - z} f(\xi).$$

Indeed, this is analytic by inspection, vanishes at infinity, $G(z) = O(1/z)$ as $z \to \infty$, and the jump discontinuity (3.58) follows from Plemelj formulae (3.57). The uniqueness can be shown in the usual way. Suppose there are two solutions $G_1(z)$ and $G_2(z)$, then the function $G_1(z) - G_2(z)$ is analytic everywhere including $\mathcal{P}$ and vanishes at infinity. Hence $G_1(z) - G_2(z) \equiv 0$ by the Liouville theorem.

Up to now we required that $G(z) \to 0$ as $z \to \infty$. We can, however, allow for an algebraic growth

$$G(z) = O(|z|^N) \quad \text{as} \quad z \to \infty.$$

In this case $G(z)$ is determined up to a polynomial of degree N (Liouville theorem again). That means that we can write

$$G(z) = \frac{1}{2\pi i} \int_{\mathcal{P}} \frac{d\xi}{\xi - z} f(\xi) + P_N(z),$$

where $P_N(z)$ is a polynomial of degree N. Clearly P_N does not change analytic properties around $\mathcal{P}$. This polynomial is determined from subsidiary conditions (for example growth restriction), which we impose for a given problem.

Example 3.16
Take $f(\xi) \equiv 1$ and as $\mathcal{P}$ the real axis segment $-1 < x < 1$. Then

$$F(z) = \frac{1}{2\pi i} \int_{-1}^{1} \frac{dx}{x - z}.$$

We want to determine the function $F(z)$ for all z, see Fig. 3.23a. Recall that we already have computed the principal value integral

$$F_p(x) = \frac{1}{2\pi i} \, \mathrm{P} \int_{-1}^{1} \frac{dx'}{x' - x} = \frac{1}{2\pi i} \ln\left(\frac{1 - x}{1 + x}\right), \quad z = x \in [-1, 1]$$

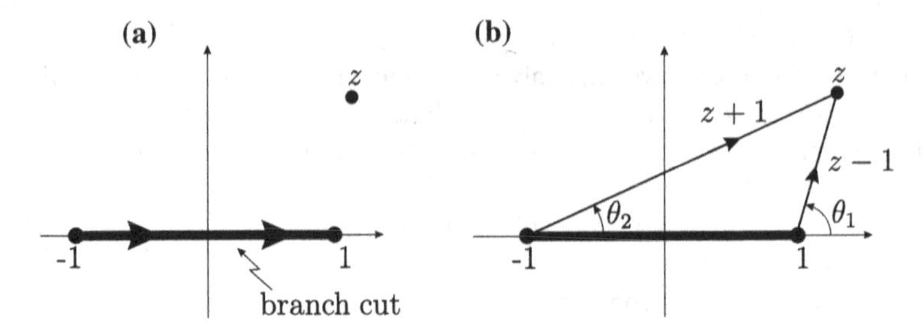

Fig. 3.23 Illustration for the calculation in Example 3.16

in Sect. 1.3.3. Next take $z = x > 1$, then

$$F(x) = \frac{1}{2\pi i}\int_{-1}^{1}\frac{dx'}{x'-x} = \frac{1}{2\pi i}\ \ln|x'-x|\Big|_{-1}^{1} = \frac{1}{2\pi i}\ \ln\left|\frac{1-x}{1+x}\right|$$
$$= \frac{1}{2\pi i}\ \ln\left(\frac{x-1}{x+1}\right) = \frac{1}{2\pi i}\left[\ln(x-1) - \ln(x+1)\right] \quad \text{for } x > 1.$$

Therefore according to the analytic continuation theorem from Sect. 1.3.6[13] the above relation holds also for complex z and we have

$$F(z) = \frac{1}{2\pi i}\left[\ln(z-1) - \ln(z+1)\right].$$

Let us now work out limiting values $F_{\pm}(x)$ for $z \to x \pm i0$. We write $z - 1 = |z-1|\, e^{i\theta_1}$ and $z + 1 = |z+1|\, e^{i\theta_2}$, see Fig. 3.23b. Then we have

$$\ln(z-1) = \ln|z-1| + i\,\theta_1,$$

where $-\pi < \theta_1 < \pi$ and

$$\ln(z+1) = \ln|z+1| + i\,\theta_2,$$

where $-\pi < \theta_2 < \pi$. Now let $z \to x + i\delta\ (-1 < x < 1)$, $\delta \to 0^+$, so that $\theta_1 \to \pi$, and $\theta_2 \to 0$, see Fig. 3.24a. Then

$$F_+(x) = \frac{1}{2\pi i}\left[\ln(1-x) + i\pi - \ln(1+x)\right].$$

Next let $z \to x - i0^+$, so that $\theta_1 \to -\pi$, and $\theta_2 \to 0$, see Fig. 3.24b. Then

[13] If two functions coincide on a line in an analyticity domain then they coincide everywhere.

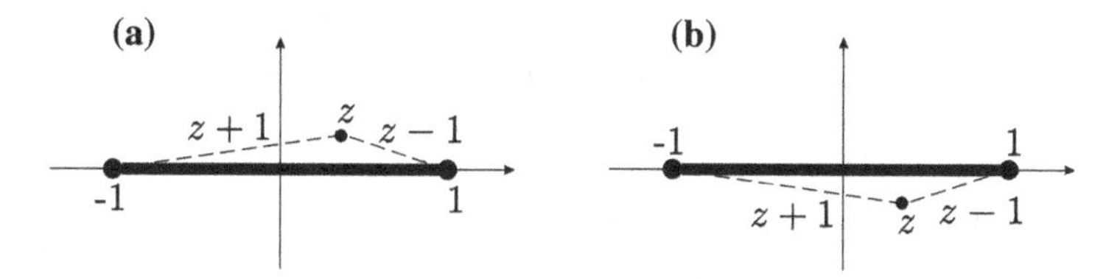

Fig. 3.24 Calculation of arguments in Example 3.16

$$F_-(x) = \frac{1}{2\pi i}\Big[\ln(1-x) - i\pi - \ln(1+x)\Big].$$

Collecting the results

$$F_\pm(x) = \frac{1}{2\pi i}\ln\left(\frac{1-x}{1+x}\right) \pm \frac{1}{2}$$

we can see that the Plemelj formulae

$$F_+(x) + F_-(x) = \frac{1}{\pi i}\ln\left(\frac{1-x}{1+x}\right) = 2F_p(x),$$
$$F_+(x) - F_-(x) = 1 = f(x),$$

hold as expected.

Now we can try to solve a singular integral equation of the first kind, one typical representative of which is the equation[14]

$$\frac{1}{\pi}\,\mathrm{P}\int_{-1}^{1}\frac{f(t)dt}{t-x} = g(x). \tag{3.59}$$

The task is to compute $f(x)$ for a given $g(x)$ which is defined for $-1 < x < 1$ and is continuous. We define an associated function

$$F(z) = \frac{1}{2\pi i}\int_{-1}^{1}\frac{f(t)\,dt}{t-z}$$

for complex z, see Fig. 3.23a. If we require that

$$\int_{-1}^{1} dt f(t)$$

converges then $F(z) = O(1/z)$ as $z \to \infty$, since

[14] This type of equations arises, for instance, in the aerofoil theory, see [24].

$$F(z)\Big|_{z\to\infty} = -\frac{1}{z}\frac{1}{2\pi i}\int_{-1}^{1} dt f(t) + O(z^{-2}).$$

Using Plemelj formulae we obtain:

$$F_+(x) - F_-(x) = f(x), \quad F_+(x) + F_-(x) = 2F_p(x), \tag{3.60}$$

where the Cauchy principal value is

$$F_p(x) = \frac{1}{2\pi i}\,\mathrm{P}\int_{-1}^{1} \frac{f(t)}{t-x}dt.$$

So we have

$$F_+(x) + F_-(x) = \frac{2}{2\pi i}\,\mathrm{P}\int_{-1}^{1} \frac{dt}{t-x} f(t) = -ig(x),$$

where the last equality is just the original integral equation. For such kind of equations, a general theory, which is referred to as the Riemann-Hilbert problem, exists and we refer the reader interested in this problem to [20]. Here we resort to a trick. We see from Eq. (3.60) that the simplest way to obtain the solution is to compute the difference of $F_\pm$ functions rather than working with the sum of them. Our problem is the minus sign between the two terms in the above equation. Thus, instead of working with the function $F(z)$ we shall work with the function

$$\omega(z) = (z^2-1)^{1/2}F(z).$$

We just multiply $F(z)$ by the function $(z^2-1)^{1/2}$ which changes sign when z is moved across the branch cut. After that we easily solve for $\omega(z)$ and then compute the desired $f(x)$.

But first we make sure that $(z^2-1)^{1/2}$ has the necessary property. As depicted on Fig. 3.23b we have

$$(z-1)^{1/2} = |z-1|^{1/2}\, e^{i\theta_1/2}$$

with $-\pi < \theta_1 < \pi$ and

$$(z+1)^{1/2} = |z+1|^{1/2}\, e^{i\theta_2/2}$$

with $-\pi < \theta_2 < \pi$. For $z \to x + i\delta$ with infinitesimal δ and for $-1 < x < 1$ the angles behave as $\theta_1 \to \pi$ and $\theta_2 \to 0$, see Fig. 3.24b. Therefore we obtain

$$(z^2-1)^{1/2} = (1-x)^{1/2}e^{i\frac{\pi}{2}}(1+x)^{1/2} = i\,(1-x^2)^{1/2}. \tag{3.61}$$

On the other hand, in the limiting case $z \to x - i\delta$, $\theta_1 \to -\pi$ and $\theta_2 \to 0$, see Fig. 3.24b, and thus

$$(z^2 - 1)^{1/2} = (1 - x)^{1/2} e^{-i\frac{\pi}{2}} (1 + x)^{1/2} = -i\,(1 - x^2)^{1/2}, \tag{3.62}$$

as required. For $-1 < x < 1$ we then have

$$\omega_{\pm}(x) = \pm i\,(1 - x^2)^{1/2}\,F_{\pm}(x).$$

So, for the new function ω we obtain a jump discontinuity

$$\omega_+(x) - \omega_-(x) = i(1 - x^2)^{1/2}[F_+(x) + F_-(x)] = (1 - x^2)^{1/2} g(x).$$

Such branch cut behaviour would be produced by the function (this is the solution of the associated integral equation)

$$\omega(z) = \frac{1}{2\pi i} \int_{-1}^{1} \frac{(1 - t^2)^{1/2}}{t - z} g(t) dt + P_N(z),$$

where $P_N(z)$ is a polynomial of degree N. To fix N we need to investigate $\omega(z)$ on infinity. Recall that $\omega(z) = (z^2 - 1)^{1/2} F(z)$ and that $F(z) = O(1/z)$ as $z \to \infty$. Then

$$\omega(z) = O(1) \quad \text{as} \quad z \to \infty.$$

Hence $N = 0$ and $P_N(z)$ is yet an arbitrary constant A_0, and the solution is

$$\omega(z) = \frac{1}{2\pi i} \int_{-1}^{1} \frac{(1 - t^2)^{1/2}}{t - z} g(t) dt + A_0.$$

Finally to get $f(x)$ we use Plemelj formula for $F(z)$

$$f(x) = F_+(x) - F_-(x) = \frac{1}{i(1 - x^2)^{1/2}} \Big[\omega_+(x) + \omega_-(x)\Big] = \frac{2}{i(1 - x^2)^{1/2}}\, \omega_p(x),$$

where the Cauchy principal part is

$$\omega_p(x) = \frac{1}{2\pi i}\, \mathrm{P} \int_{-1}^{1} \frac{(1 - t^2)^{1/2}}{t - x} g(t) dt + A_0.$$

Thus, the solution to the problem is explicitly:

$$f(x) = -\frac{1}{\pi(1-x^2)^{1/2}}\,\mathrm{P}\int_{-1}^{1}\frac{(1-t^2)^{1/2}}{t-x}g(t)dt + \frac{A}{(1-x^2)^{1/2}}, \tag{3.63}$$

where $A = -i2A_0$ is an arbitrary constant.

Example 3.17
Solve the equation

$$\frac{1}{\pi}\mathrm{P}\int_{-1}^{1}\frac{f(t)dt}{t-x} = 1 \quad \text{for} \quad -1 < x < 1. \tag{3.64}$$

We identify $g(x) = 1$, which is a special case of (3.59). We have a general solution (3.63), so essentially we need to evaluate the integral

$$I(x) = \mathrm{P}\int_{-1}^{1}\frac{(1-t^2)^{1/2}}{t-x}dt. \tag{3.65}$$

The solution will then be

$$f(x) = -\frac{I(x)}{\pi(1-x^2)^{1/2}} + \frac{A}{(1-x^2)^{1/2}}.$$

There are different ways to evaluate these type of integrals. One method is to introduce the function

$$G(z) = \frac{1}{2\pi i}\int_{-1}^{1}\frac{(1-t^2)^{1/2}}{t-z}dt,$$

for the complex variable z. For a real $z = x > 1$ outside of the segment $[-1, 1]$ we can write $t = \cos\theta$, then[15]

[15] The integral

$$\int_{0}^{\pi}\frac{d\theta}{x-\cos\theta}$$

can be very conveniently performed via residua after substitution $z = e^{i\theta}$. The integration angle can be extended to $[0, 2\pi]$, that only doubles the integral. Then the z-integration is along a unit circle around the coordinate origin. For $x > 1$ the integrand has two poles $z_{1,2} = x \pm \sqrt{x^2-1}$, of which only one lies within the circle. Finally we then obtain for the integral $\pi/\sqrt{x^2-1}$.

$$G(x) = \frac{1}{2\pi i}\int_0^\pi \frac{\sin^2\theta\, d\theta}{\cos\theta - x} = \frac{1}{2\pi i}\int_0^\pi \frac{1-\cos^2\theta}{\cos\theta - x}\, d\theta$$

$$= \frac{1}{2\pi i}\int_0^\pi \frac{(x+\cos\theta)(x-\cos\theta)-(x^2-1)}{\cos\theta - x}\, d\theta$$

$$= \frac{1}{2\pi i}\Big[-\int_0^\pi d\theta\,(x+\cos\theta) + (x^2-1)\int_0^\pi \frac{d\theta}{x-\cos\theta}\Big]$$

$$= \frac{1}{2\pi i}\Big[-\pi x + \frac{\pi(x^2-1)}{\sqrt{x^2-1}}\Big] = \frac{1}{2i}\Big[-x + \sqrt{x^2-1}\Big].$$

And therefore by analytic continuation we have for complex z

$$G(z) = \frac{1}{2i}\left(-z + \sqrt{z^2-1}\right).$$

Remembering that for $|x| < 1$, see Eqs. (3.61) and (3.62) above,

$$\sqrt{z^2-1}\Big|_{z\to x\pm i0^+} = \pm i\,\sqrt{1-x^2}, \quad -1 < x < 1,$$

we have

$$G_\pm(x) = \frac{1}{2i}\left(-x \pm i\sqrt{1-x^2}\right).$$

Therefore by Plemelj formulae we have

$$G_+(x) - G_-(x) = \sqrt{1-x^2}\, g(x) = \sqrt{1-x^2}$$

as required, whereas for the principal value we obtain

$$2G_p(x) = \frac{2}{2\pi i}\mathrm{P}\int_{-1}^{1} \frac{(1-t^2)^{1/2}}{t-x} dt = G_+(x) + G_-(x) = ix.$$

As a consequence the above integral is given by

$$I(x) = \mathrm{P}\int_{-1}^{1} \frac{(1-t^2)^{1/2}}{t-x}\, dt = -\pi x, \quad |x| < 1 \tag{3.66}$$

and for the solution of the integral equation (3.64) follows

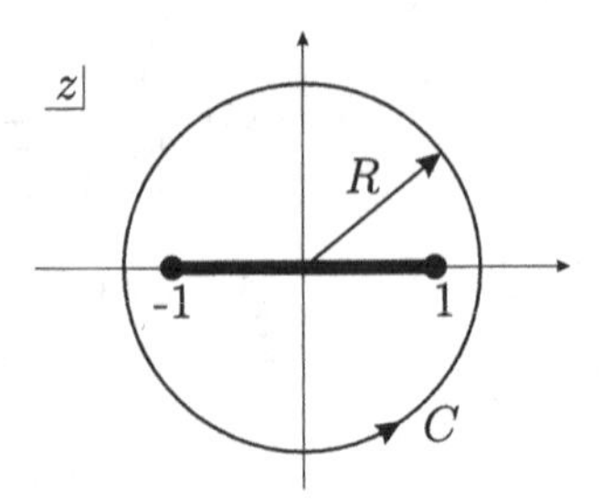

Fig. 3.25 Illustration for the calculation of the integral in (3.65)

$$f(x) = \frac{x}{(1-x^2)^{1/2}} + \frac{A}{(1-x^2)^{1/2}}. \tag{3.67}$$

The second approach for the evaluation of the integral (3.65) is to consider the integral[16]

$$F(x) = \int\limits_{C} \Big[\frac{(z^2-1)^{1/2}}{z-x} - 1\Big] dz,$$

for real $-1 < x < 1$ along the contour C encircling the branch cut of the integrand, see Fig. 3.25. Due to the subtracted unity in the integrand its behaviour at infinity is

$$\frac{(z^2-1)^{1/2}}{z-x} - 1 = \frac{\sqrt{1-\frac{1}{z^2}}}{1-\frac{x}{z}} - 1 = \frac{x}{z} + O\Big(\frac{1}{z^2}\Big).$$

On the one hand we can take the radius of the circle $R \to \infty$, then

$$F(x) = \int\limits_0^{2\pi} \Big[\frac{x}{Re^{i\varphi}} + O(1/R^2)\Big] Re^{i\varphi}\, i\, d\varphi = 2\pi i\, z_\infty = 2\pi i\, x, \tag{3.68}$$

where $z_\infty = x$ is the residue at infinity. On the other hand we can deform the contour C as shown in Fig. 3.26. On the upper/lower edge of the cut $C_\pm$ we have $z = t \pm i\delta,\ -1 < t < 1$. Then $\sqrt{z^2-1} = \pm i\sqrt{1-t^2}$ on $C_\pm$ (including $C_\rho^\pm$), so that in the limit $\rho \to 0$

[16] It can in general be used with all integrals of the type

$$\mathrm{P}\int\limits_{-1}^{1} \frac{\sqrt{1-t^2}}{t-x} g(t) dt,$$

where $g(t)$ is a rational function.

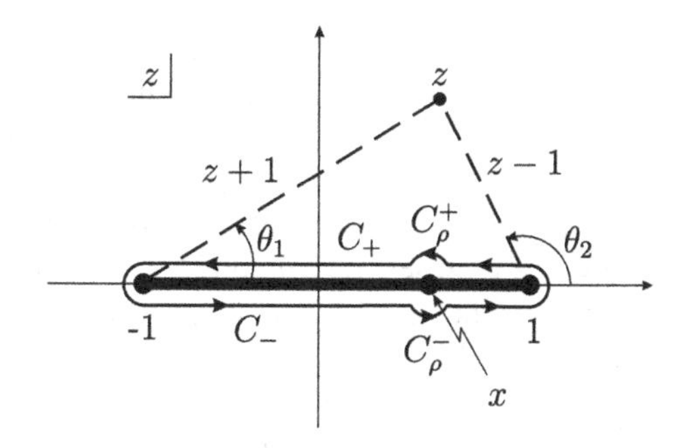

Fig. 3.26 The contour shown in Fig. 3.25 can also be "collapsed" onto the branch cut

$$S_{\text{up}} = \int\limits_{C_+ + C_\rho^+} \left[\frac{i\sqrt{1-t^2}}{t-x} - 1\right]dt = \mathrm{P}\int\limits_{1}^{-1} \frac{i\sqrt{1-t^2}}{t-x}dt + 2 + i\pi i\sqrt{1-x^2},$$

where the last term is the evaluation of the first part of the integrand along C_ρ^+ and is essentially its half-residue. Thus we obtain

$$\lim_{\rho\to 0} S_{\text{up}} = \lim_{\rho\to 0} \int\limits_{C_+ + C_\rho^+} \ldots\, dt = -i\; I(x) + 2 - \pi\,\sqrt{1-x^2}.$$

By analogy for the integrals along the lower edge of the cut we get at $\rho \to 0$

$$S_{\text{down}} = \int\limits_{C_- + C_\rho^-} \left[\frac{-i\sqrt{1-t^2}}{t-x} - 1\right]dt = -i\;\mathrm{P}\int\limits_{-1}^{1} \frac{\sqrt{1-t^2}}{t-x}dt - 2$$
$$+\, i\pi(-i\sqrt{1-x^2}) = -i\; I(x) - 2 + \pi\sqrt{1-x^2}.$$

Therefore in total the residue contributions as well as the integrals of unity cancel each other while the integrals along $C_\pm$ double and one finally obtains

$$F(x) = \int\limits_{\mathrm{C}_{\mathrm{R}\to\infty}} \left[\frac{(z^2-1)^{1/2}}{z-x} - 1\right]dz$$

$$= \int\limits_{C_+ + C_- + C_\rho^+ + C_\rho^-} \left[\frac{(z^2-1)^{1/2}}{z-x} - 1\right]dz = -2i\,I(x).$$

Comparing this result to the previous one (3.68), we find that $I(x) = -\pi x$, as before, see the result (3.66).

Let us now consider another singular integral equation

$$\frac{1}{\pi}\int_{-1}^{1} f(t)\ln|x-t|\,dt = g(x), \tag{3.69}$$

again for $-1 < x < 1$, where $g(x)$ is a given function. This equation, as we shall see shortly, is closely related to (3.59). To solve for $f(x)$, we represent the lhs in the following form:

$$\int_{-1}^{1} f(t)\ln|x-t|\,dt = \lim_{\varepsilon\to 0^+}\left\{\int_{-1}^{x-\varepsilon} dt f(t)\ln(x-t) + \int_{x+\varepsilon}^{1} dt f(t)\ln(t-x)\right\}$$

and compute a derivative with respect to x:

$$\int_{-1}^{x-\varepsilon} dt f(t)\frac{1}{x-t} + \int_{x+\varepsilon}^{1} dt f(t)\frac{(-1)}{t-x} + f(x-\varepsilon)\ln\varepsilon - f(x+\varepsilon)\ln\varepsilon. \tag{3.70}$$

Here the last two terms arise from the limits of integration. As long as $f(x)$ is sufficiently differentiable[17] the limit $\varepsilon \to 0$,

$$\begin{aligned}&\lim_{\epsilon\to 0}[f(x-\varepsilon)\ln\varepsilon - f(x+\varepsilon)\ln\varepsilon]\\ &= \lim_{\epsilon\to 0}\ln\varepsilon\,[f(x+\varepsilon) - f(x-\varepsilon)] \to 0,\end{aligned}$$

whereas the first two terms of (3.70) become a principal value integral

$$-\mathrm{P}\int_{-1}^{1} dt\frac{f(t)}{t-x}$$

after putting $\varepsilon \to 0^+$. Therefore we have shown that

$$\frac{d}{dx}\int_{-1}^{1} dt f(t)\ln|t-x| = -\mathrm{P}\int_{-1}^{1} dt\frac{f(t)}{t-x},$$

and the above integral equation reduces to

$$\frac{1}{\pi}\mathrm{P}\int_{-1}^{1}\frac{f(t)dt}{t-x} = -g'(x).$$

[17] Although this requirement could in principle be relaxed.

This is related to (3.59) the solution of which we already know.

Example 3.18
Let us solve the problem

$$\frac{1}{\pi}\int_{-1}^{1} f(t)\ln|t-x|\,dt = 3, \tag{3.71}$$

for $-1 < x < 1$.

First we differentiate with respect to x to get

$$\frac{1}{\pi}\mathrm{P}\int_{-1}^{1}\frac{f(t)dt}{t-x} = 0.$$

According to (3.63) the solution of this equation is

$$f(x) = \frac{A}{\sqrt{1-x^2}},$$

where A is a constant. To find this constant, we substitute the obtained solution into the original Eq. (3.71). We get then the equation with respect to A,

$$\frac{1}{\pi}\int_{-1}^{1}\frac{A}{(1-t^2)^{1/2}}\ln|x-t|\,dt = 3.$$

The simplest way to determine A is to take $x = 0$ since the rhs of the equation is independent of x. Then we obtain

$$\frac{A}{\pi}\int_{-1}^{1}\frac{\ln|t|}{\sqrt{1-t^2}}\,dt = \frac{2A}{\pi}\int_{0}^{1}\frac{\ln t\,dt}{\sqrt{1-t^2}} = 3.$$

In order to calculate the last integral we make the trigonometric substitution $t = \sin\theta$, leading to

$$\frac{2A}{\pi}\int_{0}^{\pi/2}\ln(\sin\theta)\,d\theta = 3.$$

For the integral we then can write

$$\int_0^{\pi/2} \ln(\sin\theta)\, d\theta = \frac{1}{2}\int_0^{\pi} \ln(\sin\theta)\, d\theta = \int_0^{\pi/2} \ln(\sin 2\theta)\, d\theta$$

$$= \int_0^{\pi/2} \Big[\ln 2 + \ln(\sin\theta) + \ln(\cos\theta)\Big] d\theta = \frac{\pi}{2}\ln 2 + 2\int_0^{\pi/2} \ln(\sin\theta)\, d\theta,$$

where we have used that $\ln(\sin\theta)$ and $\ln(\cos\theta)$ are equal after the substitution $\theta \to \pi/2 - \theta$. Hence

$$\int_0^{\pi/2} \ln(\sin\theta)\, d\theta = -\frac{\pi}{2}\ln 2$$

and therefore $A = -3/\ln 2$. The full solution of (3.71) is then

$$f(x) = -\frac{3}{\ln 2}\frac{1}{\sqrt{1-x^2}}.$$

Example 3.19
Solve the integral equation

$$\frac{1}{\pi}\int_{-1}^{1} f(t)\ln|t-x|\, dt = x, \quad -1 < x < 1 \tag{3.72}$$

for $f(x)$. We differentiate with respect to x first and obtain

$$\frac{1}{\pi}\,\mathrm{P}\int_{-1}^{1} dt\frac{f(t)}{t-x} = -1.$$

From the inversion formula (3.63) we have for the solution

$$f(x) = -\frac{1}{\pi}\frac{1}{(1-x^2)^{1/2}}\,\mathrm{P}\int_{-1}^{1}\frac{(1-t^2)^{1/2}(-1)}{t-x}\, dt + \frac{A}{(1-x^2)^{1/2}}.$$

The integral here is identical to that in (3.66), therefore we obtain

$$f(x) = -\frac{x}{(1-x^2)^{1/2}} + \frac{A}{(1-x^2)^{1/2}}.$$

Next we substitute this into the original Eq. (3.72) evaluated at $x = 0$ for simplicity:

$$\frac{1}{\pi}\int_{-1}^{1} dt\left(\frac{-t}{\sqrt{1-t^2}}+\frac{A}{\sqrt{1-t^2}}\right)\ln|t|=0.$$

The first term in the integrand is an odd function of the variable t and drops out. Therefore

$$\frac{1}{\pi}A\int_{-1}^{1}\frac{\ln|t|}{\sqrt{1-t^2}}\,dt=0.$$

It follows that $A=0$ and the solution is

$$f(x)=-\frac{x}{(1-x^2)^{1/2}}. \tag{3.73}$$

3.5 Abel's Integral Equation

Another example which falls under the category of the singular equation is *the generalized Abel's equation*[18]

$$f(x)=\int_{a}^{x} dt\,\frac{u(t)}{(x-t)^{\mu}}, \tag{3.74}$$

where $0<\mu<1$ and $f(x)$ is a given function having a continuous derivative $f'(x)$. This is a Volterra integral equation of the first kind with the weakly-singular difference kernel $k(x-t)=(x-t)^{-\mu}\to\infty$ as $t\to x$.

The Abel's equation in the form (3.74) can be solved by different methods. We encourage the reader to solve this problem using the Laplace transform (see Problem 3.10). Here we would like to employ the approach outlined in Ref. [3]. Let us define an auxiliary function

$$\varphi(x)=\int_{a}^{x} u(\xi)\,d\xi$$

---

[18] Original equation

$$f(x)=\int_{0}^{x} dt\,\frac{u(t)}{\sqrt{x-t}}.$$

was obtained by Abel in the study of the motion of a particle in the gravitational field sliding in the vertical plane along a given curve. $f(x)$ then has a meaning of the time of descent from the highest point to the lowest point on the curve, $ds=u(t)dt$ is the curve element.

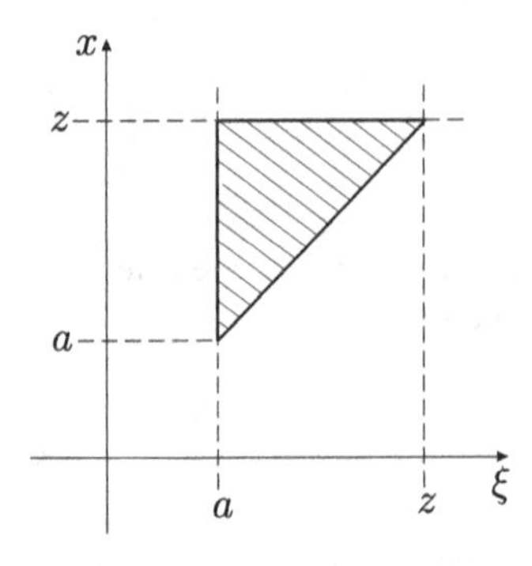

Fig. 3.27 Schematic representation of the integration region for the double integral on the page 180

and multiply both parts of the following equality[19]

$$\frac{\pi}{\sin(\mu\pi)} = \int_{\xi}^{z} \frac{dx}{(z-x)^{1-\mu}(x-\xi)^{\mu}}$$

by $u(\xi)$ and integrate with respect to ξ from a to z. Then, changing the order of integration in the double integral, see Fig. 3.27, we obtain:

$$\frac{\pi}{\sin(\mu\pi)}\Big[\varphi(z)-\varphi(a)\Big] = \int_{a}^{z} d\xi \int_{\xi}^{z} \frac{u(\xi)dx}{(z-x)^{1-\mu}(x-\xi)^{\mu}}$$

$$= \int_{a}^{z} dx \int_{a}^{x} \frac{u(\xi)d\xi}{(z-x)^{1-\mu}(x-\xi)^{\mu}} = \int_{a}^{z} \frac{f(x)dx}{(z-x)^{1-\mu}}.$$

Hence, a continuous solution of Abel's Eq. (3.74), if exists, is for $0 < \mu < 1$ given by

$$u(z) = \frac{\sin(\mu\pi)}{\pi}\frac{d}{dz}\int_{a}^{z} \frac{f(x)\,dx}{(z-x)^{1-\mu}}. \tag{3.75}$$

For example, for the equation

[19] Setting $y = (z-x)/(z-\xi)$ and using definitions (2.10) and (2.11) of the Beta function as well as the complement formula (2.6) for the Gamma function, we have:

$$\int_{\xi}^{z} \frac{dx}{(z-x)^{1-\mu}(x-\xi)^{\mu}} = \int_{0}^{1} y^{\mu-1}(1-y)^{-\mu}dy = B(\mu, 1-\mu) = \Gamma(\mu)\Gamma(1-\mu) = \frac{\pi}{\sin(\mu\pi)}.$$

$$2\sqrt{x} = \int_0^x \frac{u(t)}{\sqrt{x-t}}\,dt,$$

substituting $f(x) = 2\sqrt{x}$, $a = 0$, and $\mu = 1/2$ into the general formula (3.75), we find

$$u(x) = \frac{2}{\pi}\frac{d}{dx}\int_0^x \frac{\sqrt{t}}{\sqrt{x-t}}\,dt.$$

This can be computed with the help of a trigonometric substitution $t = x\sin^2\theta$,

$$u(x) = \frac{4}{\pi}\frac{d}{dx}\Bigg(x\underbrace{\int_0^{\pi/2} d\theta\,\sin^2\theta}_{=\pi/4}\Bigg) = 1.$$

Obviously, in this simple case the solution can be straightforwardly tested: $\int_0^x dt\,(x-t)^{-1/2} = 2\sqrt{x}$.

Let us now consider an equation of the Abel type, which additionally contains an inhomogeneity:

$$u(x) = g(x) + \lambda\int_0^x \frac{u(t)}{(x-t)^\mu}\,dt, \tag{3.76}$$

where the function $g(x)$ is assumed to be sufficiently smooth so that an unique solution to (3.76) is guaranteed; λ is a constant. To determine the solution we can use the decomposition, which can be understood as, for instance, an expansion in λ,

$$u(x) = \sum_{n=0}^{\infty} u_n(x)$$

into both sides of (3.76) to obtain

$$\sum_{n=0}^{\infty} u_n(x) = g(x) + \lambda\int_0^x \frac{1}{(x-t)^\mu}\left(\sum_{n=0}^{\infty} u_n(t)\right)dt.$$

The components $u_0(x)$, $u_1(x)$, $u_2(x)$, ... are determined upon applying the following recurrence relations

$$u_0(x) = g(x), \quad u_n(x) = \lambda\int_0^x \frac{u_{n-1}(t)}{(x-t)^\mu}\,dt, \tag{3.77}$$

for $n \geq 1$. In some problems terms with opposite signs entering the components $u_0(x)$ and $u_1(x)$ may be observed. Therefore it is sometimes convenient to use a modified decomposition method. This can be done by splitting the inhomogeneous term $g(x)$ into two parts. To illustrate that we discuss the following example.

Example 3.20
Consider the Abel type equation

$$u(x) = \sqrt{x} + \frac{\pi}{2}\,x - \int\limits_0^x \frac{u(t)}{\sqrt{x-t}}dt. \tag{3.78}$$

Using the algorithm (3.77) we set $u_0(x) = \pi x/2 + \sqrt{x}$, which gives

$$u_1(x) = -\int\limits_0^x \frac{\sqrt{t} + \pi t/2}{\sqrt{x-t}}\,dt.$$

The trigonometric substitution $t = x \sin^2 \theta$ leads to

$$u_1(x) = -2x \int\limits_0^{\pi/2} d\theta\, \sin^2\theta - \pi x^{3/2} \int_0^{\pi/2} d\theta\, \sin^3\theta = -\frac{\pi}{2}\,x - \frac{2\pi}{3}\,x^{3/2}.$$

We observe that the first term appearing in $u_0(x)$ differs from the one in $u_1(x)$ only by a sign, whereas the remaining term satisfies the Eq. (3.78). Its exact solution is thus $u(x) = \sqrt{x}$.

Alternatively, if we split the $g(x)$ into two parts and set now $u_0(x) = \sqrt{x}$, then

$$u_1(x) = \frac{\pi}{2}x - \int\limits_0^x \frac{\sqrt{t}}{\sqrt{x-t}}\,dt,$$

which gives $u_1(x) = 0$. Consequently, other components $u_2(x)$, $u_3(x)$, ... will vanish according to (3.77) and the same solution follows immediately: $u(x) = u_0(x) = \sqrt{x}$.

3.6 Problems

Problem 3.1
Solve the following integral equations:

$$\textbf{(a):}\ x^2 = 1 + \lambda \int_1^x xt\varphi(t)dt,$$

$$\textbf{(b):}\ \int_0^x e^{(x-t)}\varphi(t)dt = x.$$

Problem 3.2
Using the Laplace transform solve the following integral equations:

$$\textbf{(a):}\quad \varphi(x) = f(x) + \int_0^x (x-t)\varphi(t)dt,$$

$$\textbf{(b):}\quad \varphi(x) = e^{-|x|} + \lambda e^x \int_x^\infty e^{-t}\varphi(t)dt.$$

Problem 3.3
Perform the sum decomposition of the function

$$R(s) = \frac{1}{(s - k\cos\theta)\sqrt{(s+k)}}.$$

Similar function occurs in the diffraction problem on a half–plane, where the complex variable k has the meaning of the wave number and θ is the incident angle.

Problem 3.4
Solve the eigenvalue problem

$$f(x) = \frac{\lambda}{2}\int_\delta^\infty \frac{f(y)}{\cosh[(x-y)/2]}\,dy$$

for $\delta = 0$ using the Wiener-Hopf method. Compare the result with the one for the case $\delta = -\infty$ given in the main text, see Example 3.4.

Problem 3.5
Solve the integral equation[20]

$$D(t,t') = D_0(t-t') + \int_0^\infty dt_2\, K(t-t_1)\, D(t_1,t'),$$

for the function $D(t,t')$. The source term is given by

[20] This kind of equation emerges in the problem of transients, see for instance [25, 26].

$$D_0(t - t') = -i\Theta(t - t')\, e^{-i\Delta(t-t')}$$

and contains the Heaviside step-function $\Theta(t)$, and the kernel has the form

$$K(t - t_1) = -\Gamma\, e^{-i\Delta(t-t_1)}\, \Theta(t - t_1).$$

(a): Without solving the equation show that $D(t, t') = D(t - t')$.
(b): Solve the equation by iterations.
(c): Show that the equation can be solved by Laplace transform as defined in Sect. 1.4.3.

Problem 3.6
Solve the integral equation

$$f(x) + 4\int_0^\infty dy\, e^{-|x-y|} f(y) = e^{-x}$$

for $0 < x < \infty$ using the Wiener-Hopf method.

Problem 3.7
Solve the singular integral equation

$$\frac{1}{\pi}\, \mathrm{P}\int_{-1}^{1} \frac{f(t)dt}{t - x} = 1 - x^2$$

for $-1 < x < 1$ using the method described in Sect. 3.4.

Problem 3.8
Solve the integral equation

$$\int_0^\infty k_1(x - t) f(t)dt + \int_{-\infty}^0 k_2(x - t) f(t)dt = k_1(x)$$

for $-\infty < x < +\infty$. Here $k_1(x < 0) = (e^{3x} - e^{2x})$ and $k_2(x > 0) = -ie^{-2x}$, whereas $k_1(x > 0) = k_2(x < 0) = 0$.

Problem 3.9
Consider the integral equation

$$f(x) + \int_0^\infty (a + b|x - t|)e^{-|x-t|} f(t)dt = p(x)$$

for $x > 0$. The constants a and b ($b \neq 0$) are real, and $p(x) = 0$ for $x < 0$. Find the formal solution in terms of p using the Wiener-Hopf method.

Problem 3.10
Consider the Abel's integral equation

$$f(x) = \int_0^x \frac{u(t)}{(x-t)^\mu} dt$$

with $0 < \mu < 1$. Employ the Laplace transform in order to express u in terms of f. Find the solution $u(x)$ for $f(t) = t^\nu$.

Answers:

Problem 3.1:

$$\textbf{(a):}\ \ \varphi(x) = (x^2+1)/\lambda x^3; \qquad \textbf{(b):}\ \ \varphi(x) = 1 - x.$$

Problem 3.2:

$$\textbf{(a):}\ \ \varphi(x) = f(x) + \frac{1}{2}e^x \int_0^x e^{-t} f(t)dt - \frac{1}{2}e^{-x} \int_0^x e^t f(t)dt;$$

$$\textbf{(b):}\ \ \text{for}\ \ \lambda \neq 2: \ \ \varphi(x) = \frac{2}{2-\lambda}e^{-x} + Ce^{(1-\lambda)x}\ (x > 0),$$

$$\varphi(x) = \left(\frac{2}{2-\lambda} + C\right) e^{(1-\lambda)x}\ (x < 0),$$

where $C = 0$ for $\mathrm{Re}(1-\lambda) > 0$, $C = -\dfrac{2}{2-\lambda}$ for $\mathrm{Re}(1-\lambda) < 0$,

for $\lambda = 2$: $\varphi(x) = -2xe^{-x}\ (x > 0)$, $\varphi(x) = 0\ (x < 0)$.

Problem 3.3:

$$R_+(s) = \frac{1}{(s - k\cos\theta)}\left(\frac{1}{\sqrt{s+k}} - \frac{1}{\sqrt{k + k\cos\theta}}\right),$$

$$R_-(s) = \frac{1}{(s - k\cos\theta)\sqrt{k + k\cos\theta}}.$$

Problem 3.4:

$$f(x) = C\left[e^{i\alpha x}\frac{\Gamma(a)\,\Gamma(b)}{\Gamma(c)}\ F(a,b;c;e^{-2x}) - e^{-i\alpha x}\frac{\Gamma(\overline{a})\,\Gamma(\overline{b})}{\Gamma(\overline{c})}\ F(\overline{a},\overline{b};\overline{c};e^{-2x})\right],$$

$a = 1/4 - i\alpha/2, b = 3/4 - i\alpha/2, c = 1 - i\alpha, \overline{a} = 1/4 + i\alpha/2$ etc., $\cosh(\pi\alpha) = 2\pi\lambda$, C is an arbitrary constant.

Problem 3.5:

$$D(t - t') = -i\Theta(t - t')\, e^{-i\Delta(t-t')}\, e^{-\Gamma(t-t')}.$$

Problem 3.6:

$$f(x) = e^{-3x}/2.$$

Problem 3.7:

$$f(x) = \frac{3x/2 - x^3 + A}{\sqrt{1 - x^2}},$$

where A is an arbitrary constant.

Problem 3.8:

$$f(x) = -iCe^{-2x}, \quad x > 0,$$
$$f(x) = \left[C(e^{2x} - e^{3x}) - 4ie^{2x} + 5ie^{3x}\right], \quad x < 0,$$

where C is an arbitrary constant.

Problem 3.9:

$$f(x) = p(x) + \rho \int_0^\infty e^{-\beta|x-t|} \cos(\theta + \alpha|x - t|) p(t) dt$$
$$+ \int_0^\infty e^{-\beta(x+t)} \Big[A \cos[\alpha(x - t)] + B \cos[\psi + \alpha(x + t)]\Big] p(t) dt,$$

$$\text{where} \quad \rho e^{i\theta} = \frac{\gamma}{\beta - i\alpha}, \quad \gamma = i\frac{(\alpha + i\beta)^2(a - b) + a + b}{2\alpha\beta},$$
$$A = \frac{[\alpha^2 + (\beta - 1)^2]^2}{4\alpha^2\beta}, \quad B = \frac{R}{4\alpha^2}, \quad Re^{i\psi} = \frac{[\alpha + i(\beta - 1)]^4}{\alpha + i\beta}.$$

The positive constants α and β set the four complex roots $(\pm\alpha \pm i\beta)$ of the polynomial $M(u) = u^4 + 2(a - b + 1)u^2 + 2a + 2b + 1$.

Problem 3.10:

$$u(x) = \frac{\sin(\mu\pi)}{\pi} \frac{d}{dx} \int_0^x \frac{f(t) dt}{(x - t)^{1-\mu}}; \quad \frac{\sin(\mu\pi)}{\pi} (\nu + \mu)\, x^{\nu+\mu-1} B(\nu + 1, \mu).$$

Chapter 4
Orthogonal Polynomials

Classical orthogonal polynomials are an important class of special functions. They are intimately related to many of the problems we have discussed in previous chapters. In particular, in many cases they represent eigenfunctions of differential and integral operators. For instance, Hermite polynomials multiplied by exponentials solve the quantum harmonic oscillator problem and simultaneously represent eigenfunctions of the Fourier transform, whereas the Laguerre polynomials are used in the quantum mechanical treatment of atomic systems.

4.1 Introduction

One of the most widespread eigenfunction problem can be formulated and solved in the following way. Let us consider an ODE of the form[1]

$$\frac{d}{dx}\Big[p(x)y'(x)\Big] + q(x)y'(x) + \lambda y(x) = 0 \tag{4.1}$$

for $a \le x \le b$, where

$$p(x) = p_0 + p_1 x + p_2 x^2$$

is a polynomial of the second degree, while

$$q(x) = q_0 + q_1 x$$

is a polynomial of a first degree. λ is a constant called *eigenvalue*. Being supplemented with boundary conditions at a and b such problem is referred to as *Sturm–Liouville problem*. Here we would like to concentrate on the situation without boundary conditions. Our first goal is to find the value of $\lambda = \lambda_n$ for which there exists a

[1] Sometimes it is convenient to refer to it as a *standard form*.

A. O. Gogolin (edited by E. G. Tsitsishvili and A. Komnik), *Lectures on Complex Integration*,
Undergraduate Lecture Notes in Physics, DOI: 10.1007/978-3-319-00212-5_4,

polynomial solution of degree n, which can be written down as

$$y(x) = \sum_{r=0}^{n} a_r x^r \quad \text{with} \quad a_n \neq 0.$$

For the constituents of (4.1) we then obtain

$$\begin{aligned}
\frac{d}{dx}\Big[p(x)y'(x)\Big] &= \sum_{r=2}^{n} p_0 a_r r(r-1)x^{r-2} + \sum_{r=1}^{n} p_1 a_r r^2 x^{r-1} \\
&\quad + \sum_{r=1}^{n} p_2 a_r r(r+1)x^r, \\
q(x)y'(x) &= \sum_{r=1}^{n} q_0 a_r r x^{r-1} + \sum_{r=1}^{n} q_1 a_r r x^r, \\
\lambda_n y(x) &= \sum_{r=0}^{n} \lambda_n a_r x^r.
\end{aligned}$$

Substituting this back into the original ODE leads to

$$\begin{aligned}
&\sum_{r=2}^{n} p_0 r(r-1)a_r x^{r-2} + \sum_{r=1}^{n} p_1 r^2 a_r x^{r-1} + \sum_{r=1}^{n} p_2 r(r+1)a_r x^r \\
&\quad + \sum_{r=1}^{n} q_0 r a_r x^{r-1} + \sum_{r=1}^{n} q_1 r a_r x^r + \lambda_n \sum_{r=0}^{n} a_r x^r = 0.
\end{aligned}$$

We now rearrange the summations to make all powers of x the same,[2]

$$\begin{aligned}
&\sum_{r=0}^{n-2} p_0(r+2)(r+1)a_{r+2}x^r + \sum_{r=0}^{n-1} p_1(r+1)^2 a_{r+1}x^r + \sum_{r=0}^{n} p_2 r(r+1)a_r x^r \\
&\quad + \sum_{r=0}^{n-1} q_0(r+1)a_{r+1}x^r + \sum_{r=1}^{n} q_1 r a_r x^r + \lambda_n \sum_{r=0}^{n} a_r x^r = 0.
\end{aligned}$$

For $0 \leq r \leq n-2$, equating powers of x^r yields a recurrence relation

$$a_{r+2}p_0(r+2)(r+1) + a_{r+1}(r+1)\Big[p_1(r+1) + q_0\Big] + a_r\Big[p_2 r(r+1) + q_1 r + \lambda_n\Big] = 0, \tag{4.2}$$

which gives a_{r+2} for known a_r and a_{r+1}. Further, for the x^n term we obtain

[2] Special care is to be taken with the upper and lower bounds of the sums.

$$a_n\Big[p_2n(n+1)+q_1n+\lambda_n\Big]=0.$$

Since we require that $a_n \neq 0$ this leads to

$$\lambda_n = -p_2n(n+1) - q_1n. \tag{4.3}$$

This is the solution of the eigenvalue problem as it shows the exact value λ needs to take for the solution of the ODE to be a polynomial of degree n. The set of all possible λ_ns is then called the *spectrum of eigenvalues*.

To reconstruct the eigenfunction one starts with an *arbitrary* $a_n \neq 0$ and then uses the following relation (which is a descendant of the coefficient in front of the x^{n-1} term)

$$a_{n-1}\Big[p_2n(n-1)+q_1(n-1)+\lambda_n\Big]+a_n\Big[p_1n^2+q_0n\Big]=0, \tag{4.4}$$

in order to compute a_{n-1}. After that one uses (4.2) in order to calculate the remaining coefficients $a_{n-2}, a_{n-3}, \ldots, a_1, a_0$.

Example 4.1
Consider *Legendre equation*

$$(1-x^2)y'' - 2xy' + \lambda y = 0 \tag{4.5}$$

for $-1 \leq x \leq 1$. It arises for example in a solution of the Laplace equation in a three-dimensional case. We write this in a standard form

$$\frac{d}{dx}\Big[(1-x^2)y'\Big] + \lambda y = 0. \tag{4.6}$$

So in this case

$$\begin{aligned} p(x) = 1 - x^2 : &\quad p_0 = 1,\ p_1 = 0,\ p_2 = -1,\\ q(x) = 0 \quad : &\quad q_0 = 0,\ q_1 = 0. \end{aligned}$$

With the help of (4.3) we immediately identify the whole eigenvalue spectrum of the problem:

$$\lambda = \lambda_n = n(n+1)\,.$$

For the eigenfunctions we then obtain:

- $n = 0$: then $\lambda = 0$, the equation takes the form

$$\frac{d}{dx}\Big[(1-x^2)y'\Big] = (1-x^2)y'' - 2xy' = 0,$$

  and the polynomial solution is just a constant $y(x) = a_0$.
- $n = 1$: in this case $\lambda = 2$ and one has to solve the equation

$$(1-x^2)y'' - 2xy' + 2y = 0.$$

We try $y = x + a$, then the derivatives are trivially $y' = 1$ and $y'' = 0$, so

$$-2x + 2(x + a) = 0,$$

leading to $a = 0$. Thus the solution is $y(x) = x$ or, more generally, $y_1 = a_1 x$ with an arbitrary constant $a_1 \neq 0$.

- $n = 2$: then $\lambda_2 = 6$ and the corresponding equation is

$$(1-x^2)y'' - 2xy' + 6y = 0.$$

We seek a solution of the form $y(x) = x^2 + ax + b$. After the substitution into the above equation we obtain

$$+2 - 2x^2 - 4x^2 - 2xa + 6x^2 + 6xa + 6b = 0\,.$$

From the comparison of the coefficients we conclude that $a = 0$ and $b = -1/3$. Thus the solution up to some constant factor is $y_2(x) = x^2 - 1/3$. It is a convention to use the notation $P_n(x)$ for the Legendre polynomials, so $P_2(x) = y_2(x) = x^2 - 1/3$.

4.2 Orthogonality Relations

Let us consider two different polynomial solutions y_m and y_n with $m \neq n$. Then y_m satisfies the equation

$$(py_m')' + qy_m' + \lambda_m y_m = 0$$

and similarly y_n satisfies the equation

$$(py_n')' + qy_n' + \lambda_n y_n = 0.$$

We now multiply the former equation by $y_n w$ and the latter by $y_m w$, where $w = w(x)$ is some function to be specified in a moment, subtract the results from each other and integrate over the interval $a \leq x \leq b$. Then one obtains

$$\int_a^b dx w(x) \left[y_n(py_m')' - y_m(py_n')' + q\left(y_n y_m' - y_m y_n'\right) + \left(\lambda_m - \lambda_n\right) y_n y_m \right] = 0.$$

We evaluate the first term in the brackets by parts:

$$\int_a^b dx\, w\left[y_n(py_m')' - y_m(py_n')'\right] = w\left[y_n p y_m' - y_m p y_n'\right]\Big|_a^b$$

$$-\int_a^b dx\, w \underbrace{\left[y_n' p y_m' - y_m' p y_n'\right]}_{=0} - \int_a^b dx\, w' \left[y_n p y_m' - y_m p y_n'\right],$$

then substitute back and group the terms,

$$wp\left[y_n y_m' - y_m y_n'\right]\Big|_a^b + \int_a^b dx \left[y_n y_m' - y_m y_n'\right]\left[qw - pw'\right]$$
$$+ (\lambda_m - \lambda_n) \int_a^b w y_m y_n dx = 0.$$

Next we chose the function $w(x)$ to satisfy the condition

$$q(x)w(x) - p(x)w'(x) = 0,$$

that is

$$w(x) = \exp\Big(\int \frac{q(x)}{p(x)}\, dx\Big). \tag{4.7}$$

We also need the integrated part to vanish, therefore we require at the ends of the 'interesting' interval $[a, b]$, that

$$w(a)p(a) = w(b)p(b) = 0. \tag{4.8}$$

We then have

$$(\lambda_m - \lambda_n) \int_a^b dx\; w(x) y_m(x) y_n(x) = 0,$$

and in this way we arrive at the orthogonality relation ($\lambda_n \neq \lambda_m$)

$$\int_a^b dx\; w(x) y_n(x) y_m(x) = 0 \quad \text{for} \quad n \neq m. \tag{4.9}$$

It defines a scalar product in the space of functions $y(x)$ and $w(x)$ for obvious reasons is called a *weight function*. Using the above scalar product any arbitrary polynomial $F_n(x)$ of degree n can obviously be expanded as a linear combination of $y_0, y_1, \ldots, y_n$, that is

$$F_n(x) = \sum_{r=0}^{n} c_r\, y_r(x).$$

Let us first assume a and b to be finite. If we want to satisfy (4.8) by the properties of $p(x)$, then the minimal solution is a second degree polynomial:

$$p(x) = p_0 + p_1 x + p_2 x^2 = -p_2 \underbrace{(b-x)}_{\geq 0} \underbrace{(x-a)}_{\geq 0}.$$

This constitutes a restriction on the polynomial $p(x)$, by virtue of Vieta's formulas:

$$\frac{p_1}{p_2} = -(a+b), \quad \frac{p_0}{p_2} = ab.$$

Then the integrand in (4.7) is given by

$$\frac{q(x)}{p(x)} = \frac{q_0 + q_1 x}{-p_2(b-x)(x-a)} = \frac{\alpha}{x-a} - \frac{\beta}{b-x},$$

where, of course,

$$\alpha = -\frac{q_0 + q_1 a}{p_2(b-a)}, \quad \beta = \frac{q_0 + q_1 b}{p_2(b-a)}. \tag{4.10}$$

By a formal integration we then arrive at

$$\int \frac{q(x)}{p(x)}\, dx = \alpha \ln(x-a) + \beta \ln(b-x).$$

In connection with (4.7) this yields for the weight function the following result:

$$w(x) = c_0\,(x-a)^{\alpha}(b-x)^{\beta}, \tag{4.11}$$

where c_0 is some rather unimportant numerical constant. For the product $p(x)w(x)$ to vanish and not diverge at the ends $x = a, b$ we must additionally require

$$\alpha > -1, \quad \beta > -1.$$

Polynomials with the weight function (4.11) are called *Jacobi polynomials* $P_n^{(\alpha,\,\beta)}(x)$. They are solutions of the following differential equation,

$$(1-x^2)y'' + [\beta - \alpha - (\alpha+\beta+2)x]y' + n(n+\alpha+\beta+1)y = 0 \tag{4.12}$$

on the interval $[-1, 1]$.

Example 4.2
For the *Legendre polynomials* on the interval $[-1, 1]$ according to (4.6) we have

$$\frac{d}{dx}\left[(1-x^2)y'\right] + n(n+1)y = 0.$$

Here $p(x) = 1 - x^2$, so that $a = -1$ and $b = 1$ are indeed zeros of $p(x)$. Since $q = 0$ it immediately follows that $\alpha = \beta = 0$. This is obviously a special case of Jacobi polynomials with the simplest possible weight function

$$w(x) = e^{\int 0\, dx} = 1. \tag{4.13}$$

Hence the orthogonality relations for Legendre polynomials $P_n^{(0,\,0)}(x) = P_n(x)$ are simply (remember $n \neq m$)

$$\int_{-1}^{1} dx\ P_n(x)P_m(x) = 0.$$

Let us now turn to the case of the semi-infinite interval $[a, b]$, that is $a = 0$ and $b \to +\infty$. It is clear that now the boundary condition (4.8) cannot be satisfied only by the properties of the polynomial $p(x)$, a special form of the weight function $w(x)$ is required as well.

Example 4.3
For instance, we can set $p(x) = p_1 x$ with $p_1 > 0$. While at $x = 0$ $p = 0$, at infinity we have to require $p(x)w(x) \to 0$, which means having $xw(x) \to 0$ at $x \to \infty$. One possibility to fulfill it is to choose $q(x) = -Qx$, so that $q_0 = 0$, $q_1 = -Q$ with $Q > 0$.[3] Then from (4.7) one obtains

$$w(x) = \exp\left(\int dx\, \frac{q}{p}\right) = \exp\left(\int dx\, \frac{-Qx}{p_1 x}\right) = e^{-\frac{Q}{p_1}x}. \tag{4.15}$$

This weight function is exponentially small for $x \to +\infty$ and thus meets all necessary requirements. The polynomials produced by this constellation of parameters are referred to as *Laguerre polynomials* and are denoted by $L_n(x)$. They emerge, for example, as solutions for an electron wave function in a Coulomb potential, see for example Ref. [11]. Setting $Q = p_1 = 1$, the Laguerre equation reads

$$xy'' + (1 - x)y' + ny = 0. \tag{4.16}$$

Finally we consider the whole real axis $a \to -\infty$, $b \to +\infty$. Obviously, in this case both the 'right' and the 'left' boundary conditions (4.8) can only be satisfied by correctly chosen weight function $w(x)$.

---

[3] A finite real q_0 would spawn more general polynomials, called the *associated Laguerre polynomials* $L_n^{\alpha}(x)$ with $\alpha = q_0$, see Problem 4.4. The respective weight function $w(x) = x^{\alpha} e^{-\frac{Q}{p_1}x}$ immediately reduces to (4.15) at $q_0 = 0$, so that $L_n^0(x) = L_n(x)$. These polynomials satisfy the following differential equation:

$$xy'' + (\alpha + 1 - x)y' + ny = 0. \tag{4.14}$$

Example 4.4
Let us set $p = p_0 > 0$ and $q = -Qx$. Then, using (4.7) for the weight function we obtain

$$w(x) = \exp\left(\int dx\, \frac{q}{p}\right) = \exp\left(\int dx\, \frac{-Qx}{p_0}\right) = e^{-\frac{Qx^2}{2p_0}}.$$

All necessary requirements are met if we set $Q/p_0 > 0$. The polynomials generated by this weight function are called *Hermite polynomials* with the standard notation $H_n(x)$. They emerge as solutions of a Schrödinger equation for a harmonic oscillator, see, for instance [11]. In the most widespread incarnations $Q/p_0 = 2$ and $w(x) = e^{-x^2}$. Then the *Hermite equation* reads

$$y'' - 2xy' + 2ny = 0. \tag{4.17}$$

4.3 Uniqueness

As we have seen in Sect. 4.1 our set of polynomials which is generated by the recurrence relations is not unique as we are allowed to start with an arbitrary coefficient a_n. We can remove this ambiguity by fixing it. In most cases one encounters one of three different normalization prescriptions. For instance we can require that the coefficient of x^n term in $y_n(x)$ is unity. The second option is to demand that every polynomial has a fixed value at the boundary, for example $y_n(a) = 1$. By now we also have a prescription for a scalar product so that we can fix it by requiring the polynomials to be normalized in the following way:

$$\int_a^b dx\, w(x) y_n^2(x) = 1. \tag{4.18}$$

It turns out that then the set of *any* orthogonal polynomials $\{y_n(x)\}$ with a given weight function $w(x)$ becomes unique.

In order to show that let us suppose that there is another orthogonal set $\{Q_m(x)\}$ with the same weight function $w(x)$ over the interval $[a, b]$. We shall show that $Q_n = c\, y_n$, where c is a constant. The proof is based on the orthogonality relation (4.9) and the representation (4.20) of any arbitrary polynomial as a linear combination of a set of other polynomials. So we take two different integers m and n, $m \neq n$, and consider the cases of $m < n$ and $m > n$ separately.

For the case $m < n$, we expand Q_m as a linear combination of the y_n's

$$Q_m(x) = \sum_{r=0}^{m} c_r\, y_r(x)$$

and then obtain

$$\int_a^b dx\, w(x) y_n(x) Q_m(x) = \int_a^b dx\, w(x) y_n(x) \sum_{r=0}^{m} c_r y_r(x) = 0,$$

because of the orthogonality of y_n's for $r \le m < n$. That immediately entails the orthogonality of y_n and Q_m at $m < n$:

$$\int_a^b dx\, w(x) y_n(x) Q_m(x) = 0. \tag{4.19}$$

Quite similar orthogonality relation emerges for the case $m > n$. Here we expand y_n in terms of Q_m's and use their orthogonality.

Now we fix an integer n and consider the expansion

$$y_n(x) = \sum_{r=0}^{n} a_r Q_r(x). \tag{4.20}$$

By multiplying both sides of this relation with $w(x)Q_m(x)$ and integrating over x we obtain

$$\int_a^b dx w(x) Q_m(x) y_n(x) = \sum_{r=0}^{n} a_r \int_a^b dx w(x) Q_r(x) Q_m(x) = a_m N_m,$$

where

$$N_m = \int_a^b dx\, w(x) Q_m^2(x) > 0$$

is definitely positive. So if $m \neq n$ we get $a_m N_m = 0$ due to the orthogonality (4.19). But $N_m > 0$ and therefore $a_m = 0$. From the expression (4.20) then follows

$$y_n(x) = a_n Q_n(x),$$

which proves the uniqueness.

4.4 Gram-Schmidt Orthonormalisation Procedure

Interestingly, the knowledge of the weight function is fully sufficient for the calculation of y_n and no information about either $p(x)$ or $q(x)$ is required. The corresponding procedure is the iterative Gram-Schmidt orthonormalisation, which works as follows.

One starts by defining y_0—a constant consistent with the normalization procedure.[4] In the next step we find the two coefficients of $y_1(x)$ from the condition

$$\int_a^b dx\, w(x)\, y_0\, y_1(x) = 0 \tag{4.21}$$

and from the normalisation requirement. The three coefficients of $y_2(x)$ are computed from the conditions

$$\int_a^b dx\, w(x)\, y_0\, y_2(x) = 0\,, \quad \int_a^b dx\, w(x)\, y_1(x)\, y_2(x) = 0,$$

as well as from the normalization prescription etc. In general $y_n(x)$ is found from the normalization and orthogonality conditions

$$\int_a^b dx w(x) y_r(x) y_n(x) = 0 \quad \text{for} \quad r = 0, 1, \ldots, n-1.$$

Example 4.5
Let us consider Legendre equation (4.5)–(4.6) and find a few first polynomial solutions $P(x)$ using the above Gram-Schmidt procedure. We already know the respective weight function $w = 1$ from (4.13).

We normalize by setting the coefficient in front of x^n to unity, thus $P_0 = 1$. For $P_1(x)$ we propose $P_1(x) = x + a$, where a is found from (4.21), resulting in $a = 0$. Therefore we obtain $P_1(x) = x$ as before, see Example 4.1.

For the second order polynomial we make a substitution $P_2(x) = x^2 + ax + b$, where a and b are determined from

$$\int_{-1}^{1} dx\; 1\; (x^2 + ax + b) = 0 \quad \text{and} \quad \int_{-1}^{1} dx\; x\; (x^2 + ax + b) = 0.$$

As expected it leads to $a = 0$ and $b = -1/3$ and hence to $P_2(x) = x^2 - 1/3$, which is consistent with the previous result found in Example 4.1.

Example 4.6
The Laguerre equation

$$x\, y'' + (1 - x)\, y' + n\, y = 0 \quad \text{for} \quad 0 \leq x < \infty, \tag{4.22}$$

[4] As before we want $\int_a^b dx\, y_n^2(x) w(x) = 1$. Alternatively we set the coefficient in front of highest exponent term x^n equal to 1.

being written in the standard form

$$(xy')' - xy' + ny = 0$$

has $p(x) = x$ and $q(x) = -x$. In accordance with (4.15) it leads to the weight function $w(x) = e^{-x}$. We chose the lowest order Laguerre polynomial to be $L_0 = 1$. For the next one we again substitute $L_1(x) = x + a$, where a is determined from

$$\int_0^\infty dx\, e^{-x}\, 1\, (x + a) = 0.$$

Recalling that $\int_0^\infty dx\, x^n\, e^{-x} = n!$ we immediately find $a = -1$. Therefore $L_1(x) = x - 1$.

For $L_2(x)$ we again suggest $L_2(x) = x^2 + ax + b$, where a and b are determined from

$$\int_0^\infty dx\, e^{-x}\, 1\, (x^2 + ax + b) = 0 \quad \text{and} \quad \int_0^\infty dx\, e^{-x}\, (x - 1)\, (x^2 + ax + b) = 0.$$

Evaluation of the integrals leads to $a = -4$ and $b = 2$. Hence $L_2(x) = x^2 - 4x + 2$.

4.5 Rodrigues' Formula

There is a far more explicit way of generating orthogonal polynomials with the help of *Rodrigues' formula*:

$$y_n(x) = \frac{1}{K_n\, w(x)}\, \frac{d^n}{dx^n}\, \left\{ w(x)\, [p(x)]^n \right\}, \tag{4.23}$$

which yields the nth order polynomial. K_n is the normalization factor, which we shall specify later.

In order to prove (4.23) in a finite interval $[a, b]$ we first verify that $y_n(x)$ as defined by the above formula is indeed a polynomial of degree n for the case of the 'minimal' weight (4.11). We have

$$w(x)[p(x)]^n = c_0\, (x - a)^{n+\alpha} (b - x)^{n+\beta}.$$

So, by the Leibniz rule for the nth derivative of a product of two factors we obtain

$$\frac{d^n}{dx^n}\Big[(x-a)^{n+\alpha}(b-x)^{n+\beta}\Big] = \sum_{r=0}^{n} \frac{n!}{r!(n-r)!} \frac{d^{n-r}}{dx^{n-r}}(x-a)^{n+\alpha} \frac{d^r}{dx^r}(b-x)^{n+\beta}$$
$$= \sum_{r=0}^{n} C_{r,n}\,(x-a)^{r+\alpha}(b-x)^{n-r+\beta},$$

where $C_{r,n}$ is some numerical coefficient. Hence

$$\frac{1}{w(x)}\frac{d^n}{dx^n}\left(wp^n\right) = \frac{\sum_{r=0}^{n} C_{r,n}\,(x-a)^{r+\alpha}(b-x)^{n-r+\beta}}{(x-a)^{\alpha}(b-x)^{\beta}}$$
$$= \sum_{r=0}^{n} C_{r,n}\,(x-a)^{r}\,(b-x)^{n-r}.$$

This is unmistakably a polynomial of degree n.

In the next step we show the orthogonality property

$$\int_a^b dx\, w(x)\, y_n(x) y_m(x) = 0$$

for all $m < n$ (the case $n < m$ follows by symmetry). This is the same as to show that

$$I_m = \int_a^b dx\, w(x)\, y_n(x)\, x^m = 0$$

for all $m = 0, 1, 2, \ldots, n-1$. We substitute Rodrigues' formula into the above expression (obviously, the factor $1/K_n$ can be omitted):

$$I_m = \int_a^b dx\, x^m \frac{d^n}{dx^n}(wp^n)$$

and integrate by parts. Then we obtain

$$I_m = x^m \frac{d^{n-1}}{dx^{n-1}}(wp^n)\Big|_a^b - m\int_a^b dx x^{m-1}\frac{d^{n-1}}{dx^{n-1}}(wp^n).$$

At the boundary $x = a$ we have

$$wp^n = \text{constant} \times (x-a)^{n+\alpha} \quad \text{as} \quad x \to a.$$

Note that for the 'minimal' weight we assume $\alpha > -1$. Consequently, for the involved derivatives we find

$$\frac{d^{n-1}}{dx^{n-1}}\left(wp^n\right) \sim (x-a)^{1+\alpha} \to 0$$

as $x \to a$. Moreover, this conclusion holds for all $r = 0, 1, \ldots, n-1$:

$$\left.\frac{d^r}{dx^r}\left(wp^n\right)\right|_{x\to a} \to 0.$$

In the similar way we can show that the terms on the other boundary $x = b$ vanish as well. Therefore we keep integrating by parts:

$$I_m = -m\int_a^b dx x^{m-1}\frac{d^{n-1}}{dx^{n-1}}\left(wp^n\right) = +m(m-1)\int_a^b dx x^{m-2}\frac{d^{n-2}}{dx^{n-2}}\left(wp^n\right)$$

$$= \cdots = (-1)^m m!\int_a^b dx\frac{d^{n-m}}{dx^{n-m}}\left(wp^n\right) = (-1)^m m!\left.\frac{d^{n-m-1}}{dx^{n-m-1}}\left(wp^n\right)\right|_a^b = 0,$$

provided $n \geq m+1$, that is $m = 1, 2, \ldots, n-1$. Thus $y_n(x)$ defined by (4.23) is indeed a polynomial degree n and is orthogonal to all $1, x, x^2, \ldots, x^{n-1}$, that is to all $y_0, y_1, y_2, \ldots, y_{n-1}$. Hence Rodrigues' formula defines a set of orthogonal polynomials. Due to the uniqueness property, after normalization this set of polynomials is exactly the same as the one obtained from the Gram–Schmidt procedure.

Example 4.7
Use the Rodrigues' formula to find the Legendre polynomial $P_3(x)$. We recall that in that case $p(x) = 1 - x^2$ and $w = 1$. Then up to the normalization formula (4.23) yields

$$P_n(x) = \frac{d^n}{dx^n}\left[(1-x^2)^n\right]. \tag{4.24}$$

Therefore

$$P_3(x) = \frac{d^3}{dx^3}\left[(1-x^2)^3\right] = 24(-5x^3 + 3x).$$

We want to normalise to the highest coefficient, that is set the coefficient in front of x^3 equal to 1, then we obtain

$$P_3(x) = x^3 - \frac{3}{5}\,x.$$

Obviously, using the Gram–Schmidt procedure to calculate this polynomial results in much more work.

The polynomials given in (4.24) are unnormalized. The most widespread normalization is

$$P_n(x) = \frac{1}{K_n} \frac{d^n}{dx^n} \left[(1 - x^2)^n \right], \tag{4.25}$$

with $K_n = (-1)^n/(2^n n!)$. This corresponds to the second normalization prescription, according to which all polynomials have one and the same value at the boundary, see Example 4.10 in the next section.

Example 4.8
The *Chebyshev equation* of the first kind reads

$$(1 - x^2)y'' - xy' + n^2 y = 0, \quad -1 \le x \le 1. \tag{4.26}$$

Let us compute the respective eigenfunction of the second order—the *Chebyshev polynomial* $T_2(x)$. First we rewrite the equation in the standard form

$$\frac{d}{dx}\left[(1 - x^2)\frac{dy}{dx}\right] + x\frac{dy}{dx} + n^2 y = 0,$$

in order to identify $p(x) = 1 - x^2$ and $q(x) = x$. Then the weight function can be calculated according to the relation (4.7),

$$w(x) = \exp\left[\int dx\, \frac{x}{1 - x^2}\right] = \exp\left[-\frac{1}{2}\ln(1 - x^2)\right] = \frac{1}{\sqrt{1 - x^2}}. \tag{4.27}$$

We immediately observe that the eigenfunctions of the Chebyshev Eq. (4.26) are a special case of the Jacobi polynomials $P_n^{(\alpha,\,\beta)}(x)$ with $\alpha = \beta = -1/2$.[5] It becomes clear comparing for example the result (4.27) with the expression (4.11) for the 'minimal' weight (for both cases $p(x) = 1 - x^2$).

By virtue of the Rodrigues' formula:

$$\begin{aligned} T_n(x) &= \frac{1}{K_n}\sqrt{1 - x^2}\, \frac{d^n}{dx^n}\left[\frac{1}{\sqrt{1 - x^2}} \cdot (1 - x^2)^n\right] \\ &= \frac{1}{K_n}(1 - x^2)^{\frac{1}{2}}\, \frac{d^n}{dx^n}\,(1 - x^2)^{n - \frac{1}{2}}. \end{aligned}$$

Therefore for $T_2(x)$ we obtain

$$T_2(x) = \frac{1}{K_2}(1 - x^2)^{1/2}\frac{d^2}{dx^2}\,(1 - x^2)^{3/2} = \frac{1}{K_2}3(2x^2 - 1).$$

---

[5] One distinguishes the Chebyshev polynomials of the second kind denoted by $U_n(x)$. They satisfy the equation

$$(1 - x^2)y'' - 3xy' + n^2 y = 0,$$

obey the weight function $w(x) = \sqrt{1 - x^2}$ and, consequently, reduce to the Jacobi polynomials $P_n^{(1/2,\,1/2)}(x)$.

Normalizing the highest exponent to unity leads to $T_2(x) = x^2 - 1/2$. However, the conventional normalization is $K_n = (-1)^n (2n)!/n!$ [8], yielding $T_n(1) = 1$, see Problem 4.10.

Rodrigues' formula is not restricted to the polynomials defined on finite intervals but also works on the full real axis or semiaxes. For instance, for the Laguerre polynomials, which are defined on the semiaxis $a = 0,\ b = \infty$ with the weight $w(x) = e^{-x}$ and $p(x) = x$ we obtain

$$L_n(x) = \frac{1}{K_n} e^x \frac{d^n}{dx^n} \left(e^{-x} x^n \right),$$

where for the normalization prescription (4.18) one has to set $K_n = n!$. Hermite polynomials for the full real axis with $a = -\infty,\ b = +\infty$ and $w(x) = e^{-x^2}$ can be generated by

$$H_n(x) = (-1)^n e^{x^2} \frac{d^n}{dx^n} \left(e^{-x^2} \right). \tag{4.28}$$

This formula produces the polynomials in their canonical and most widespread form, which is unnormalized.

4.6 Generating Functions

Instead of using the Rodrigues' formula the orthogonal polynomials could be computed using a *generating function* $G(x,t)$ defined by the infinite series

$$G(x,t) = \sum_{n=0}^{\infty} \frac{y_n(x)}{n!} t^n, \tag{4.29}$$

which we assume to be convergent for sufficiently small t. Then if we have a simple expression for $G(x,t)$ $y_n(x)$ is just the coefficient in front of $t^n/n!$ in the Taylor expansion of the generating function.

Similarly to the Rodrigues' formula, $G(x,t)$ can be derived for any polynomial set with the knowledge of the respective $p(x)$ and $w(x)$. Let us define a new variable z by the equation

$$z = x + tp(z). \tag{4.30}$$

Since $p(x)$ is quadratic in x, an equation for z as a function of x has two roots. Later we shall need the root such that $z \to x$ as $t \to 0$. According to the above definition z is a function of both x and t and its derivative satisfies

$$\frac{\partial z}{\partial x} = 1 + tp'(z) \frac{\partial z}{\partial x} \quad \text{and therefore} \quad \frac{\partial z}{\partial x} = \frac{1}{1 - tp'(z)}.$$

One can show that the generating function can be found to be given by

$$G(x,t) = \frac{w(z)}{w(x)} \frac{\partial z}{\partial x}, \tag{4.31}$$

or, alternatively, by

$$G(x,t) = \frac{w(z)}{w(x)} \frac{1}{1 - tp'(z)}. \tag{4.32}$$

In order to verify this formula we consider the integral

$$I_m(t) = \int_a^b dx\, w(x)\, G(x,t)\, y_m(x)$$

and substitute into it the expansion (4.29). Then, using the orthogonality relation (4.9), we obtain

$$I_m(t) = \frac{t^m}{m!} \int_a^b dx\, w(x)\, y_m^2(x) = \frac{t^m}{m!} N_m,$$

where

$$N_m = \int_a^b dx\, w(x)\, y_m^2(x) > 0$$

is nothing else but the normalization constant of the polynomial $y_m(x)$.

We need to show that the same holds for the generating function (4.32). Then (4.29) and (4.32) are equivalent. Using expression (4.31) we obtain

$$I_m = \int_a^b dx\, w(x) \frac{w(z)}{w(x)} \frac{\partial z}{\partial x}\, y_m(x) = \int_a^b dx\, w(z) \frac{\partial z}{\partial x}\, y_m(x).$$

We observe that for small t the mapping $x \to z$ is one–to–one. Moreover, since $p(a) = p(b) = 0$ we have $z = a$ when $x = a$ and $z = b$ when $x = b$, see Fig. 4.1. Therefore we can change the parametrisation of the integration domain,

$$I_m(t) = \int_a^b dz\, w(z) y_m\Big(x(z)\Big) = \int_a^b dz\, w(z) y_m\Big(z - tp(z)\Big).$$

In the next step we perform a Taylor expansion of y_m around z for small t. As y_m is a polynomial this expansion terminates at the mth term:

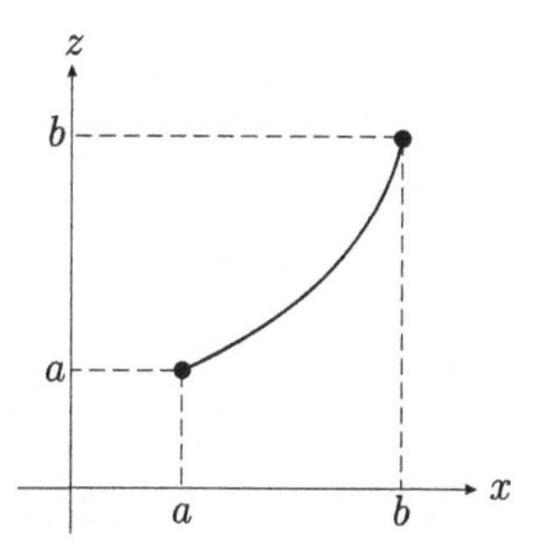

Fig. 4.1 Schematic representation of $z - x$ dependence

$$I_m(t) = \sum_{r=0}^{m} \int_a^b dz\, w(z) \frac{1}{r!} (-t)^r [p(z)]^r \frac{d^r}{dz^r} y_m(z).$$

The rth term for $r > 0$ can be integrated by parts,

$$\frac{(-t)^r}{r!} \int_a^b dz\, w(z)[p(z)]^r \frac{d^r}{dz^r}[y_m(z)]$$

$$= \frac{(-t)^r}{r!} \left\{ w(z)[p(z)]^r \frac{d^{r-1}}{dz^{r-1}} y_m(z) \Big|_a^b - \int_a^b dz\, (wp^r)' \frac{d^{r-1}}{dz^{r-1}} y_m(z) \right\}.$$

Now we observe that $w(z)[p(z)]^r = 0$ at the ends $z = a, b$ for all $r > 0$.[6] Therefore the boundary term vanishes. We continue the partial integration:

$$= \frac{(-t)^r}{r!} \left\{ - \int_a^b dz\, (wp^r)' \frac{d^{r-1}}{dz^{r-1}} y_m(z) \right\}$$

$$= \frac{(-t)^r}{r!} \left\{ \underbrace{- (wp^r)' \frac{d^{r-2}}{dz^{r-2}} y_m(z) \Big|_a^b}_{=0} + \int_a^b dz\, (wp^r)'' \frac{d^{r-2}}{dz^{r-2}} y_m(z) \right\} = \dots$$

$$= \frac{(-t)^r}{r!} \left\{ (-1)^{r-1} \underbrace{(wp^r)^{(r-1)} y_m(z) \Big|_a^b}_{=0} + (-1)^r \int_a^b dz\, (wp^r)^{(r)} y_m(z) \right\}$$

[6] The situation with $r = 0$ is simple—this term does not contribute as its integral corresponds to the scalar product of y_m and y_0, which is equal $\delta_{m0} N_m$ by definition.

$$= \frac{t^r}{r!}\int_a^b dz\, \frac{d^r(wp^r)}{dz^r} y_m(z) = \frac{t^r}{r!}\int_a^b dz\, w(z)y_r(z)y_m(z) = \delta_{rm}\, \frac{t^m}{m!}\, N_m.$$

Hence the result is

$$I_m(t) = \frac{y^m}{m!}\, N_m,$$

and therefore (4.29) and (4.32) are indeed equivalent. Note that, as a rule, the expansion (4.29) produces unnormalized polynomials.

Example 4.9
Compute the generating function for the Laguerre polynomials.

From Example 4.6 we need $p(x) = x$ and $w(x) = e^{-x}$. The solution of (4.30) is

$$z = \frac{x}{1-t}.$$

(We observe that indeed $z = 0$ when $x = 0$.) According to (4.31) the generating function is then

$$G(x,t) = \frac{w(z)}{w(x)}\,\frac{\partial z}{\partial x} = \frac{e^{-x/(1-t)}}{e^{-x}}\,\frac{1}{1-t} = \frac{e^{-xt/(1-t)}}{1-t}. \tag{4.33}$$

In order to verify the polynomials we expand for small t:

$$\begin{aligned} G(x,t) &= (1+t+t^2+\ldots)\,(1 - \frac{xt}{1-t} + \frac{1}{2}\,\frac{x^2t^2}{(1-t)^2} + \ldots) \\ &= 1 + (1-x)t + \left[1 + \frac{1}{2}(x^2 - 2x) - x\right]t^2 + \ldots \end{aligned}$$

Therefore we obtain $L_0 = 1$, $L_1(x) = 1-x$, $L_2(x) = x^2 - 4x + 2$ etc. as expected.[7] Observe that at $x = 0$

$$G(0,t) = \frac{1}{1-t} = 1 + t + t^2 + \ldots,$$

whence, according to (4.29), follows that $L_n(0) = n!$. The expansion (4.29) produces unnormalized $L_n(x)$ as one would also obtain using the Rodrigues' formula. We observe however, that we can also produce normalized polynomials with a fixed value at the boundary $a = 0$ using an alternative definition of the generating function:

$$\Psi(x,t) = \sum_{n=0}^{\infty} L_n(x)\, t^n.$$

[7] Just set the coefficient in front of x to unity, $L_1(x) = 1 - x \to x - 1$.

Example 4.10
Calculate the generating function for the Legendre polynomials defined by (4.24).

From Example 4.2 we know $p(x) = 1 - x^2$ and $w = 1$. Of the two solutions of the equation

$$z = x + t\,p(z) = x + t(1 - z^2)$$

only the solution

$$z = \frac{-1 + \sqrt{1 + 4t(x + t)}}{2t}$$

has the property $z \to x$ as $t \to 0$. Thus the generating function is

$$G(x,t) = \frac{w(z)}{w(x)}\frac{\partial z}{\partial x} = \frac{1}{1 - tp'(z)} = \frac{1}{1 + 2tz} = \frac{1}{\sqrt{1 + 4xt + 4t^2}}. \tag{4.34}$$

Now we expand for small t to access the individual polynomials. For that purpose we use

$$\frac{1}{\sqrt{1+\delta}} = 1 - \frac{1}{2}\,\delta + \frac{3}{8}\,\delta^2 + \ldots$$

with $\delta = 4xt + 4t^2$, so that

$$\begin{aligned} G(x,t) &= 1 - 2xt - 2t^2 + \frac{3}{8}\,16x^2t^2 + \ldots \\ &= 1 - 2xt + (6x^2 - 2)t^2 + \ldots = P_0 + \frac{t}{1!}\,P_1 + \frac{t^2}{2!}\,P_2 + \ldots \end{aligned}$$

Hence $P_0 = 1$, $P_1(x) = -2x$, $P_2(x) = 12x^2 - 4$ and after the normalization: $P_1(x) = x$, $P_2(x) = x^2 - 1/3$ etc. as we already know. So we get the same result as we found in Example 4.1.

The generating function for the traditional normalization [see (4.25)] can be produced in the same way as in the previous example. An additional prefactor $(-1/2)^n$ we achieve by a simple variable substitution $x \to -x/2$ while $1/n!$ simply becomes the part of the polynomial itself. Thus we obtain

$$\Psi(x,t) = \sum_{n=0}^{\infty} P_n(x)\,t^n \tag{4.35}$$

with[8]

$$\Psi(x,t) = \frac{1}{\sqrt{1 - 2xt + t^2}}. \tag{4.36}$$

---

[8] A familiar use of this generating function for the Legendre polynomials is to compute the multipole expansion useful in classical field theory or electrostatics. An expansion of the reciprocal distance between two charges situated at the points $\mathbf{r}_1$ and $\mathbf{r}_2$ is, see e. g. [27],

Using the generating function (4.36) we can evaluate the Legendre polynomials at the endpoints and the origin. Thus at $x = 1$ we have[9]

$$\Psi(1,t) = \frac{1}{1-t} = \sum_{n=0}^{\infty} t^n = \sum_{n=0}^{\infty} P_n(1)\, t^n \;\Rightarrow\; P_n(1) = 1,$$

so that the Legendre polynomials defined by (4.25) have a fixed value at the upper boundary. For the endpoint $x = -1$ one obtains

$$\Psi(-1,t) = \frac{1}{1+t} = \sum_{n=0}^{\infty} (-1)^n t^n = \sum_{n=0}^{\infty} P_n(-1)\, t^n \;\Rightarrow\; P_n(-1) = (-1)^n.$$

At the origin $x = 0$ we have:

$$\Psi(0,t) = \frac{1}{\sqrt{1+t^2}} = \sum_{m=0}^{\infty} (-1)^m t^{2m} \frac{\Gamma\left(\frac{2m+1}{2}\right)}{m!\Gamma\left(\frac{1}{2}\right)} = \sum_{n=0}^{\infty} P_n(0)\, t^n,$$

where Γ is the Gamma function. Hence for odd $n = 2k+1$ $P_{2k+1}(0) = 0$, whereas for even $n = 2k$

$$P_{2k}(0) = \frac{(-1)^k \Gamma\left(k + \frac{1}{2}\right)}{k!\sqrt{\pi}}.$$

Example 4.11

Calculate the generating function of the Hermite polynomials. Then, using the obtained generating function compute their normalization constant.

From Example 4.4 we know that $w(x) = e^{-x^2}$ and $p = 1$. Then from (4.31) we immediately obtain

$$G(x,t) = e^{-2xt-t^2}. \tag{4.37}$$

The traditional normalization as given in (4.28) is achieved by a simple sign change: $G(x,t) \to G(x,-t)$, since now $p = -1$.

In order to find the normalization constant we first evaluate the following product,

---

(Footnote 8 continued)

$$\frac{1}{|\mathbf{r}_1 - \mathbf{r}_2|} = \frac{1}{\sqrt{r_1^2 - 2r_1r_2\cos\theta + r_2^2}} = \frac{1}{r_1}\sum_{n=0}^{\infty} P_n(\cos\theta)\left(\frac{r_2}{r_1}\right)^n, \text{ for } r_1 > r_2.$$

[9] The same result readily follows from the Rodrigues' formula (4.25). Indeed, in the limit $x \to 1$ we have

$$\frac{d^n}{dx^n}(x^2-1)^n = \frac{d^n}{dx^n}\left((x-1)^n(x+1)^n\right) \to \frac{d^n}{dx^n}(x+1)^n\Big|_{x\to 1} = n!(1+1)^n.$$

$$\int_{-\infty}^{\infty} dx\, G(x, y)G(x, z)\, e^{-x^2} = e^{-(y^2+z^2)} \int_{-\infty}^{\infty} dx\, e^{-x^2-2x(y+z)}$$
$$= \sqrt{\pi}\, e^{2yz} = \sqrt{\pi} \sum_{m=0}^{\infty} \frac{(2yz)^m}{m!}.$$

On the other hand, using the definition (4.29) we get for the product of two generating functions

$$\sum_{n,k=0}^{\infty} \frac{y^n z^k}{n!k!} \int_{-\infty}^{\infty} dx\, H_n(x)\, H_k(x)\, e^{-x^2} = \sum_{n,k=0}^{\infty} \frac{y^n z^k}{n!k!} N_n\, \delta_{nk}$$
$$= \sum_{n=0}^{\infty} \frac{(yz)^n}{(n!)^2} N_n.$$

From the comparison of both last relations we obtain[10]

$$N_n = n!\, 2^n \sqrt{\pi}. \tag{4.38}$$

4.7 Recurrence Relations

As we have seen in Sect. 4.1 the coefficients of the orthogonal polynomials satisfy recurrence relations, see for instance (4.2). Therefore it is an interesting question, whether there are any recurrence relations for whole polynomials. It turns out that corresponding expressions not only exist but are in many cases also quite compact. In fact, all classical orthogonal polynomials satisfy the three-term recurrence relation of the form:

$$y_{n+1}(x) = \Big(A_n\, x + B_n\Big)\, y_n(x) + C_n\, y_{n-1}(x) \tag{4.39}$$

for $n \geq 1$, where A_n, B_n, and C_n are some numerical coefficients. In what follows we try to prove this statement.

Let $\Big\{y_n(x)\Big\}$ be an orthogonal set. Since $y_n(x)$ is obviously a polynomial of degree n, $x\, y_n(x)$ is a polynomial of degree $n+1$ and can be expanded in terms of the $y_n(x)$s as

$$x\, y_n(x) = \sum_{r=0}^{n+1} b_{n,r}\, y_r(x), \tag{4.40}$$

where $b_{n,\, r}$ are numerical coefficients such that $b_{n,\, n+1} \neq 0$. Next we single out y_{n+1} on the rhs and write:

---

[10] Evidently, the same result holds for the canonical form (4.28).

$$y_{n+1}(x) = \alpha_n x y_n(x) + \sum_{r=0}^{n} a_{n,\,r} y_r(x), \tag{4.41}$$

where

$$\alpha_n = \frac{1}{b_{n,n+1}} \quad \text{and} \quad a_{n,r} = -\frac{b_{n,r}}{b_{n,n+1}},$$

which is fine as $b_{n,\,n+1} \neq 0$. We shall show that the expansion (4.41) terminates after the third term, that is $a_{n,n} \neq 0$ and $a_{n,n-1} \neq 0$ but all others vanish: $a_{n,n-2} = a_{n,n-3} = \cdots = a_{n,0} = 0$. Let us take $y_s(x)$ with $s < n + 1$ and multiply both sides of (4.41) by $w(x)\, y_s(x)$ and integrate. Then for the lhs we obtain

$$\int_a^b dx\; w(x) y_s(x) y_{n+1}(x) = 0.$$

For the rhs we get

$$\alpha_n \int_a^b dx\; w(x) x y_s(x)\, y_n(x) + \sum_{r=0}^{n} a_{n,r} \int_a^b dx\; w(x)\, y_s(x)\, y_r(x) = 0.$$

In the next step we use the expansion (4.40) to obtain

$$\sum_{r=0}^{s+1} \alpha_n b_{s,r} \int_a^b dx\; w(x)\, y_r(x)\, y_n(x) + \sum_{r=0}^{n} a_{n,r} \int_a^b dx\; w(x)\, y_s(x)\, y_r(x) = 0 \tag{4.42}$$

for all $s < n + 1$. Suppose $n \geq 2$ [otherwise the expansion (4.41) terminates naturally]. As long as $0 \leq s \leq n - 2$ the first term in (4.42) vanishes due to the orthogonality (the summation is extended over $0 \leq r \leq n - 1$). For the same reason only the term with $r = s$ survives in the second part, so that we obtain

$$a_{n,s} \int_a^b dx\; w(x)\, y_s^2(x) = a_{n,s} N_s = 0.$$

The integral is just the normalization of $y_s(x)$ and therefore nonzero so that $a_{n,s} = 0$ follows immediately.

We can not continue further to larger s because for $s = n - 1$ the first term in (4.42) is not zero any more (that is $s + 1$ hits n). Therefore $a_{n,0} = 0$, $a_{n,1} = 0$, …, $a_{n,n-2} = 0$, but $a_{n,n-1}$ and $a_{n,n}$ is generally finite. Thus (4.41) indeed terminates and becomes

$$y_{n+1}(x) = \alpha_n \, x \, y_n(x) + a_{n,n} \, y_n(x) + a_{n,n-1} \, y_{n-1}(x).$$

Therefore orthogonal polynomials satisfy a three-term recurrence relation of the form (4.39).

Example 4.12
One can use the Rodrigues' formula (4.28) for the Hermite polynomials in order to derive a recurrence relation. Using the Leibniz rule we obtain

$$\begin{aligned} H_{n+1}(x) &= (-1)^{n+1} e^{x^2} \frac{d^n}{dx^n} \frac{d}{dx} \, (e^{-x^2}) = 2 \, (-1)^n e^{x^2} \frac{d^n}{dx^n} \, (xe^{-x^2}) \\ &= 2(-1)^n e^{x^2} \Big[x \frac{d^n}{dx^n} \, (e^{-x^2}) + n \frac{d^{n-1}}{dx^{n-1}} \, (e^{-x^2}) \Big]. \end{aligned}$$

From this result we can read off the recurrence relation for the Hermite polynomials in its canonical form,

$$H_{n+1}(x) = 2x \, H_n(x) - 2n \, H_{n-1}(x). \tag{4.43}$$

Finally we observe that taking the derivative of (4.28) with respect to x we obtain

$$\begin{aligned} \frac{dH_n(x)}{dx} &= 2x(-1)^n e^{x^2} \frac{d^n}{dx^n} \, (e^{-x^2}) + (-1)^n e^{x^2} \frac{d^{n+1}}{dx^{n+1}} \, (e^{-x^2}) \\ &= 2x H_n(x) - H_{n+1}(x) = 2n \, H_{n-1}(x), \end{aligned}$$

where in the last step we have used the recurrence relation (4.43). Thus the Hermite polynomials defined by (4.28) satisfy the following differential equation:

$$\frac{dH_n(x)}{dx} = 2n \, H_{n-1}(x). \tag{4.44}$$

4.8 Integral Representation of Orthogonal Polynomials

With the help of Rodrigues' formula (4.23) one can derive a very useful expression for the orthogonal polynomials as contour integrals. Being an nth order derivative of some function the polynomial $y_n(x)$ can thus be understood as a coefficient in a Taylor expansion of some other function at the point x (we shall see in a moment that this is, in fact, the generating function). On the other hand, according to (1.4) any such coefficient can be written down as a contour integral along C enclosing the point x. In this way one obtains the *Schläfli integral*

$$y_n(x) = \frac{1}{w(x)} \frac{n!}{2\pi i} \int_C dz \, \frac{w(z)[p(z)]^n}{(z-x)^{n+1}}. \tag{4.45}$$

A necessary prerequisite is the analyticity of $w(z)\,[p(z)]^n$ within and on the contour C.

As an example, consider the properties of the Schläfli integral for the Legendre polynomials $P_n(x)$. Writing $P_n(x)$ in a standard form (4.25) with the normalization $P_n(1) = 1$,

$$P_n(x) = \frac{1}{n!\,2^n}\frac{d^n}{dx^n}[(x^2-1)^n],$$

the Schläfli integral is given by

$$P_n(x) = \frac{1}{2\pi i}\int_C dz\,\frac{(z^2-1)^n}{2^n(z-x)^{n+1}}. \tag{4.46}$$

Let the contour C be a circle of radius $|x^2-1|^{1/2}$ centered at the point x. Using the parametrization $z = x + (x^2-1)^{1/2}e^{i\theta}$ $(-\pi \le \theta \le \pi)$, we obtain

$$P_n(x) = \frac{1}{2\pi}\int\limits_{-\pi}^{\pi}\left\{\frac{[x-1+\sqrt{x^2-1}e^{i\theta}][x+1+\sqrt{x^2-1}e^{i\theta}]}{2\sqrt{x^2-1}e^{i\theta}}\right\}^n d\theta.$$

After some algebra we arrive at the Laplace integral formula or the first Laplace integral[11]

$$P_n(x) = \frac{1}{\pi}\int\limits_0^{\pi} d\theta\left(x+\sqrt{x^2-1}\cos\theta\right)^n. \tag{4.47}$$

From the Laplace formula we can derive an important inequality satisfied by the Legendre polynomials. Let x be a real number such that $-1 \le x \le 1$. Then

$$|x+\sqrt{x^2-1}\cos\theta| = \sqrt{x^2+(1-x^2)\cos^2\theta} \le 1$$

and hence

$$|P_n(x)| \le 1, \qquad -1 \le x \le 1.$$

Note also that in a particular case of $x = 1$,

$$P_n(1) = \frac{1}{2^{n+1}\pi i}\int\limits_C \frac{(t+1)^n}{(t-1)}dt,$$

and using the residue theorem, we have $P_n(1) = 2^{-n}(t+1)^n|_{t=1} = 1$, as expected.

Let us now merge the definition of the generating function (4.35) with the Schläfli formula, then

[11] The choice of a branch is not important. The right hand-side of the formula (4.47) contains no odd degrees $\sqrt{x^2-1}$.

$$\Psi(x,t) = \sum_{n=0}^{\infty} t^n P_n(x) = \frac{1}{2\pi i}\int_C dz \frac{1}{z-x}\sum_{n=0}^{\infty}\left[\frac{t(z^2-1)}{2(z-x)}\right]^n$$

$$= \frac{1}{2\pi i}\int_C dz \frac{1}{z-x} \underbrace{\left[1 - \frac{t(z^2-1)}{2(z-x)}\right]^{-1}}_{2(z-x)/(-t z^2+2z+t-2x)}$$

$$= -\frac{1}{\pi i t}\int_C \frac{dz}{(z-z_+)(z-z_-)},$$

where $z_\pm$ satisfy

$$z^2 - \frac{2z}{t} + \frac{2x}{t} - 1 = 0 \;\Rightarrow\; z_\pm = \frac{1}{t}\left[1 \pm \sqrt{1-2xt+t^2}\right].$$

Now we see that as $t \to 0$, $z_+ \simeq 2/t \to \infty$ and $z_- \to x$. Thus only the pole $z = z_-$ is within the (small) contour encircling the point x and by the residue theorem the generating function is thus

$$\Psi(x,t) = -\frac{2}{t}\frac{1}{z_- - z_+} = \frac{1}{\sqrt{1-2xt+t^2}},$$

which coincides with the previous result (4.36).

Let us now verify that the Schläfli integral for the Legendre polynomials $P_n(x)$ is a solution to the original Eq. (4.5). Writing the differential equation for $P_n(x)$ (generally speaking, x may be a complex variable) and inserting the integral (4.46) into its lhs after some algebra we find

$$(1-z^2)\frac{d^2 P_n(z)}{dz^2} - 2z\frac{dP_n(z)}{dz} + n(n+1)P_n(z)$$
$$= \frac{n+1}{2^n 2\pi i}\int_C \frac{d}{dt}\left[\frac{(t^2-1)^{n+1}}{(t-z)^{n+2}}\right]dt = \frac{n+1}{2^n 2\pi i}\left[\frac{(t^2-1)^{n+1}}{(t-z)^{n+2}}\right]\Big|_C.$$

So, the Schläfli integral is a solution to the Eq. (4.5) under the following condition: the function

$$(t^2-1)^{n+1}(t-z)^{-n-2} \tag{4.48}$$

must return to its initial value after going around the contour C. Clearly, this condition is satisfied for integer n.[12]

---

[12] When n is non-integer, a solution of the Legendre equation may still be given by the Schläfli integral. The latter defines then a regular function of z ($\mathrm{Re}z > -1$) in the complex plane with the branch cut from -1 to $-\infty$. This function is usually called the *Legendre function* of the first kind and denoted by $P_n(z)$. Indeed, the function (4.48) has now three branch

4.9 Relation to Homogeneous Integral Equations

In Sect. 4.6 we have seen the advantages of the generating function in calculating different properties of orthogonal polynomials. It turns out that $G(x, t)$ is also useful in the field of integral equations. Here we give some examples.

Let us consider the following integral equation:

$$g(x) = \lambda \int_{-\infty}^{\infty} dy\, e^{-(x+y)^2}\, f(y), \tag{4.49}$$

which we would like to solve for $f(y)$. First we observe that the kernel of this equation resembles the generating function of Hermite polynomials (4.37), therefore we can write

$$g(x) = \lambda \sum_{n=0}^{\infty} \frac{x^n}{n!} \int_{-\infty}^{\infty} dy\, e^{-y^2}\, H_n(y)\, f(y).$$

Furthermore, assuming $f(y)$ to fulfill all necessary prerequisites we can replace it by an expansion in terms of Hermite polynomials, $f(y) = \sum_{m=0}^{\infty} f_m\, H_m(y)$. Taking into account the normalization of $H_m(x)$ given by (4.38) we then end up with

$$g(x) = \lambda\sqrt{\pi} \sum_{n=0}^{\infty} 2^n x^n\, f_n.$$

Obviously it is solved for f_n in the following way:

$$f_n = \frac{1}{\lambda\sqrt{\pi}\, 2^n\, n!} \left[\frac{d^n}{dx^n} g(x)\right]\Bigg|_{x\to 0}. \tag{4.50}$$

Let us consider $g(x) = (2x)^m$. Then the nth order derivative is obviously zero unless $n = m$, when it is equal $2^n n!$. In this situation we obtain for the solution

$$f(y) = \frac{1}{\lambda\sqrt{\pi}} H_n(y).$$

In view of the above one can ask a question whether it is possible to have a Hermite polynomial as a solution of (4.49) when the lhs itself is $H_n(x)$. To investigate that we consider the following integral

(Footnote 12 continued)
points at $t = z$ and $t = \pm 1$. Taking the branch cut from -1 to $-\infty$ and assuming that it does not contain the point z, the suitable contour C may begin on the real axis at any point $t > 1$ and must enclose the points $t = z$ and $t = 1$. Then the term $(t - z)^{-n-2}$ acquires the factor $e^{2\pi i(-n-2)}$, the term $(t^2 - 1)^{n+1}$ acquires the factor $e^{2\pi i(n+1)}$, so that the function (4.48) returns to the original value: $(t^2 - 1)^{n+1}(t - z)^{-n-2} \to e^{-2\pi i}(t^2 - 1)^{n+1}(t - z)^{-n-2} \equiv (t^2 - 1)^{n+1}(t - z)^{-n-2}$. For further details see [3].

$$\int_{-\infty}^{\infty} dy\, e^{-(x+y)^2}\, G(\gamma y, t) = \int_{-\infty}^{\infty} dy\, e^{-(x+y)^2}\, e^{-2\gamma y t - t^2}$$

$$= \sqrt{\pi} e^{2\gamma x t - t^2(1-\gamma^2)} = \sqrt{\pi} \sum_{n=0}^{\infty} \frac{\left(t\sqrt{1-\gamma^2}\right)^n}{n!}\, H_n\left(-\frac{\gamma x}{\sqrt{1-\gamma^2}}\right).$$

On the other hand, for the lhs of the first line we have

$$\int_{-\infty}^{\infty} dy\, e^{-(x+y)^2}\, G(\gamma y, t) = \int_{-\infty}^{\infty} dy\, e^{-(x+y)^2} \sum_{n=0}^{\infty} \frac{t^n}{n!} H_n(\gamma y).$$

Comparing both expansions we conclude that the following identity holds,

$$\sqrt{\pi}\left(\sqrt{1-\gamma^2}\right)^n H_n\left(-\frac{\gamma x}{\sqrt{1-\gamma^2}}\right) = \int_{-\infty}^{\infty} dy\, e^{-(x+y)^2}\, H_n(\gamma y). \tag{4.51}$$

This has obviously the form of (4.49) with $\lambda = 1/\sqrt{\pi}$, $f(y) = H_n(\gamma y)$ and

$$g(x) = \left(\sqrt{1-\gamma^2}\right)^n H_n\left(-\frac{\gamma x}{\sqrt{1-\gamma^2}}\right). \tag{4.52}$$

Using this result we can derive the *multiplication theorem* for the Hermite polynomials. To do that we start with the formal solution of (4.51), which is due to (4.50) given by

$$f(y) = \sum_{n=0}^{\infty} \frac{g^{(n)}(0)}{2^n n!} H_n(y).$$

According to (4.44) for the kth order derivative of (4.52) with $k \leq n$ we obtain[13]

$$g^{(k)}(0) = 2^k \frac{n!}{(n-k)!} H_{n-k}(0)\, \gamma^k\, (1-\gamma^2)^{(n-k)/2},$$

which leads to

$$f(y) = \sum_{k=0}^{n} \frac{n!}{k!(n-k)!} H_k(y)\, H_{n-k}(0)\, \gamma^k\, (1-\gamma^2)^{(n-k)/2}.$$

[13] For the considered here nonclassical form of the Hermite polynomials the differential Eq. (4.44) reads

$$\frac{dH_n(x)}{dx} = -2n H_{n-1}(x).$$

From (4.37) we immediately realize that $H_k(0)$ is non-zero only for even $n-k=2p$ and equal to $H_{2p}(0)=(-1)^p(2p)!/p!$. Therefore we obtain

$$H_n(\gamma y)=\sum_{p=0}^{[n/2]}\frac{n!}{(n-2p)!p!}H_{n-2p}(y)\,\gamma^{n-2p}\,(\gamma^2-1)^p,$$

which is the multiplication theorem we were looking for. Note that very same form has the multiplication theorem in the case of traditional normalization (4.28).

Let us now turn to the following version of the homogeneous Fourier equation (3.3),

$$g(x)=\lambda\int_{-\infty}^{\infty}\frac{dy}{\sqrt{2\pi}}\,e^{ixy}\,f(y), \tag{4.53}$$

and try to evaluate the rhs for the special choice

$$f(y)=G(y,t)\,e^{-y^2/2}=e^{-2yt-t^2-y^2/2}.$$

Then we obtain

$$g(x)=\lambda\,e^{-2ixt+t^2-x^2/2}=\lambda\,G(x,it)\,e^{-x^2/2}.$$

Inserting the last two relations into the original Eq. (4.53), expanding in t and comparing term by term we conclude that functions

$$f_n(x)=H_n(x)\,e^{-x^2/2}$$

are eigenfunctions of the Fourier integral equation. There are only four different eigenvalues $\lambda_n=(-i)^n=\pm 1,\pm i$, corresponding to the even and odd values of n, respectively.

4.10 Orthogonal Polynomials and Hypergeometric Functions

According to the definition (2.34) and (2.35) the hypergeometric series $F(a,b;c;z)$ breaks up when either parameter a or b is a negative integers $-n$. Then it is some polynomial of degree n. It turns out that under certain conditions it coincides with one of the Jacobi orthogonal polynomials. An important observation is that both a and b enter the last, linear in F term of the hypergeometric Eq. (2.51). The eigenvalue (4.3) happens to be factorizable (at least under the conditions of Sect. 4.1),

$$\lambda_n=-ab=-n\,[p_2\,(n+1)-q_1].$$

Comparing the hypergeometric Eq. (2.51) with (4.1) and consulting (4.10) for the precise relation between the parameters p_i and q_i with α and β we find the following identification[14]

$$P_n^{(\alpha,\beta)}(x) = C\, F\left(-n, n+\alpha+\beta+1; \alpha+1; \frac{1-x}{2}\right), \tag{4.54}$$

where the constant $C = P_n^{(\alpha,\beta)}(1) = \Gamma(n+\alpha+1)/n!\Gamma(1+\alpha)$ is related to the normalization constant, see Problem 4.2. This identity leads to a number of interesting results. For instance, for the special case $\alpha = \beta = 0$ the above identity defines the Legendre polynomials. In our normalization convention[15]

$$P_n(x) = F\left(-n, n+1; 1; \frac{1-x}{2}\right). \tag{4.55}$$

Since the hypergeometric function is symmetric with respect to exchange $a \leftrightarrow b$ of two first parameters we, on the one hand, immediately obtain a definition of polynomials of negative order,

$$P_{-n-1}(x) = P_n(x).$$

On the other hand, by comparison with Eq. (4.47) we obtain a *second Laplace integral*,

$$P_n(x) = \frac{1}{\pi}\int_0^{\pi} d\theta\, \frac{1}{\left(x+\sqrt{x^2-1}\cos\theta\right)^{n+1}}. \tag{4.56}$$

Even more straightforward is the relation between the Laguerre polynomials and the confluent hypergeometric function. From the direct comparison of (2.57) and (4.22) we immediately realize that

$$L_n(x) = F(-n, 1; x). \tag{4.57}$$

---

[14] We assume the interval $[a, b] = [-1, 1]$.

[15] For an arbitrary $n = \nu$ (and the complex variable z as well) this equality serves as a definition of the Legendre function $P_\nu(z)$ which we have already encountered on page 211. It is absolutely convergent inside a circle of radius 2 with a center at the point $z = 1$. Here we can draw an analogy to the theory of Gamma function, which is an extension of the factorial function to real and complex numbers.

4.11 Problems

Problem 4.1
Show that the equation

$$6(1-x^2)y'' - (5+11x)y' + \lambda y = 0$$

has polynomial solutions for $\lambda = 6n^2 + 5n$. Show that the first three polynomials are

$$y_0 = 1\,, \quad y_1 = 11x + 5\,, \quad \text{and} \quad y_2 = 391x^2 + 170x - 113.$$

Derive the Rodrigues' formula for this particular family of polynomials and verify the above results.

Problem 4.2
We have seen above that the Jacobi polynomials satisfy Eq. (4.12). Derive the corresponding Rodrigues' formula and compute the generating function $\Psi(x,t) = \sum_{n=0}^{\infty} P_n^{(\alpha,\beta)}(x)t^n$. Evaluate Jacobi polynomials at the endpoints $x = \pm 1$.

Problem 4.3
Gegenbauer polynomials are special cases of Jacobi polynomials $P_n^{(\alpha,\beta)}(x)$ with $\alpha = \beta = \lambda - 1/2$ and the following normalization:

$$C_n^{\lambda}(x) = \frac{\Gamma(n+2\lambda)}{\Gamma(n+\lambda+\frac{1}{2})}\,\frac{\Gamma(\lambda+\frac{1}{2})}{\Gamma(2\lambda)}\,P_n^{(\lambda-\frac{1}{2},\,\lambda-\frac{1}{2})}(x).$$

They are related to the hypergeometric series:

$$\begin{aligned} C_n^{\lambda}(x) &= \frac{\Gamma(2\lambda+n)}{n!\Gamma(2\lambda)} F\left(2\lambda+n,\, -n;\, \lambda+\frac{1}{2};\, \frac{1-x}{2}\right) \\ &= \frac{2^n\Gamma(\lambda+n)}{n!\Gamma(\lambda)} x^n F\left(-\frac{n}{2},\, \frac{1-n}{2};\, 1-\lambda-n;\, \frac{1}{x^2}\right). \end{aligned}$$

Often these polynomials are called the ultraspherical polynomials and denoted $P_n^{\lambda}(x)$.

(a): Evaluate the normalization constant

$$N_n = \int_{-1}^{1} dx\, [C_n^{\lambda}(x)]^2 (1-x^2)^{\lambda-\frac{1}{2}}.$$

(b): Following the procedure outlined in Sect. 4.7 obtain the three-term recurrence relation.

(c): Find the generating function $\Psi(x,t) = \sum_{n=0}^{\infty} C_n^{\lambda}(x)t^n$.

Problem 4.4
The *associated Laguerre polynomials* are given by

$$L_n^{\alpha}(x) = e^x x^{-\alpha} \frac{d^n}{dx^n} \left(e^{-x} x^{n+\alpha} \right).$$

Show that their generating function is

$$G^{\alpha}(x, y) = (1 - y)^{-\alpha-1} e^{-xy/(1-y)},$$

and compute the first three polynomials. Show that the generating function satisfies the following ODE:

$$(1 - y)^2 \frac{dG^{\alpha}(x, y)}{dy} + [x - (1 + \alpha)(1 - y)] G^{\alpha}(x, y) = 0.$$

Using it derive a recurrence relation for $L_n^{\alpha}(x)$.

Problem 4.5
Taking advantage of the generating function for the Legendre polynomials:

$$\frac{1}{\sqrt{1 - 2xt + t^2}} = \sum_{n=0}^{\infty} P_n(x) t^n,$$

(a): derive an explicit formula for $P_n(x)$,
(b): find the three–term recurrence relation,
(c): determine the normalisation constant $N_n = \int_{-1}^{1} dx\, P_n^2(x)$.

Problem 4.6
Show that there exists an nth order polynomial $S_n(x)$ such that

$$\int_{-1}^{1} dx\, S_n(x) K(x) = K(1)$$

for any polynomial $K(x)$ of order n or less.
Hint: take advantage of the recurrence relation and of the normalization constant for the Legendre polynomials and use that $P_0 = 1$, $P_1 = x$ and $P_n(1) = 1$, $P_n(-1) = (-1)^n$.

Problem 4.7
Hermite polynomials satisfy the Eq. (4.17) and their traditional normalization is given by (4.28).

(a): Use the Gram–Schmidt process to find the first four polynomials.
(b): Use the Schläfli integral to obtain the standard series representation.
(c): Verify the real axis integral representation

$$H_n(x) = \frac{2^n}{\sqrt{\pi}} \int_{-\infty}^{\infty} dt e^{-t^2} (x + it)^n.$$

Problem 4.8
Consider the Legendre equation

$$\frac{d}{dz}\Big[(1 - z^2)\frac{du}{dz}\Big] + \nu(\nu + 1)u = 0,$$

where ν is an arbitrary number. Show that the solution is given in terms of hypergeometric functions,

$$P_\nu(z) = A_\nu\, z^\nu\, F\left(-\frac{\nu}{2}, \frac{1-\nu}{2}; \frac{1}{2} - \nu\,; \frac{1}{z^2}\right)$$
$$+ B_\nu\, z^{-\nu-1}\, F\left(\frac{\nu+1}{2}, \frac{\nu+2}{2}; \nu + \frac{3}{2}; \frac{1}{z^2}\right),$$

where A_ν and B_ν are some constants. Recover the Legendre polynomials for integer $\nu = n = 0, 1, 2, \ldots.$.

Problem 4.9
Show that the Hermite polynomials, defined by (4.28), are related to the confluent hypergeometric series in the following way:

$$H_{2n}(x) = (-1)^n \frac{(2n)!}{n!} F\Big(-n, \frac{1}{2}; x^2\Big),$$
$$H_{2n+1}(x) = 2(-1)^n z \frac{(2n+1)(2n)!}{n!} F\Big(-n, \frac{3}{2}; x^2\Big).$$

Derive the relation between these polynomials and the associated Laguerre polynomials given by the Rodrigues formula:

$$L_n^\alpha(x) = \frac{x^{-\alpha} e^x}{n!} \frac{d^n}{dx^n} (e^{-x} x^{n+\alpha}).$$

Problem 4.10
As we have seen in Example 4.8, the Chebyshev polynomials of the first kind satisfy Eq. (4.26). They are usually normalized to obey the condition $T_n(1) = 1$.

(a): Show that these polynomials are explicitly defined by the formula

$$T_n(x) = \cos\big(n \arccos x\big) = \frac{1}{2}\Big[(x + i\sqrt{1-x^2})^n + (x - i\sqrt{1-x^2})^n\Big].$$

(b): Derive the corresponding three–terms recurrence relation.
(c): Obtain the following generating functions:

$$\Psi(x, s) = \sum_{n=0}^{\infty} T_n(x)\, s^n \quad \text{and} \quad G(x, s) = \sum_{n=0}^{\infty} \frac{T_n(x)}{n!}\, s^n.$$

Problem 4.11
Using the method developed in Sect. 4.9 solve the integral equation

$$\int_{-1}^{1} \frac{y(t)dt}{\sqrt{1 + x^2 - 2xt}} = x^m, \quad m > 0.$$

Answers:

Problem 4.1:

$$y_n(x) = \frac{6^n}{(1-x)^{1/3}(x+1)^{-1/2}} \frac{d^n}{dx^n}\Big[(1-x)^{n+1/3}(x+1)^{n-1/2}\Big].$$

Problem 4.2:

$$P_n^{(\alpha,\beta)}(x) = \frac{(-1)^n}{2^n n!}(1-x)^{-\alpha}(1+x)^{-\beta}\frac{d^n}{dx^n}[(1-x)^{n+\alpha}(1+x)^{n+\beta}].$$

$$\Psi(x,t) = \sum_{n=0}^{\infty} P_n^{(\alpha,\beta)}(x)\, t^n = 2^{\alpha+\beta}R^{-1}(1-t+R)^{-\alpha}(1+t+R)^{-\beta},$$

where $R = \sqrt{1 - 2xt + t^2}$.

$$P_n^{(\alpha,\beta)}(1) = \frac{\Gamma(n+1+\alpha)}{n!\Gamma(1+\alpha)}, \qquad P_n^{(\alpha,\beta)}(-1) = (-1)^n \frac{\Gamma(n+1+\beta)}{n!\Gamma(1+\beta)}.$$

Problem 4.3:

$$\textbf{(a)}: \quad N_n = \frac{2^{1-2\lambda}\pi\Gamma(n+2\lambda)}{n!(\lambda+n)[\Gamma(\lambda)]^2};$$

$$\textbf{(b)}: \quad nC_n^{\lambda}(x) - 2(n-1+\lambda)xC_{n-1}^{\lambda}(x) + (n+2\lambda-2)C_{n-2}^{\lambda}(x) = 0;$$

$$\textbf{(c)}: \quad \Psi(x,t) = \sum_{n=0}^{\infty} C_n^{\lambda}(x)t^n = \frac{1}{(1-2xt+t^2)^{\lambda}}.$$

Problem 4.4:

$$1,\ (1+\alpha-x),\ [(1+\alpha)(2+\alpha) - 2(2+\alpha)x + x^2],$$
$$[(1+\alpha)(2+\alpha)(3+\alpha) - 3(2+\alpha)(3+\alpha)x + 3(3+\alpha)x^2 - x^3].$$
$$L_{n+1}^{\alpha}(x) + (x - \alpha - 1 - 2n)\, L_n^{\alpha}(x) + n\,(n+\alpha)\, L_{n-1}^{\alpha}(x) = 0.$$

Problem 4.5:

$$\textbf{(a)}:\quad P_n(x) = \sum_{r=0}^{[\frac{n}{2}]} \frac{(-1)^r (2n-2r)!}{2^n r!(n-r)!(n-2r)!} x^{n-2r};$$

$$\textbf{(b)}:\quad (n+1)P_{n+1}(x) - (2n+1)xP_n(x) + nP_{n-1}(x) = 0;$$

$$\textbf{(c)}:\quad N_n = \frac{2}{2n+1}.$$

Problem 4.6:

$$\textbf{(a)}:\quad H_0 = 1,\, H_1(x) = x,\, H_2(x) = x^2 - 1/2,\, H_3(x) = x^3 - 3x/2;$$

$$\textbf{(b)}:\quad H_n(x) = n! \sum_{l=0}^{[\frac{n}{2}]} (-1)^l \frac{(2x)^{n-2l}}{l!(n-2l)!}.$$

Problem 4.7:

$$P_n(x) = \frac{(2n)!}{2^n (n!)^2} x^n F\left(-\frac{n}{2}, \frac{1-n}{2}; \frac{1}{2} - n; \frac{1}{x^2}\right).$$

Problem 4.8:

$$H_{2n}(x) = (-1)^n 2^{2n} n!\, L_n^{-1/2}(x^2), \quad H_{2n+1}(x) = (-1)^n 2^{2n+1} n!\, x\, L_n^{1/2}(x^2).$$

Problem 4.9:

$$\textbf{(b)}:\quad T_{n+1}(x) + T_{n-1}(x) - 2\, T_n(x)\, x = 0;$$

$$\textbf{(c)}:\quad \Psi(x,s) = \frac{1-xs}{1-2xs+s^2},\quad G(x,s) = \frac{1}{2}\left[e^{(x-\sqrt{x^2-1})s} + e^{(x+\sqrt{x^2-1})s}\right].$$

Problem 4.10:

$$y(x) = \frac{2m+1}{2} P_m(x),$$

where $P_m(x)$ is the Legendre polynomial.

Chapter 5
Solutions to the Problems

5.1 Chapter 1

Problem 1.1
In polar coordinates $z = re^{i\theta}$ we have: $f(z) = f(r,\theta) = u(r,\theta) + iv(r,\theta)$. If we take the limit $\Delta z = [(r+\Delta r)e^{i(\theta+\Delta\theta)} - re^{i\theta}] \to 0$ 'radially', i.e. $\Delta\theta = 0$ first and then $\Delta r \to 0$, by definition of the derivative we obtain

$$f'(z) = \lim_{\Delta r\to 0} \frac{u(r+\Delta r,\theta) + iv(r+\Delta r,\theta) - u(r,\theta) - iv(r,\theta)}{\Delta r e^{i\theta}} = e^{-i\theta}\Big[\frac{\partial u}{\partial r} + i\frac{\partial v}{\partial r}\Big].$$

On the other hand, taking the limit $\Delta r = 0$ first and subsequently computing $\Delta\theta \to 0$, we find

$$f'(z) = \lim_{\Delta\theta\to 0} \frac{u(r,\theta+\Delta\theta) + iv(r,\theta+\Delta\theta) - u(r,\theta) - iv(r,\theta)}{ire^{i\theta}\Delta\theta} = \frac{e^{-i\theta}}{r}\Big[\frac{\partial v}{\partial\theta} - i\frac{\partial u}{\partial\theta}\Big].$$

Equating these two expressions results in Cauchy–Riemann equations we are looking for:

$$\frac{\partial u}{\partial r} = \frac{1}{r}\frac{\partial v}{\partial\theta}, \qquad \frac{\partial v}{\partial r} = -\frac{1}{r}\frac{\partial u}{\partial\theta}.$$

We can combine these into one equation for $f(r,\theta)$:

$$\frac{\partial f}{\partial r} = \frac{1}{ir}\frac{\partial f}{\partial\theta}.$$

Let us now consider $f(z) = \ln z = \ln r + i \arg z$. Then at $r \neq 0$ we have

$$\frac{\partial \ln z}{\partial r} = \frac{1}{r}, \qquad \frac{\partial \ln z}{\partial\theta} = i,$$

whence the Cauchy–Riemann equation for the $\ln z$:

A. O. Gogolin (edited by E. G. Tsitsishvili and A. Komnik), *Lectures on Complex Integration*, 221
Undergraduate Lecture Notes in Physics, DOI: 10.1007/978-3-319-00212-5_5,

$$\frac{\partial \ln z}{\partial r} = \frac{1}{ir}\frac{\partial \ln z}{\partial \theta}$$

immediately follows. Therefore the function $\ln z$ is analytic for all finite z, except for $z = 0$.

Problem 1.2
In all cases we substitute $z = e^{i\varphi}$ and determine the residues of all poles inside the unit circle.

$$(\mathbf{a}): \int_0^{2\pi} \frac{d\varphi}{(a + b\cos\varphi)^2} = \frac{4}{ib^2} \int_{|z|=1} \frac{z dz}{(z^2 + 2az/b + 1)^2}.$$

There is only one pole inside $|z| = 1$ at $z_0 = (\sqrt{a^2 - b^2} - a)/b$ with the residue

$$a_{-1} = \left[\frac{d}{dz}\frac{z}{(z + (\sqrt{a^2 - b^2} + a)/b)^2}\right]_{z=z_0} = \frac{b^2 a}{4(a^2 - b^2)^{3/2}},$$

hence the answer given.

$$(\mathbf{b}): \int_0^{2\pi} \frac{\cos^2(3\varphi)d\varphi}{1 - 2p\cos(2\varphi) + p^2} = \frac{1}{4i} \int_{|z|=1} \frac{(z^6 + 1)^2 dz}{z^5(z^2 - p)(1 - pz^2)}.$$

The poles inside the unit circle are located at $0, -p^{1/2}, p^{1/2}$. The residues are

$$-\frac{1 + p^2 + p^4}{p^3}, \quad \frac{(p^3 + 1)^2}{2p^3(1 - p^2)}, \quad \frac{(p^3 + 1)^2}{2p^3(1 - p^2)},$$

respectively. Note that while $z = \pm p^{1/2}$ are simple poles and the residues are easily read off, $z = 0$ is a 5th degree pole and thus the following Laurent expansion (which is a product of two geometric series) must be used as $z \to 0$:

$$-\frac{1}{pz^5}\frac{1 + O(z^6)}{(1 - z^2/p)(1 - pz^2)}$$

$$= -\frac{1}{pz^5}[1 + O(z^6)][1 + z^2/p + z^4/p^2 + O(z^6)][1 + pz^2 + p^2z^4 + O(z^6)]$$

$$= -\frac{1}{pz^5} - \frac{1 + p^2}{p^2 z^3} - \frac{1 + p^2 + p^4}{p^3 z} + O(z).$$

A massive algebraic simplification occurs when residues are added up leading to the final answer.

(c): We have

$$\int_0^{2\pi} \frac{(1+2\cos\varphi)^n e^{in\varphi} d\varphi}{1-a-2a\cos\varphi} = \frac{1}{i} \int_{|z|=1} \frac{(1+z+z^2)^n dz}{(1-a)z - a(1+z^2)}.$$

The poles of the integrand are the zeros of the denominator. There are two roots,

$$z_{1,2} = \frac{1-a \pm \sqrt{1-2a-3a^2}}{2a},$$

which satisfy $z_1 z_2 = 1$. Therefore one of the roots is inside the unit circle and the other one is outside. For $0 < a < 1/3$, the root

$$z_1 = \frac{1-a-\sqrt{1-2a-3a^2}}{2a}$$

is inside and the residue is:

$$\frac{(1+z_1+z_1^2)^n}{-a(z_1-z_2)} = \frac{1}{\sqrt{1-2a-3a^2}} \left(\frac{z_1}{a}\right)^n,$$

where the quadratic equation was used again. As the above expression is real, the answer follows.

Problem 1.3

We need to calculate the following integral:

$$\int_{-\infty}^{\infty} \frac{1}{1+x'^2} \frac{dx'}{1+x^2+x'^2-xx'}.$$

As the integrand decays fast enough in infinity, we close the contour in the upper half-plane. One pole is at $x' = i$ and the other one is at $x' = x/2 + i\sqrt{1+3x^2/4}$. The residues are

$$\frac{1}{2ix} \frac{x+i}{1+x^2}, \quad \frac{1}{2ix\sqrt{1+3x^2/4}} \frac{-x/2 - i\sqrt{1+3x^2/4}}{1+x^2},$$

respectively. Plugging this in and simplifying, we obtain

$$\int_{-\infty}^{\infty} \frac{1}{1+x'^2} \frac{dx'}{1+x^2+x'^2-xx'} = \frac{\pi}{2} \frac{1}{1+x^2} \left[2 - \frac{1}{\sqrt{1+3x^2/4}}\right].$$

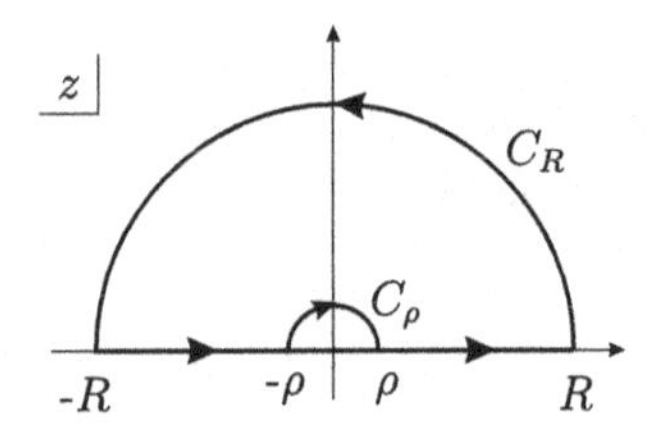

Fig. 5.1 Integration contour used in Problem 1.4(a) and 1.6(b)

Consequently $f(x) = 1/(1 + x^2)$ is indeed the solution with $\lambda = 2$.

Problem 1.4

(a): A useful contour here consists of two segments of the real axis $[-R, -\rho]$ and $[\rho, R]$, completed by two semi-circles C_ρ and C_R with radii ρ and R, in the limit $\rho \to 0$ and $R \to \infty$, see Fig. 5.1. The auxiliary function is $f(z) = e^{iz}/z$. As this function is analytic in the domain enclosed by the contour, the Cauchy theorem reads

$$\int_{-R}^{-\rho} + \int_{C_\rho} + \int_{\rho}^{R} + \int_{C_R} = 0.$$

Around $z = 0$ we have $f(z) = 1/z + g(z)$, where $g(z)$ is continuous at $z = 0$, so

$$\lim_{\rho\to 0} \int_{C_\rho} f(z)dz = \int_{\pi}^{0} \frac{\rho i e^{i\varphi} d\varphi}{\rho e^{i\varphi}} = -i\pi.$$

The C_R integral evidently tends to zero, so the Cauchy theorem takes the form

$$\int_{-\infty}^{0} \frac{e^{ix}}{x} dx + \int_{0}^{\infty} \frac{e^{ix}}{x} dx = i\pi.$$

Changing $x \to -x$ in the first integral and combining it with the second one results in

$$\int_{0}^{\infty} \frac{e^{ix} - e^{-ix}}{x} dx = i\pi,$$

leading to the $\pi/2$ result.

(b): We take the rectangular contour consisting of four segments I, II, III and IV, see Fig. 5.2. Note that the real part of the function $f(z) = e^{-az^2}$ evaluated on the upper segment III, $f(x + ih) = e^{ah^2} e^{-ax^2} [\cos(2ahx) - i \sin(2ahx)]$, is proportional to the integrand when $h = b/(2a)$. For $a > 0$, the integrand is decaying exponentially

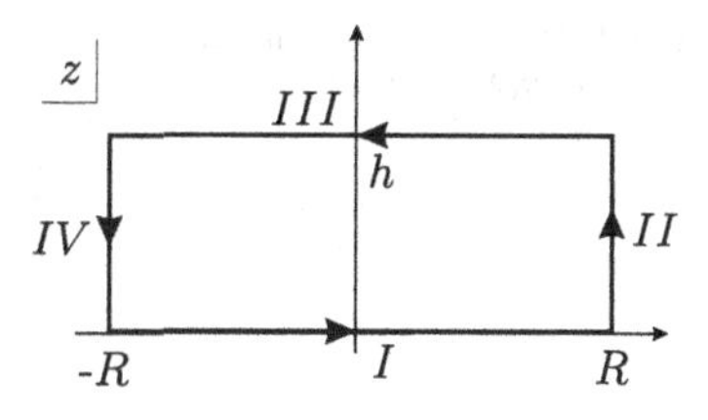

Fig. 5.2 Integration contour used in Problem 1.4(b)

on the side segments so that

$$\lim_{R\to\infty} \int\limits_{\text{II, IV}} = 0.$$

Furthermore,

$$\lim_{R\to\infty} \int\limits_{\text{I}} = \int\limits_{-\infty}^{\infty} e^{-ax^2} dx = \sqrt{\frac{\pi}{a}},$$

and therefore the Cauchy theorem takes the form

$$\sqrt{\frac{\pi}{a}} - e^{b^2/(4a)} \int\limits_{-\infty}^{\infty} e^{-ax^2} \cos(bx) dx = 0,$$

from the real part of which we read off the answer.

In order to compute the integral

$$I = \int\limits_{-\infty}^{\infty} e^{-ax^2} dx$$

we can employ the following trick: square the original integral and go over to polar coordinates in 2D,

$$I^2 = \int\limits_{-\infty}^{\infty} dx \int\limits_{-\infty}^{\infty} dy e^{-a(x^2+y^2)} = \int\limits_{0}^{2\pi} d\varphi \int\limits_{0}^{\infty} r dr e^{-ar^2} = \pi \int\limits_{0}^{\infty} du e^{-au} = \frac{\pi}{a}.$$

(c): Here we need the contour consisting of the segment $[0, R]$ on the real axis, C_R, which is $1/8$ of a circle reaching point A $(R/\sqrt{2}, R/\sqrt{2})$, and returning to the origin 0 via the straight line $A0$, see Fig. 5.3. The function $f(z) = e^{iz^2}$ is analytic inside this triangular domain. On C_R substitute $z^2 = \xi$:

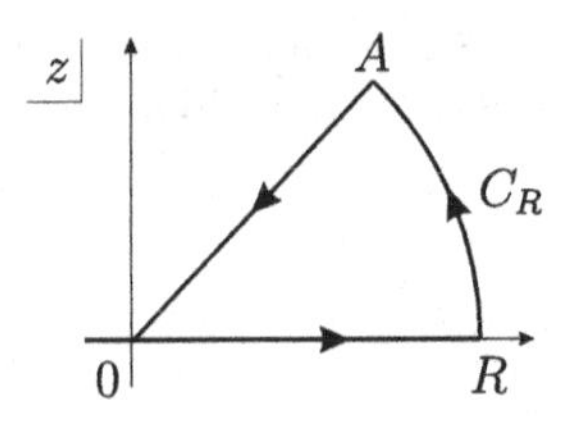

Fig. 5.3 Integration contour used in Problem 1.4(c)

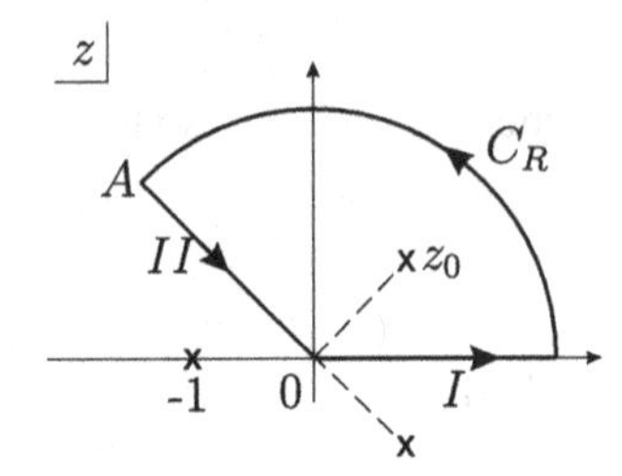

Fig. 5.4 Integration contour used in Problem 1.5(a)

$$\int_{C_R} f(z)dz = \frac{1}{2}\int_{C'_R} \frac{e^{i\xi}d\xi}{\sqrt{\xi}},$$

which tends to zero when $R \to \infty$ (C'_R is $1/4$ of a circle). Furthermore, on the line $A0$ substitute $z = \sqrt{it}$, so that the Cauchy theorem becomes

$$\int_0^\infty e^{ix^2}dx = -\int_{A0} f(z)dz = -\sqrt{i}\int_\infty^0 e^{-t^2}dt = \sqrt{i}\frac{\sqrt{\pi}}{2}.$$

Separating the real and imaginary parts results in

$$\int_0^\infty \cos(x^2)dx = \int_0^\infty \sin(x^2)dx = \frac{1}{2}\sqrt{\frac{\pi}{2}}.$$

Problem 1.5

(a): We use the contour shown in Fig. 5.4. It consists of the segment $[0, R]$ on the real axis, C_R, which is $3/8$ of a circle reaching point $A(-R/\sqrt{2}, R/\sqrt{2})$, and returning to the origin 0 via the straight line A0. The auxiliary function $f(z) = 1/(1+z^3)$ has three poles at $z = -1, e^{-i\pi/3}, e^{+i\pi/3}$, one of which $z_0 = e^{i\pi/3}$ is inside the contour with the residue

$$-\frac{e^{i\frac{\pi}{3}}}{3}.$$

The Cauchy theorem reads

$$\int_I + \int_{II} + \int_{C_R} = -2\pi i \frac{e^{i\frac{\pi}{3}}}{3}.$$

In the limit $R \to \infty$ the C_R integral tends to zero and the real axis integral is

$$\lim_{R\to\infty} \int_I = \int_0^\infty \frac{dx}{1+x^3} = I.$$

Substituting $z = e^{2\pi i/3}x$ on the straight line $A0$, the corresponding integral is found to be equal to

$$\lim_{R\to\infty} \int_{II} = \int_\infty^0 \frac{e^{\frac{2\pi i}{3}}dx}{1+\left(e^{\frac{2\pi i}{3}}x\right)^3} = -e^{\frac{2\pi i}{3}} I.$$

Combining the above expressions results in

$$\lim_{R\to\infty} \left(\int_I + \int_{II} \right) = (1 - e^{\frac{2\pi i}{3}})I = e^{\frac{\pi i}{3}}(e^{-\frac{\pi i}{3}} - e^{\frac{\pi i}{3}})I = -2iIe^{\frac{\pi i}{3}} \sin\frac{\pi}{3} = -i\sqrt{3}e^{\frac{\pi i}{3}} I$$

leading to the $2\pi/3\sqrt{3}$ result.
(b): This example makes use of the asymptotic properties of the trigonometric functions in the complex plane. First of all we observe that the integrand is an even function and the integration can be extended to the whole real axis:

$$I(a,b) = \frac{1}{2}\int_{-\infty}^{\infty} \frac{\sin ax}{\sin bx}\frac{1}{1+x^2}\,dx.$$

The integrand has poles along the real axis at $x = n\pi/b, n = 0, \pm 1, \pm 2, \ldots$. To avoid them, we use the principal value prescription:

$$I(a,b) = \frac{1}{2}\lim_{\varepsilon\to 0} \int_{-\infty+i\varepsilon}^{\infty+i\varepsilon} \frac{\sin az}{\sin bz}\frac{1}{1+z^2}\,dz.$$

Taking $\varepsilon > 0$, we consider a contour consisting of the segment parallel to the real axis and the semicircle C_R, as is shown in Fig. 5.5. (The sign of ε is not important: at $\varepsilon < 0$ the contour must be closed in the lower half-plane.) The chosen contour contains the pole at $z = +i$ with the residue

$$\frac{\sinh a}{\sinh b}\frac{1}{2i},$$

Fig. 5.5 Integration contour used in Problem 1.5(b)

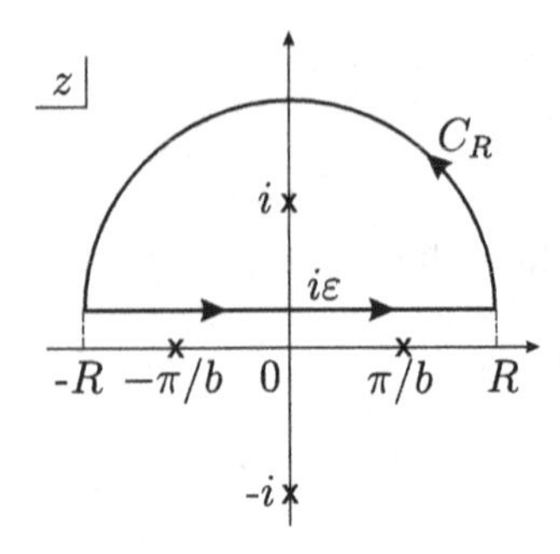

so by virtue of the Cauchy theorem we obtain

$$\int_{-R+i\varepsilon}^{R+i\varepsilon} + \int_{C_R} = \pi \frac{\sinh a}{\sinh b}.$$

In the limit $R \to \infty$ the C_R integral tends to zero if $|b| > |a|$:

$$\lim_{R\to\infty} \frac{\sin |a|z}{\sin |b|z} \to e^{-i(|a|-|b|)\text{Re}z} e^{-(|b|-|a|)\text{Im}z} \to 0$$

and the first integral is $2I(a, b)$ leading to the final answer.

Problem 1.6
(a): From Eq. (1.14) we have

$$I = \frac{\pi}{\sin(\pi p)} \sum \text{Res} f$$

with

$$f(z) = \frac{(-z)^{-p}}{1 + 2z\cos\lambda + z^2} = \frac{(-z)^{-p}}{(z + e^{i\lambda})(z + e^{-i\lambda})}$$

(when $a - 1 \to -p$, the sine changes sign twice). Poles are at $z = -e^{\pm i\lambda}$ with the residues being given by

$$\frac{e^{\mp i\lambda p}}{-e^{\pm i\lambda} + e^{\mp i\lambda}},$$

respectively. Adding everything up yields

$$\frac{\sin(p\lambda)}{\sin\lambda}.$$

Hence the result.
(b): Here we use the same contour as in Problem 1.4 (a). The function

$$f(z) = \frac{\ln z}{(z^2+1)^2}$$

has a pole at $z = i$ with the residue

$$a_{-1} = \frac{\pi + 2i}{8}.$$

Integrals over the small and large semi-circles vanish as $\rho \to 0$ and $R \to \infty$. On the negative half of the real axis substitute $z = -x$ and use $\ln(-x) = \ln x + \pi i$ (upper edge of the branch cut) to obtain:

$$\int_{-\infty}^{0} f(z)dz = \int_{0}^{\infty} \frac{\ln x + \pi i}{(x^2+1)^2} dx.$$

Consequently the Cauchy theorem yields

$$2\int_{0}^{\infty} \frac{\ln x dx}{(x^2+1)^2} + \pi i \int_{0}^{\infty} \frac{dx}{(x^2+1)^2} = \frac{\pi^2 i}{4} - \frac{\pi}{2}.$$

Comparing the real parts leads to the answer. We observe that from the comparison of the imaginary parts one obtains the elementary integral

$$\int_{0}^{\infty} \frac{dx}{(x^2+1)^2} = \frac{\pi}{4}.$$

(c): Let us define the function

$$f(z) = \frac{1}{(z-1)^{1/3}(z+1)^{2/3}},$$

and integrate it over the circle C_R ($R > 1$). There are two ways to calculate this. On the one hand, $f(z) \sim (1/z)$ as $z \to \infty$, so the residue at infinity is equal to 1 and therefore at $R \to \infty$ one gets:

$$\int_{C_R} f(z)dz = 2\pi i.$$

On the other hand, we can collapse the contour on the branch cut. According to the arguments given in the main text, on the upper edge of the cut $\theta_1 \to \pi$ and $\theta_2 \to 0$ (arguments of $z \mp 1$), so we shall have

$$f(z) = e^{-\pi i/3} f(x),$$

while on the lower edge of the branch cut

$$f(z) = e^{\pi i/3} f(x),$$

with

$$f(x) = \frac{1}{[(1-x)(1+x)^2]^{1/3}}.$$

Given the directions of integration we conclude that

$$\int_{C_R} f(z)dz = 2i \sin\left(\frac{\pi}{3}\right) \int_{-1}^{1} f(x)dx.$$

Comparing the two results one finds

$$\int_{-1}^{1} \frac{dx}{[(1-x)(1+x)^2]^{1/3}} = \frac{\pi}{\sin(\pi/3)} = \frac{2\pi}{\sqrt{3}},$$

as required.

Problem 1.7
(a): Here we have

$$G(k) = -\int_{-\infty}^{\infty} d\lambda \frac{e^{i\lambda k}\, e^{-\lambda}}{1+e^{-2\lambda}}.$$

It can be computed going over to the complex plane $\lambda \to z$ and closing the contour in the upper half-plane. Obviously, the resulting integral exists in the strip $-1 < \operatorname{Im} k < 1$ only. The integrand has the poles $\lambda_n = i\pi(n + 1/2)$, $n > 0$ with the residua

$$\operatorname{Res} \frac{e^{i\lambda k}\, e^{-\lambda}}{1+e^{-2\lambda}}\bigg|_{\lambda \to \lambda_n} = -\frac{i}{2}(-1)^n\, e^{-\pi k(n+1/2)}.$$

The final summation over n is then trivial and yields

$$G(k) = 2\pi i \sum_{n=0}^{\infty} \frac{i}{2}(-1)^n e^{-\pi k(n+1/2)} = -\frac{\pi}{2\cosh(\pi k/2)}. \qquad (5.1)$$

(b): Is solved in the similar way. In the first step we rewrite the integral as

$$K(k) = -\frac{4U}{\pi}\frac{1}{1+U^2}\int_{-\infty}^{\infty} d\lambda \frac{e^{(ik-1)\lambda}}{e^{-2\lambda}+2e^{-\lambda}\cos\gamma+1}$$
$$= -\frac{4U}{\pi}\frac{1}{1+U^2}\int_{0}^{\infty} dz \frac{z^{(ik-1)}}{(1/z+e^{i\gamma})(1/z+e^{-i\gamma})},$$

where $\cos\gamma = (1-U^2)/(1+U^2)$ and in the last line $z = e^{\lambda}$. Finding the roots of the denominator of the first line we can perform elementary fraction decomposition and then read the result off the solution of part **(a)** or the solution of Problem 1.6 (a). Alternatively, we can bring the second line to the integrals of the form discussed in Sect. 1.3.2.

Problem 1.8

Let $F(p)$ be the image of $f(t)$, $f(t) \equiv 0$ for $t < 0$. Let us consider the shifted function

$$f_\tau(t) = \begin{cases} 0,\ t < \tau,\ \tau > 0\,, \\ f(t-\tau),\ t \geq \tau\,. \end{cases}$$

The respective image is

$$F_\tau(p) = \int_{\tau}^{\infty} e^{-pt} f(t-\tau)dt = e^{-p\tau}\int_{0}^{\infty} e^{-pt} f(t)dt = e^{-p\tau}F(p)\,,$$

hence $f(t-\tau) \leftrightarrow e^{-p\tau}F(p)$. For the periodic function $f_1(t) = f_1(t+\tau)$ (with a period τ), this property leads to the geometric series:

$$F_1(p) = \int_0^{\infty} e^{-pt} f_1(t)dt$$
$$= \int_0^{\tau} e^{-pt} f_1(t)dt + \int_{\tau}^{2\tau} e^{-pt} f_1(t)dt + ... + \int_{n\tau}^{(n+1)\tau} e^{-pt} f_1(t)dt + ...$$
$$= \int_0^{\tau} e^{-pt} f_1(t)dt + e^{-p\tau}\int_0^{\tau} e^{-pt} f_1(t)dt + ... + e^{-pn\tau}\int_0^{\tau} e^{-pt} f_1(t)dt + ...$$
$$= \int_0^{\tau} e^{-pt} f_1(t)dt \sum_{n=0}^{\infty}(e^{-p\tau})^n = \frac{\int_0^{\tau} e^{-pt} f_1(t)dt}{1-e^{-p\tau}}.$$

A similar formula can be obtained for the image of the function $f_2(t) = -f_2(t+\tau)$,

$$F_2(p) = \int_0^{\tau} e^{-pt} f_2(t)dt \sum_{n=0}^{\infty} (-1)^n e^{-pn\tau} = \frac{\int_0^{\tau} e^{-pt} f_2(t)dt}{1 + e^{-p\tau}}.$$

Next for a particular case of $f_1(t) = |f_2(t)|$ and $|f_2(t)| = f_2(t)$ at $t \leq \tau$, from the above equalities follows:

$$F_1(p) = \frac{1 + e^{-p\tau}}{1 - e^{-p\tau}} F_2(p).$$

Let now $f_2(t) = \sin t$ and $f_1(t) = |\sin t|$, so that $\tau = \pi$ and $F_2(p) = 1/(p^2 + 1)$. Therefore

$$F_1(p) = \frac{1 + e^{-p\pi}}{1 - e^{-p\pi}} \frac{1}{p^2 + 1}$$

and finally

$$|\sin t| \quad \leftrightarrow \quad \frac{1}{p^2 + 1} \coth\left(\frac{p\pi}{2}\right).$$

Problem 1.9
The original function is given by the Mellin formula

$$f(t) = \frac{1}{2\pi i} \int_{a-i\infty}^{a+i\infty} e^{pt} F(p)\, dp, \; a > 0.$$

In both cases $F(p)$ is generally the multivaled function. Analytical continuation of $F(p)$ from the region Re$p > 0$ to the region Re$p < 0$ has the branch points

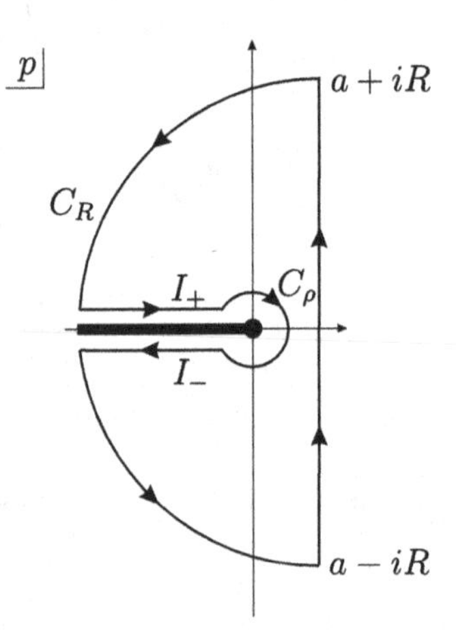

Fig. 5.6 Integration contour used in Problem 1.9

$p = 0$ and $p = \infty$. A suitable contour is shown in Fig. 5.6. It consists of a segment $[a - iR, a + iR]$ parallel to the imaginary axis, the upper and lower edges of the branch cut I_+, I_- and the closing arcs C_ρ, $|p| = \rho$, and C_R, $|p - x| = R$, in the limit $\rho \to 0$ and $R \to \infty$. By Jordan lemma the C_R–integrals tend to zero in the limit $R \to \infty$. Also on the upper edge of the cut $\arg p = \pi$, which gives $p = \xi e^{i\pi}$, and on the lower edge of the cut $\arg p = -\pi$, which gives $p = \xi e^{-i\pi}$. As a result the Cauchy theorem leads to

$$f(t) = \frac{1}{2\pi i}\left\{\lim_{\rho\to 0}\int_{C_\rho} e^{pt}F(p)\,dp + \int_0^\infty e^{-t\xi}\left[F\left(\xi e^{-i\pi}\right) - F\left(\xi e^{i\pi}\right)\right]d\xi\right\}.$$

(a): By setting $p = \rho e^{i\varphi}$, the C_ρ–integral tends to zero in the limit $\rho \to 0$:

$$\left|\frac{1}{2\pi i}\int_{C_\rho} e^{pt}\frac{dp}{p^{\alpha+1}}\right| \le \frac{1}{2\pi\rho^\alpha}\int_{-\pi}^{\pi} e^{t\rho\cos\varphi}d\varphi \to 0 \quad \text{as } \rho \to 0,\ \alpha < 0.$$

Therefore we obtain

$$f(t) = \frac{1}{2\pi i}\int_0^\infty e^{-t\xi}\xi^{-\alpha-1}\left(e^{i\pi(\alpha+1)} - e^{-i\pi(\alpha+1)}\right)d\xi = t^\alpha\,\frac{\sin(-\pi\alpha)}{\pi}\int_0^\infty e^{-s}\,s^{-\alpha-1}ds,$$

where the integral in the rhs determines the Gamma function $\Gamma(-\alpha)$, which satisfies the complement formula $\Gamma(-\alpha)\Gamma(1+\alpha) = \pi/\sin(-\pi\alpha)$. Hence the quoted answer.
(b): For the C_ρ–integral in the limit $\rho \to 0$ we have

$$\frac{1}{2\pi i}\lim_{\rho\to 0}\int_{C_\rho} \ldots dp = \frac{1}{2\pi}\lim_{\rho\to 0}\int_{-\pi}^{\pi} e^{\rho t e^{i\varphi}}e^{-\alpha\sqrt{\rho}e^{i\varphi/2}}\,d\varphi = 1,$$

whence

$$f(t) = 1 - \frac{1}{\pi}\int_0^\infty e^{-t\xi}\,\frac{\sin(\alpha\sqrt{\xi})}{\xi}\,d\xi = 1 - \frac{1}{\pi}\int_0^\alpha d\beta\int_{-\infty}^\infty e^{-ts^2}\,\cos(\beta s)ds.$$

The double integral in the rhs is obtained in the following way: We substitute $\xi = s^2$, use the identity $\sin(\alpha s)/s = \int_0^\alpha \cos(\beta s)\,d\beta$ and change the order of integrations. Next in the inner integral we rewrite $\cos(\beta s) = (e^{i\beta s} + e^{-i\beta s})/2$, substitute $s = \eta - i\beta/2t$ in the first exponent, $s = \eta + i\beta/2t$ in the second exponent and obtain

$$\int_{-\infty}^{\infty} ...ds = e^{-\frac{\beta^2}{4t}} \int_{-\infty}^{\infty} e^{-t\eta^2}\, d\eta = \sqrt{\frac{\pi}{t}}\, e^{-\frac{\beta^2}{4t}},$$

where we have used $\int_{-\infty}^{\infty} e^{-t\eta^2}\, d\eta = \sqrt{\pi/t}$, see Problem 1.4 (b). Finally, the answer is given by

$$f(t) = 1 - \frac{1}{\sqrt{t\pi}} \int_0^{\alpha} e^{-\frac{\beta^2}{4t}}\, d\beta = 1 - \frac{2}{\sqrt{\pi}} \int_0^{\alpha/(2\sqrt{t})} e^{-\gamma^2}\, d\gamma = 1 - \operatorname{erf}\left(\frac{\alpha}{2\sqrt{t}}\right),$$

where $\operatorname{erf}(x)$ is the error function defined in (2.62).

5.2 Chapter 2

Problem 2.1
From (2.2) we have $\Gamma(1-z) = -z\Gamma(-z)$. Then from (2.6) follows that $\Gamma(z)\Gamma(-z) = -\pi/(z \sin \pi z)$. Substituting $z = iy$ results in

$$|\Gamma(iy)|^2 = \frac{\pi}{y \sinh \pi y}.$$

Therefore for $y \to +\infty$ the Gamma function actually decays according to

$$|\Gamma(iy)| = \sqrt{\frac{2\pi}{y}}\, e^{-\pi y/2}\, [1 + o(y)]$$

and similarly for $y \to -\infty$. On the other hand, substituting $z = iy$ into the Stirling's formula $\Gamma(iy) = \sqrt{2\pi} e^{-\pi y/2 + iy \ln y - iy}[1 + o(y)]$, where $\ln iy = \ln y + i\pi/2$ was used, yields the same result for the modulus. Finally, the $y \to 0+$ limiting form is $|\Gamma(iy)| = 1/y + O(y^0)$, which is consistent with $\Gamma(z)$ having a simple pole at $z = 0$.

Problem 2.2
We are interested in the value of the product

$$P = \Gamma\left(\frac{1}{n}\right)\Gamma\left(\frac{2}{n}\right)...\Gamma\left(\frac{n-1}{n}\right).$$

We write this out in the opposite order

$$P = \Gamma\left(\frac{n-1}{n}\right)\Gamma\left(\frac{n-2}{n}\right)...\Gamma\left(\frac{1}{n}\right)$$

and multiply the above expressions combining every pair of factors with the help of identity (2.6):

$$P^2 = \frac{\pi}{\sin(\pi/n)} \frac{\pi}{\sin(2\pi/n)} \cdots \frac{\pi}{\sin((n-1)\pi/n)} = \frac{\pi^{n-1}}{\prod_{k=1}^{n-1} \sin(\pi k/n)}.$$

To simplify the product of sine functions that has emerged here, we recall the nth order roots of unity and use them in the following identity:

$$z^n - 1 = (z-1) \prod_{k=1}^{n-1} (z - e^{2\pi i k/n}),$$

which leads to

$$\lim_{z \to 1} \frac{z^n - 1}{z - 1} = \frac{dz^n}{dz}|_{z=1} = n = \prod_{k=1}^{n-1} (1 - e^{2\pi i k/n}).$$

The rhs contains pairs of complex conjugate roots (plus possibly the real root $z = -1$ depending on the parity of n), so the modulus can be taken which is:

$$n = 2^{n-1} \prod_{k=1}^{n-1} \sin(\pi k/n).$$

The Euler formula

$$\prod_{k=1}^{n-1} \Gamma\left(\frac{k}{n}\right) = \frac{1}{\sqrt{n}} (2\pi)^{(n-1)/2}$$

then follows immediately. Note that for $n = 2$ this formula is consistent with the result $\Gamma(1/2) = \sqrt{\pi}$ obtained previously.

Problem 2.3

We can use the following identity:

$$I = \int_0^1 \ln \Gamma(1-x) dx \,,$$

which is easily proven by the substitution $x \to 1 - x$. Then

$$2I = \int_0^1 \ln \Gamma(x)dx + \int_0^1 \ln \Gamma(1-x)dx = \int_0^1 \ln[\Gamma(x)\Gamma(1-x)]dx$$

$$= \int_0^1 \ln \frac{\pi}{\sin(\pi x)} dx = \ln \pi - \frac{1}{\pi} I',$$

where

$$I' = \int_0^\pi \ln \sin x dx = 2 \int_0^{\pi/2} \ln \sin(2x) dx$$

$$= \pi \ln 2 + 2 \int_0^{\pi/2} \ln \sin x dx + 2 \int_0^{\pi/2} \ln \sin x dx = \pi \ln 2 + 2I'.$$

So $I' = -\pi \ln 2$ and therefore $I = \ln \sqrt{2\pi}$.

Problem 2.4

First we observe that $I_{1,3}(x, \alpha) = -I_{1,3}(x, -\alpha)$ and $I_{2,4}(x, \alpha) = I_{2,4}(x, -\alpha)$. Furthermore, all integrals in question are expressed in terms of the following one:

$$I(x, \alpha) = \int_0^\infty t^{x-1} e^{-te^{-i\alpha}} dt = e^{i\alpha x} \, \Gamma(x).$$

Namely,

$$\begin{aligned}
(\mathbf{a}): \quad & I_1 = \lambda^{-x} \, \mathrm{Im} I(x, \alpha) = \lambda^{-x} \sin(\alpha x) \, \Gamma(x), \\
(\mathbf{b}): \quad & I_2 = \lambda^{-x} \, \mathrm{Re} I(x, \alpha) = \lambda^{-x} \cos(\alpha x) \, \Gamma(x), \\
(\mathbf{c}): \quad & I_3 = \mathrm{sgn}(\alpha)|\alpha|^{x-1} \, \mathrm{Im} I(1-x, \frac{\pi}{2}) \\
& \quad = \mathrm{sgn}(\alpha)|\alpha|^{x-1} \Gamma(1-x) \overbrace{\sin\left(\frac{\pi}{2}(1-x)\right)}^{\cos(\pi x/2)}, \\
(\mathbf{d}): \quad & I_4 = |\alpha|^{x-1} \, \mathrm{Re} I(1-x, \frac{\pi}{2}) = |\alpha|^{x-1} \Gamma(1-x) \sin\left(\frac{\pi}{2} x\right).
\end{aligned}$$

We would like to point out that for **(c)** and **(d)** the results can be rewritten with a help of the complement formula $\Gamma(x)\Gamma(1-x) = \pi / \sin(\pi x)$ and the duplication formula $\sin(2x) = 2 \sin x \cos x$ in the following way:

$$(\mathbf{c}):\quad I_3 = \operatorname{sgn}(\alpha)|\alpha|^{x-1}\cos\Big(\frac{\pi}{2}x\Big)\frac{\pi}{\Gamma(x)\sin(\pi x)} = \operatorname{sgn}(\alpha)\frac{|\alpha|^{x-1}\,\pi}{2\Gamma(x)\sin(\frac{\pi}{2}x)},$$

$$(\mathbf{d}):\quad I_4 = \frac{|\alpha|^{x-1}\,\pi}{2\Gamma(x)\cos(\frac{\pi}{2}x)}.$$

Problem 2.5

We substitute $t = 1/(\cosh x)^2$ in the integrand of $I_{\mu\nu}$ first. Then $x = 0 \to t = 1$, $x = \infty \to t = 0$. Moreover,

$$dx = -\frac{1}{2t}\frac{1}{\sqrt{1-t}}dt,\quad (\sinh x)^{\mu} = \frac{(1-t)^{\mu/2}}{t^{\mu/2}},\quad (\cosh x)^{\nu} = \frac{1}{t^{\nu/2}},$$

and, according to the definition (2.10) of the Beta function and its symmetry property, we obtain

$$I_{\mu\nu} = \frac{1}{2}\int_0^1 t^{-\frac{\mu}{2}-\frac{\nu}{2}-1}\underbrace{(1-t)^{\frac{\mu}{2}-\frac{1}{2}}}_{(1-t)^{\frac{\mu}{2}+\frac{1}{2}-1}}\,dt = \frac{1}{2}B\Big(\frac{\mu+1}{2},\,\frac{-\nu-\mu}{2}\Big).$$

Problem 2.6

According to the definition (2.34) of the hypergeometric function, the numerator is given by the series

$$\begin{aligned}F(a+1,b;c;z) - F(a,b;c;z) &= \sum_{n=0}^{\infty}\frac{z^n}{n!}\frac{(b)_n}{(c)_n}\Big[(a+1)_n - (a)_n\Big]\\ &= \sum_{n=1}^{\infty}\frac{z^n}{(n-1)!}\frac{(b)_n}{(c)_n}\underbrace{\left[\frac{(a+1)_n}{a+n}\right]}_{\Gamma(a+n)/\Gamma(a+1)}.\end{aligned}$$

Setting now $n-1 \to n$, we obtain the result:

$$\begin{aligned}F(a+1,b;c;z) - F(a,b;c;z) &= z\sum_{n=0}^{\infty}\frac{z^n}{n!}\frac{\overbrace{(b)_{n+1}}^{b(b+1)_n}}{\underbrace{(c)_{n+1}}_{c(c+1)_n}}\underbrace{\left[\frac{\Gamma(a+1+n)}{\Gamma(a+1)}\right]}_{(a+1)_n}\\ &= z\,\frac{b}{c}\,F(a+1,b+1;c+1;z).\end{aligned}$$

Problem 2.7

(a): By substitution $t = 1-x$ in the Pochhammer integral Eq. (2.49) we immediately obtain the first identity:

$$F(a, b; c; z) = (1-z)^{-a} \underbrace{\frac{\Gamma(c)}{\Gamma(b)\Gamma(c-b)} \int_0^1 x^{c-b-1}(1-x)^{b-1}\left(1 - x\frac{z}{z-1}\right)^{-a} dx}_{F\left(a,\ c-b;\ c;\ \frac{z}{z-1}\right)} .$$

From this formula the second identity follows by virtue of the symmetry property $a \leftrightarrow b$,

$$F(a, b; c; z) = F(b, a; c; z) = (1-z)^{-b} F\left(b,\ c-a;\ c;\ \frac{z}{z-1}\right).$$

(b): This relation is obtained by combining the upper two formulae:

$$F(a, b; c; z) = (1-z)^{-a} F\left(c-b, a; c; \frac{z}{z-1}\right) = (1-z)^{c-b-a} F(c-b, c-a; c; z).$$

(c): Setting $2z \to (1-z)$ in the hypergeometric Eq. (2.51), the function $F(-\nu, \nu+1; 1; , \frac{1-z}{2}) \equiv F$ in the lhs obeys the equation

$$(1-z^2)F'' - 2zF' + \nu(\nu+1)F = 0.$$

Let us now show that the functions $F(-\frac{\nu}{2}, \frac{\nu+1}{2}; \frac{1}{2}; z^2)$ and $zF(-\frac{\nu-1}{2}, \frac{\nu+2}{2}; \frac{3}{2}; z^2)$ on the rhs also satisfy this equation. Consider the series $F(a, b; c; z) = \sum_{n=0}^{\infty} A_n z^n$ and remember that the coefficients A_n obey the recurrence relation

$$\frac{A_{n+1}}{A_n} = \frac{(a+n)(b+n)}{(c+n)(n+1)}.$$

Our goal is to find the appropriate set of parameters a, b, and c. To that end, we substitute F, $F' = \sum_{n=0}^{\infty} nA_n z^{n-1}$, and $F'' = \sum_{n=0}^{\infty} n(n-1)A_n z^{n-2} = \sum_{n=0}^{\infty}(n+1)(n+2)A_{n+2}z^n$ into the above equation. We then obtain

$$\sum_{n=0}^{\infty} z^n \left\{(n+1)(n+2)A_{n+2} - n(n-1)A_n - 2nA_n + \nu(\nu+1)A_n\right\} = 0,$$

whence

$$\frac{A_{n+2}}{A_n} = \frac{(n-\nu)(n+1+\nu)}{(n+1)(n+2)}.$$

Consider the case of even and odd n separately.

- Even $n = 2k$. Take $A_{2k} = \overline{A}_k$, $A_{2k+2} = \overline{A}_{k+1}, \ldots,$ $k = 0, 1, 2, \ldots$. Then $F(a, b; c; z) = \sum_{k=0}^{\infty} \overline{A}_k z^{2k}$ and $\overline{A}_k$ satisfies the relation

$$\frac{\overline{A}_{k+1}}{\overline{A}_k} = \frac{(2k-\nu)(2k+1+\nu)}{(2k+1)(2k+2)} = \frac{(k-\frac{\nu}{2})(k+\frac{\nu+1}{2})}{(k+\frac{1}{2})(k+1)},$$

whence the appropriate parameters are: $a = -\nu/2,\ b = (\nu+1)/2,\ c = 1/2$. Therefore the required solution is the function $F_{\text{even}} = F(-\frac{\nu}{2}, \frac{\nu+1}{2}; \frac{1}{2}; z^2)$.

• Odd $n = 2k+1$. Similarly, taking $A_{2k+1} = \overline{A}_k,\ A_{2k+2} = \overline{A}_{k+1}, \dots\ (k = 0, 1, 2, \dots.)$, we obtain

$$\frac{\overline{A}_{k+1}}{\overline{A}_k} = \frac{(2k+1-\nu)(2k+2+\nu)}{(2k+2)(2k+3)} = \frac{(k+\frac{1-\nu}{2})(k+\frac{\nu+2}{2})}{(k+\frac{3}{2})(k+1)}.$$

Hence now $a = -(\nu-1)/2$, $b = (\nu+2)/2$, $c = 3/2$, and the required solution is $F_{\text{odd}} = zF(-\frac{\nu-1}{2}, \frac{\nu+2}{2}; \frac{3}{2}; z^2)$. Consequently, general solution has the form

$$F\Big(-\nu, \nu+1; 1; \frac{1-z}{2}\Big) = AF\Big(-\frac{\nu}{2}, \frac{\nu+1}{2}; \frac{1}{2}; z^2\Big) + BzF\Big(-\frac{\nu-1}{2}, \frac{\nu+2}{2}; \frac{3}{2}; z^2\Big),$$

where the constant coefficients A and B are still to be determined. Clearly, the constant A is equal to

$$A = F\Big(-\nu, \nu+1; 1; \frac{1-z}{2}\Big)\Big|_{z=0} = F\Big(-\nu, \nu+1; 1; \frac{1}{2}\Big),$$

since $F(a, b; c; z)|_{z=0} = 1$. To determine B let us differentiate both sides of the obtained identity and take into account that $F'(a, b; c; z) = (ab/c)\ F(a+1, b+1; c+1; z)$. One obtains then that the constant B is given by

$$B = -\frac{1}{2}\ F'\Big(-\nu, \nu+1; 1; \frac{1-z}{2}\Big)\Big|_{z=0} = \frac{\nu(\nu+1)}{2}\ F\Big(-\nu+1, \nu+2; 2; \frac{1}{2}\Big).$$

Finally, note that coefficients A and B can be expressed in terms of the Gamma functions,

$$A = \frac{\sqrt{\pi}}{\Gamma(\frac{1-\nu}{2})\Gamma(\frac{2+\nu}{2})}, \qquad B = \frac{\sqrt{\pi}\nu}{\Gamma(\frac{2-\nu}{2})\Gamma(\frac{\nu+1}{2})},$$

see Problem 2.8 (a) below.

Problem 2.8

(a): After setting $t \to (1-e^{-2x})$ in the integral representation (2.49) for $F(a, b; c; z)$ at $z = 1/2$, we find

$$F\Big(a, b; c; \frac{1}{2}\Big) = \frac{\Gamma(c)\ 2^b}{\Gamma(b)\Gamma(c-b)} \int_0^\infty e^{-(2c-b-a-1)x}\ (\sinh x)^{b-1}(\cosh x)^{-a} dx\ .$$

At $2c = a + b + 1$, the above integral represents the Beta function, see Problem 2.5, therefore

$$F\Big(a, b; \frac{a+b+1}{2}; \frac{1}{2}\Big) = \frac{\Gamma(\frac{a+b+1}{2})}{\Gamma(b)\Gamma(\frac{a-b+1}{2})}\, 2^{b-1}\, B\Big(\frac{b}{2}, \frac{a-b+1}{2}\Big).$$

Finally, using the relation (2.11) of the Beta function to Gamma functions, the result is

$$F\Big(a, b; \frac{a+b+1}{2}; \frac{1}{2}\Big) = \frac{\sqrt{\pi}\,\Gamma(\frac{a+b+1}{2})}{\Gamma(\frac{a+1}{2})\,\Gamma(\frac{b+1}{2})}.$$

(b): Using the transformation formula of Problem 2.7 (a),

$$F(a, b; a-b+1; -1) = 2^{-a} F\Big(a, \overbrace{a+1-2b}^{\overline{b}};\ \underbrace{a+1-b}_{(a+\overline{b}+1)/2}; \frac{1}{2}\Big),$$

we return to the above case **(a)** and the result can easily be read off.

Problem 2.9

The problem can be easily solved by means of the representation of the F–functions as the series in the variable z by (2.34), see Problem 2.6 where a similar example is considered. The involved Pochhammer symbols satisfy the recursion relation of the type $(a+n)(a)_n = a(a+1)_n$ which is related to the property $\Gamma(z+1) = z\Gamma(z)$.

(a): Let us express the first F–function in the lhs of the equation in terms of the second one. We write out

$$F(a, b; c; z) = \sum_{n=0}^{\infty} \frac{(a)_n (b)_n}{(c)_n n!}\, z^n = \sum_{n=0}^{\infty} \frac{c+n}{c} \frac{(a)_n (b)_n}{(c+1)_n n!}\, z^n$$

and then obtain

$$F(a, b; c; z) = F(a, b; c+1; z) + \frac{1}{c} A(z),$$

where

$$A(z) = \sum_{n=1}^{\infty} \frac{(a)_n (b)_n}{(c+1)_n (n-1)!}\, z^n.$$

Similarly

$$F(a+1, b; c+1; z) = F(a, b; c+1; z) + \frac{1}{a} A(z).$$

Combining these two results we obtain the desired relation.

(b): Here it is convenient to consider the difference

$$
\begin{aligned}
F(a, b-1; c; z) &- F(a-1, b; c; z) \\
&= \sum_{n=0}^{\infty} \frac{1}{(c)_n n!} z^n \Big[(a)_n (b-1)_n - (a-1)_n (b)_n\Big] \\
&= \sum_{n=0}^{\infty} \frac{1}{(c)_n n!} z^n \frac{\Gamma(a-1+n)\Gamma(b-1+n)}{\Gamma(a)\Gamma(b)} \\
&\quad \times \overbrace{\Big[(b-1)(a+n-1) - (a-1)(b+n-1)\Big]}^{(b-a)\,n} \\
&= (b-a) \sum_{n=1}^{\infty} \frac{1}{(c)_n (n-1)!} \frac{\Gamma(a-1+n)\Gamma(b-1+n)}{\Gamma(a)\Gamma(b)} z^n .
\end{aligned}
$$

Substituting $n-1 \to n$ in the last line yields

$$
F(a, b-1; c; z) - F(a-1, b; c; z) = \frac{b-a}{c} z\, F(a, b; c+1; z).
$$

Problem 2.10
Let us start with the hypergeometric equation

$$
z(1-z) f'' + [c - (a+b+1)z] f' - abf = 0.
$$

We make the following variable substitution:

$$
z = \frac{4\xi}{(1+\xi)^2} \quad \to \quad \frac{d}{dz} = \frac{(1+\xi)^3}{4(1-\xi)} \frac{d}{d\xi}
$$

and introduce a new function $g = (1+\xi)^{2a} f$. After some algebra we find that $g(\xi)$ satisfies the differential equation of the form

$$
\xi(1-\xi^2)\frac{d^2 g}{d\xi^2} + [c - (4b-2c)\xi + (c-4a-2)\xi^2]\frac{dg}{d\xi} - 2a[2b - c + (2a-c+1)\xi]g = 0.
$$

(a): Let us assume that $b = a + \frac{1}{2}$. The above equation then rewrites as

$$
\xi(1-\xi)g'' + [c - (4a-c+2)\xi]g' - 2a(2a+1-c)g = 0,
$$

which is nothing but a hypergeometric equation with new parameters $\overline{a} = 2a, \overline{b} = 2a + 1 - c, \overline{c} = c$ (note that $\overline{a} + \overline{b} + 1 = 4a + 2 - c$). Therefore we have:

$$F(2a, 2a+1-c; c; z) = (1+z)^{-2a} F\left(a, a+\frac{1}{2}; c; \frac{4z}{(1+z)^2}\right).$$

(b): Let us now set $c = 2b$. Then the original equation for g acquires the form

$$\xi(1-\xi^2)g'' + [2b + (2b - 4a - 2)\xi^2]g' - 2a(2a - c + 1)\xi g = 0,$$

the solution of which is $F(a, a-b+\frac{1}{2}; b+\frac{1}{2}; z^2)$. Indeed, substituting $z = u^2$ yields the following equation for $f = F(a, b; c; u^2)$:

$$u(1-u^2)f'' + [2c - 1 - (2a + 2b + 1)u^2]f' - 4abf = 0.$$

It has the same form as the equation for g but with the new parameters $\overline{c} = b+\frac{1}{2}$, $\overline{b} = a - b + \frac{1}{2}$, $\overline{a} = a$. So we conclude that

$$F\left(a, a-b+\frac{1}{2}; b+\frac{1}{2}; z^2\right) = (1+z)^{-2a} F\left(a, b; 2b; \frac{4z}{(1+z)^2}\right).$$

Problem 2.11
(a): Mellin transform is given by

$$F_M(s) = \int_0^\infty \frac{t^{s-1}}{(1+t)^a}\, dt.$$

Changing the variable $t = x(1-x)^{-1}$ and thus $(1+t)^{-1} = 1-x$, $dt = (1-x)^{-2}dx$, the transform $F_M(s)$ is given by the Euler integral (2.10) with $z = s$ and $w = a - s$:

$$F_M(s) = \int_0^1 x^{s-1}(1-x)^{a-s-1}dx = B(s, a-s) = \frac{\Gamma(s)\Gamma(a-s)}{\Gamma(a)}, \quad 0 < \mathrm{Re}(s) < a.$$

(b): In this case we perform the partial integration first,

$$F_M(s) = \int_0^\infty dt\, t^{s-1} \ln(1+t) = \frac{t^s}{s} \ln(1+t)\Big|_0^\infty - \frac{1}{s}\int_0^\infty dt\, \frac{t^s}{1+t}$$
$$= -\frac{1}{s}\Gamma(1+s)\,\Gamma(-s) = \frac{\pi}{s\, \sin(\pi s)},$$

where we have used the result from **(a)** and the identity (2.6). The above integrals are, of course, only convergent for $-1 < \mathrm{Re} s < 0$.

(c): First we prove an interesting identity

$$I = \int_0^\infty \frac{t^{s-1}}{1-t}\, dt = \cos(\pi s) \int_0^\infty \frac{t^{s-1}}{1+t}\, dt\,, \tag{5.2}$$

from which the result follows immediately. By making the substitution $t = x^2$ we can transform I to an integral over the whole real axis,

$$I = \int_{-\infty}^\infty dx\, \frac{x^{2s-1}}{1-x^2}\,.$$

In the next step we go into the whole complex plane and deform the integration contour into $C_- \cup C_\rho \cup C_+$, as shown in Fig. 5.7 An integral along C_ρ can be shown to vanish, while the integrals along $C_\pm$ can be reparametrised as $x = i\xi$. Since the function x^{2s-1} has a cut along the positive imaginary semi-axis, it has different values on the edges of the cut, leaving us with

$$\begin{aligned} I &= i \int_0^\infty d\xi\, \frac{\xi^{2s-1}}{1+\xi^2} \left(e^{-i(\pi/2)(2s-1)} - e^{i(\pi/2)(2s-1)} \right) \\ &= 2\cos(\pi s) \int_0^\infty d\xi\, \frac{\xi^{2s-1}}{1+\xi^2}\,. \end{aligned}$$

By a substitution $t = \xi^2$ we recover the identity (5.2). Using this identity, the result from **(a)**, and the complement formula for Γ we obtain the answer.

Problem 2.12

We rewrite the given function as a double series

$$f(x) = \sum_{m=0}^\infty (-1)^m \sum_{n=1}^\infty e^{-n(2m+1)x}\,.$$

Then the Mellin transform can be performed in the way very similar to that of the example in Sect. 2.2,

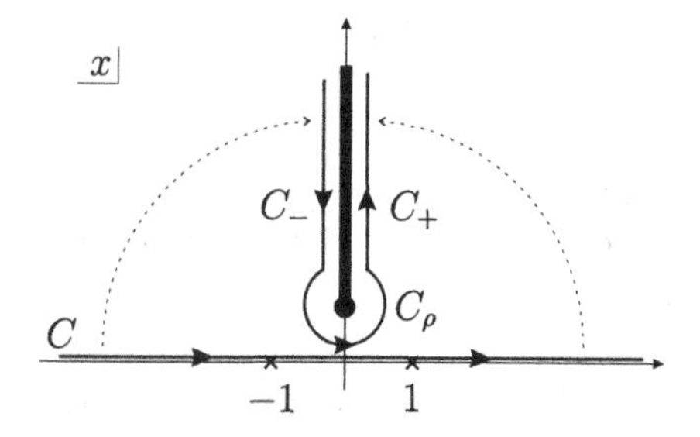

Fig. 5.7 Integration contour used in Problem 2.11(c)

$$F_M(s) = \Gamma(s)\,\zeta(s)\,g(s)\,,$$

where $g(s) = \sum_{m=0}^{\infty}(-1)^m/(2m+1)^s$ and the fundamental strip is Re $s > 0$. According to the inversion formula we then obtain

$$f(x) = \frac{1}{2\pi i}\int_{3/2-i\infty}^{3/2+i\infty} ds\, x^{-s}\,\Gamma(s)\,\zeta(s)\,g(s)\,.$$

We shift the integration line into the 'forbidden' region sweeping over the two poles $s = 0, 1$ of the integrand. Taking into account that $g(0) = 1/2$ and $g(1) = \pi/4$ we can calculate the residua and obtain

$$f(x) = \frac{\pi}{4x} - \frac{1}{4} + \frac{1}{2\pi i}\int_{-1/2-i\infty}^{-1/2+i\infty} ds\, x^{-s}\,\Gamma(s)\,\zeta(s)\,g(s)\,.$$

Using the duplication formula (2.12) it can be transformed to

$$f(x) = \frac{\pi}{4x} - \frac{1}{4} + \frac{2^{s-1}}{2\pi^{3/2} i}\int_{-1/2-i\infty}^{-1/2+i\infty} ds\, x^{-s}\,\Gamma(s/2)\,\Gamma(s/2+1/2)\,\zeta(s)\,g(s)\,.$$

In the next step we use the functional relation (2.25) as well as the identity

$$\Gamma(1-s/2)\,g(1-s) = 2^{2s-1}\,\pi^{1/2-s}\Gamma(s/2+1/2)\,g(s)\,.$$

Performing then the substitution $s \to 1-s$ leads to

$$f(x) = \frac{\pi}{4x} - \frac{1}{4} + \frac{1}{2\pi i}\int_{3/2-i\infty}^{3/2+i\infty} ds\, \pi^{1-2s}\, x^{s-1}\,\Gamma(s)\,\zeta(s)\,g(s)\,,$$

thus proving the identity.

Problem 2.13
(a): We first observe that the integral in (2.82) is an even function of z and can be rewritten as

$$\int_0^{\pi} \cosh(z\cos\varphi)\sin^{2\nu}\varphi d\varphi \equiv \int_{-1}^{1} e^{\pm zt}(1-t^2)^{\nu+\frac{1}{2}}dt$$

after the substitution $t = \cos\varphi$. Expanding the integrand in powers of z and using the definition (2.10) of the Beta function, we find:

$$I_\nu(z) = \frac{z^\nu}{2^\nu\Gamma(\nu+\frac{1}{2})\Gamma(\frac{1}{2})}\sum_{k=0}^{\infty}\frac{z^{2k}}{(2k)!}B\Big(k+\frac{1}{2}, \nu+\frac{1}{2}\Big).$$

Further, we express the Beta function in terms of Gamma functions, formula (2.11), and note that $\Gamma(k+1/2) = \sqrt{\pi}(2k)!/2^{2k}k!$ [consult the duplication formula (2.12)]. We obtain then the series (2.66) for the modified Bessel function.
(b): One possible way here is to use the series representation (2.66). For $\nu = 1/2$ we then obtain [we need again the duplication formula (2.12)]

$$I_{\frac{1}{2}}(z) = \sqrt{\frac{z}{2}} \sum_{k=0}^{\infty} \frac{z^{2k}}{2^{2k}\Gamma(k+1)\Gamma(k+\frac{3}{2})} = \sqrt{\frac{2}{\pi z}} \sum_{k=0}^{\infty} \frac{z^{2k+1}}{(2k+1)!} = \sqrt{\frac{2}{\pi z}} \sinh z,$$

see also (2.70) in Sect. 2.6.3. Similarly,

$$I_{-\frac{1}{2}}(z) = \sqrt{\frac{2}{z}} \sum_{k=0}^{\infty} \frac{z^{2k}}{2^{2k}k!\Gamma(k+\frac{1}{2})} = \sqrt{\frac{2}{\pi z}} \sum_{k=0}^{\infty} \frac{z^{2k}}{(2k)!} = \sqrt{\frac{2}{\pi z}} \cosh z.$$

Consequently, since

$$K_\nu(z) = \frac{\pi}{2\sin(\nu\pi)} \Big[I_{-\nu}(z) - I_\nu(z) \Big]$$

we immediately obtain

$$K_{\frac{1}{2}}(z) = K_{-\frac{1}{2}}(z) = \sqrt{\frac{\pi}{2z}} e^{-z}.$$

Problem 2.14
Let us first find the differential equation satisfied by the integral $\mathcal{I}_{i\alpha}(\xi)$. The equation for the function $F(\nu+1, -\nu; 1; \frac{1-t}{2}) \equiv P_\nu(t)$ can be derived using (2.51), one finds[1]

$$\frac{d}{dt}\left[(1-t^2)\frac{dP_\nu(t)}{dt} \right] + \nu(\nu+1)P_\nu(t) = 0.$$

Consequently, the original integral rewrites as

$$\mathcal{I}_{i\alpha}(\xi) = \frac{4}{1+4\alpha^2} \int_1^\infty \frac{d}{dt}\left[(1-t^2)\frac{dP_{-\frac{1}{2}+i\alpha}(t)}{dt} \right] e^{-\xi t} dt.$$

Integrating the rhs by parts, the required equation has the form

$$\xi^2 \mathcal{I}''_{i\alpha}(\xi) + 2\xi \mathcal{I}'_{i\alpha}(\xi) + \Big(\frac{1}{4} + \alpha^2 - \xi^2\Big)\mathcal{I}_{i\alpha}(\xi) = 0.$$

[1] In fact, this is the Legendre equation, see (4.6) on page 189.

In the next step we set $\mathcal{I}_{i\alpha}(\xi) = \xi^{-1/2} F_{i\alpha}(\xi)$. Then the corresponding derivatives are:

$$\mathcal{I}'_{i\alpha}(\xi) = \xi^{-1/2}\Big(-\frac{1}{2\xi} F_{i\alpha}(\xi) + F'_{i\alpha}(\xi)\Big),$$
$$\mathcal{I}''_{i\alpha}(\xi) = \xi^{-1/2}\Big(\frac{3}{4\xi^2} F_{i\alpha}(\xi) - \frac{1}{\xi} F'_{i\alpha}(\xi) + F''_{i\alpha}(\xi)\Big).$$

Substitution of these functions into the above equation for $I_{i\alpha}(\xi)$ results in

$$F''_{i\alpha}(\xi) + \frac{1}{\xi} F'_{i\alpha}(\xi) + \Big(\frac{\alpha^2}{\xi^2} - 1\Big) F_{i\alpha}(\xi) = 0,$$

so that the function $F_{i\alpha}(\xi)$ is a solution of the modified Bessel equation, see footnote on page 96. Since the integral $\mathcal{I}_{i\alpha}(\xi)$ is a decaying function of $\xi > 0$, among two linearly independent solutions we must choose the Macdonald function $K_{i\alpha}(\xi)$, that is $F_{i\alpha}(\xi) = \xi^{1/2}\mathcal{I}_{i\alpha}(\xi) = C K_{i\alpha}(\xi)$. To determine the constant C, we consider a special case $i\alpha = 1/2$. Then $\mathcal{I}_{\frac{1}{2}}(\xi) = e^{-\xi}/\xi$ (the integrand now reduces to the exponential function) and $K_{\frac{1}{2}}(\xi) = \sqrt{\frac{\pi}{2\xi}} e^{-\xi}$ [see Problem 2.13 (b)], so that $C = \sqrt{2/\pi}$. Hence, the result is:

$$\mathcal{I}_{i\alpha}(\xi) = \sqrt{\frac{2}{\pi\xi}} K_{i\alpha}(\xi).$$

Problem 2.15

Plugging the integral representation into the integral in question we obtain

$$g(\nu) = \frac{\sqrt{\pi}}{\Gamma(\nu + \frac{1}{2})} \int_0^\infty e^{-at} t^{\mu-1} \Big(\frac{\beta t}{2}\Big)^\nu \int_1^\infty e^{-\beta\xi t} (\xi^2 - 1)^{\nu-1/2}\, d\xi dt.$$

Now we change the order of integrations and observe that the t–integral reduces to the Gamma function:

$$g(\nu) = \frac{\sqrt{\pi}\beta^\nu}{2^\nu \Gamma(\nu + \frac{1}{2})} \Gamma(\nu + \mu) \int_1^\infty (a + \beta\xi)^{-\nu-\mu} (\xi^2 - 1)^{\nu-1/2}\, d\xi.$$

Next we substitute $\xi = (1 + z)/(1 - z)$ and obtain

$$g(\nu) = \frac{\sqrt{\pi}\beta^\nu \Gamma(\nu + \mu) 2^{2\nu}}{2^\nu \Gamma(\nu + \frac{1}{2})} (a + \beta)^{-\nu-\mu} \int_0^1 z^{\nu-\frac{1}{2}} (1 - z)^{\mu-\nu-1} \Big(1 - \frac{a - \beta}{a + \beta} z\Big)^{-\nu-\mu} dz,$$

where in the rhs we recognise the hypergeometric function [see formula (2.49)]

$$\int_0^1 \dots dz = \frac{\Gamma(\nu+\frac{1}{2})\Gamma(\mu-\nu)}{\Gamma(\mu+\frac{1}{2})} F\left(\nu+\mu, \nu+\frac{1}{2}; \mu+\frac{1}{2}; \frac{a-\beta}{a+\beta}\right).$$

5.3 Chapter 3

Problem 3.1
The integral equations in question may be solved by various approaches.
First method: We first observe that in both cases the kernel is factorised: $k(x,t) = k_1(x)\,k_2(t)$. Let us now consider a Volterra equation

$$g(x) = \lambda \int_a^x k(t)\varphi(t)dt,$$

where the kernel is independent of x. Then the solution

$$\varphi(x) = \frac{1}{\lambda k(x)} \frac{dg(x)}{dx}$$

follows immediately.
(a): Here we have:

$$g(x) = \frac{x^2-1}{x} \quad \rightarrow \quad g'(x) = \frac{x^2+1}{x^2}$$

and $k(x) = x$. Therefore the solution is

$$\varphi(x) = \frac{1}{\lambda x^3}(x^2+1).$$

(b): Here we have $\lambda = 1$,

$$g(x) = xe^{-x} \quad \rightarrow \quad g'(x) = e^{-x}(1-x),$$

and $k(x) = e^{-x}$, whence the solution is: $\varphi(x) = 1 - x$.
Second method: The considered equations can be presented also in the form

$$g(x) = \lambda k_1(x) \int_a^x k_2(t)\varphi(t)dt.$$

Let

$$F(x) = \int_a^x k_2(t)\varphi(t)dt \quad \rightarrow \quad F(x) = \frac{g(x)}{\lambda k_1(x)}.$$

Now take the derivative of both sides of the above equation:

$$g'(x) = \lambda k_1(x)k_2(x)\varphi(x) + \lambda k_1'(x)F(x).$$

Then the solution is

$$\varphi(x) = \left[g'(x) - g(x)\frac{k_1'(x)}{k_1(x)}\right]\frac{1}{\lambda k_1(x)k_2(x)}.$$

(a): We have $g(x) = x^2 - 1$ and $k_1(x) = k_2(x) = x$, whence the solution is

$$\varphi(x) = \left(2x - \frac{x^2-1}{x}\right)\frac{1}{\lambda x^2} = \frac{(x^2+1)}{\lambda x^3}.$$

(b): Here one has: $g(x) = x, k_1(x) = e^x, k_2(x) = e^{-x}$ and $\lambda = 1$. Hence $k_1(x)k_2(x) = 1$, $k_1'(x) = k_1(x)$ and the result $\varphi(x) = 1 - x$ immediately follows. We would like to remark that for the case **(b)**, where the kernel is an exponent, the Laplace transformation method is also applicable. Indeed, the Laplace transformation of both sides of this equation [the rhs is a convolution of e^t and $\varphi(t)$] yields

$$\frac{\varphi_L(s)}{s-1} = \frac{1}{s^2}.$$

According to the Mellin formula (1.51), the solution then is written as

$$\varphi(x) = \frac{1}{2\pi i}\int_{\sigma-i\infty}^{\sigma+i\infty} \frac{s-1}{s^2}\, e^{sx} ds.$$

For $x > 0$ take $\sigma > 0$ and close the contour in the left half–plane (Jordan lemma is then fulfilled). The integrand has a double pole at $s = 0$, the respective residue is $1 - x$ leading to the same answer as before.

Problem 3.2

(a): Here we use the resolvent method, where the solution is written in the form [see Eq. (3.2)]:

$$\varphi(x) = f(x) + \int_0^x R(x-t)f(t)dt.$$

Performing the Laplace transformation of the both sides of this equation, we have

$$\varphi_L(s) = f_L(s) + R_L(s) f_L(s).$$

On the other hand, from the original equation [in the rhs of which we recognise a convolution of t and $\varphi(t)$] follows

$$\varphi_L(s) = f_L(s) + \frac{1}{s^2} \varphi_L(s).$$

Combining these two results we obtain that $R_L(s) = 1/(s^2 - 1)$, so that

$$R(x) = \frac{1}{2\pi i} \int_{\sigma - i\infty}^{\sigma + i\infty} \frac{e^{sx}}{s^2 - 1} ds.$$

For $x > 0$ we take $\sigma > 1$ and close the contour in the left half–plane. The integrand has two simple poles at $s = +1, -1$, the residues at which are $e^x/2$ and $-e^{-x}/2$, respectively. Consequently, for the resolvent we obtain

$$R(x) = \frac{1}{2}(e^x - e^{-x}),$$

which immediately leads to the solution.
(b): Here we perform the two–side Laplace transformation, that is:

$$\Phi_L(s) = \int_{-\infty}^{\infty} e^{-sx} \varphi(x) dx.$$

For the functions in the rhs of the equation we have:

$$e^{-|x|} \rightarrow F_L(s) = \int_{-\infty}^{0} e^{(1-s)x} dx + \int_{0}^{\infty} e^{-(1+s)x} dx = \frac{2}{1 - s^2},$$

$$e^x \int_{x}^{\infty} e^{-t} \Phi(t) dt \rightarrow \int_{-\infty}^{0} e^{(1-s)y} \int_{-\infty}^{\infty} e^{-s(x-y)} \varphi(x - y) dx\, dy = \frac{\varphi_L(s)}{1 - s}.$$

So we obtain

$$\Phi_L(s) = F_L(s) + \frac{\lambda}{1 - s} \Phi_L(s),$$

whence

$$\varphi(x) = \frac{1}{\pi i} \int_{\sigma - i\infty}^{\sigma + i\infty} \frac{e^{sx}}{(1 + s)(1 - \lambda - s)} ds.$$

Now we take $\sigma = 0$ in order to obtain a bounded solution and assume that $\mathrm{Re}(1-\lambda) \neq 0$ to have no poles on the imaginary axis. Let first $\mathrm{Re}(1-\lambda) > 0$. Then for $x > 0$ we close the contour in the left half–plane, where the integrand has a pole at $s = -1$. For $x < 0$ we close the contour in the right half–plane, where the integrand has a pole at $s = 1 - \lambda$. Hence the answer is

$$\varphi(x) = \begin{cases} \frac{2}{2-\lambda}\, e^{-x}, & x > 0, \\ \frac{2}{2-\lambda}\, e^{(1-\lambda)x}, & x < 0. \end{cases}$$

Let now $\mathrm{Re}(1-\lambda) < 0$, but $\lambda \neq 2$. Then both poles at $s = -1, 1-\lambda$ are on the left side of the imaginary axis and, clearly, the answer is

$$\varphi(x) = \begin{cases} \frac{2}{2-\lambda}\, [e^{-x} - e^{(1-\lambda)x}], & x > 0, \\ 0, & x < 0. \end{cases}$$

Finally, at $\lambda = 2$ the integrand has the double pole at $s = -1$, so that

$$\varphi(x) = \begin{cases} -2xe^{-x}, & x > 0, \\ 0\,, & x < 0. \end{cases}$$

Problem 3.3
The function $R(s)$ has the simple pole at $s = k\cos\theta$ and the branch cut at $s = -k$ extending to $-k - i\infty$ in the lower half–plane. According to (3.31), the 'plus'–function can be found from

$$R_+(s) = \frac{1}{2\pi i} \int\limits_{i\alpha-\infty}^{i\alpha+\infty} \frac{d\zeta}{(\zeta - s)(\zeta - k\cos\theta)(\zeta + k)^{\frac{1}{2}}}.$$

We complete the integration path by a semicircle at infinity in the upper half–plane that encloses the poles at $\zeta = k\cos\theta$ and $\zeta = s$ but excludes the branch cut. Integration along the semicircle vanishes, so that the application of the Cauchy theorem yields

$$R_+(s) = \mathrm{Res}[...]\big|_{\zeta=s} + \mathrm{Res}[...]\big|_{\zeta=k\cos\theta}, \quad [...] = \frac{1}{(\zeta - s)(\zeta - k\cos\theta)(\zeta + k)^{\frac{1}{2}}},$$

where the residues are

$$\mathrm{Res}[...]\big|_{\zeta=s} = \frac{1}{(s - k\cos\theta)\sqrt{s+k}},$$

$$\mathrm{Res}[...]\big|_{\zeta=k\cos\theta} = -\frac{1}{(s - k\cos\theta)\sqrt{k + k\cos\theta}}.$$

Clearly the function $R_+(s)$ is analytic for $s > k\cos\theta$, that is in the upper half-plane. The 'minus'–function is $R_-(s) = R(s) - R_+(s)$ and is analytic in the lower half-plane. Note that the function $R(s)$ is almost a 'plus'—function but for the pole $s = k\cos\theta$, so that ideas from Example 3.14 still works. Subtracting the pole off immediately yields

$$R_+(s) = \frac{1}{(s-k\cos\theta)\sqrt{s+k}} - \frac{1}{(s-k\cos\theta)\sqrt{k+k\cos\theta}}.$$

Problem 3.4

According to the prescription from Sect. 3.3.2 the respective Wiener-Hopf equation is

$$[1-\lambda k(u)]\ F_+(u) + G_-(u) = 0,$$

where

$$k(u) = \int_{-\infty}^{\infty} \frac{e^{iuz}}{\cosh(z/2)}\, dz$$

is the Fourier transform of the kernel. It is computed in Problem 1.7, see page 230 and is given by

$$k(u) = \frac{2\pi}{\cosh(\pi u)}.$$

Consequently, the Wiener-Hopf equation acquires the form

$$\left[1 - \frac{2\pi\lambda}{\cosh(\pi u)}\right] F_+(u) + G_-(u) = 0 \ \text{ or } \ K(u)F_+(u) + G_-(u) = 0.$$

By a substitution $2\pi\lambda = \cosh(\pi\alpha)$ the function $K(u)$ rewrites as

$$\begin{aligned} K(u) = 1 - \frac{2\pi\lambda}{\cosh(\pi u)} &= \frac{\cosh(\pi u) - \cosh(\pi\alpha)}{\cosh(\pi u)} \\ &= 2\,\frac{\sinh[\frac{\pi}{2}(u-\alpha)]\sinh[\frac{\pi}{2}(u+\alpha)]}{\cosh(\pi u)} = -2\,\frac{\sin[\frac{\pi i}{2}(u-\alpha)]\sin[\frac{\pi i}{2}(u+\alpha)]}{\cos(\pi i u)}. \end{aligned}$$

In order to factorise $K(u)$ we express the involved trigonometric functions in terms of the Gamma functions using the following functional relations [the first one is the complement formula (2.6), while the second one is its extension obtained by a shift $z \to z - 1/2$]:

$$z\Gamma(z)\Gamma(1-z) = \frac{\pi z}{\sin(\pi z)} = \Gamma(1+z)\Gamma(1-z)\,,$$

$$\Gamma\left(\frac{1}{2} - z\right)\Gamma\left(\frac{1}{2} + z\right) = \frac{\pi}{\cos(\pi z)}.$$

We find

$$K(u) = -2\pi \frac{\Gamma(\frac{1}{2} - iu)\Gamma(\frac{1}{2} + iu)}{\Gamma[\frac{i}{2}(u+\alpha)]\Gamma[1 - \frac{i}{2}(u+\alpha)]\Gamma[\frac{i}{2}(u-\alpha)]\Gamma[1 - \frac{i}{2}(u-\alpha)]}$$

$$= \frac{\pi}{2} \frac{\overbrace{(u^2-\alpha^2)}^{\text{'plus'}-function} \overbrace{\Gamma(\frac{1}{2} - iu)}^{\text{'plus'}-function} \overbrace{\Gamma(\frac{1}{2} + iu)}^{\text{'minus'}-function}}{\underbrace{\Gamma[1 + \frac{i}{2}(u+\alpha)]}_{\text{'minus'}-function} \underbrace{\Gamma[1 - \frac{i}{2}(u+\alpha)]}_{\text{'plus'}-function} \underbrace{\Gamma[1 + \frac{i}{2}(u-\alpha)]}_{\text{'minus'}-function} \underbrace{\Gamma[1 - \frac{i}{2}(u-\alpha)]}_{\text{'plus'}-function}}.$$

So, the kernel $K(u) = K_+(u)K_-(u)$ is factorised into a product of the 'plus'–function $K_+(u)$ and the 'minus'–function $K_-(u)$. The function $K_-(u)$ is analytic in the lower half-plane and in the upper half-plane has

- poles due to $\Gamma\left(\frac{1}{2} + iu\right)$ at $u = i(n + 1/2)$, $n = 0, 1, 2, \ldots$,
- zeros due to $\Gamma[1 + \frac{i}{2}(u \pm \alpha)]$ at $u = \mp\alpha + 2i(1+n)$, $n = 0, 1, 2, \ldots$

On the contrary, the function $K_+(u)$ is analytic in the upper half-plane and in the lower half-plane has

- poles due to $\Gamma\left(\frac{1}{2} - iu\right)$ at $u = -i(n + \frac{1}{2})$, $n = 0, 1, 2, \ldots$,
- zeros due to $\Gamma[1 - \frac{i}{2}(u \pm \alpha)]$ at $u = \mp\alpha - 2i(1+n)$, $n = 0, 1, 2, \ldots$

Also $K_+(u)$ has zeros on the real axis at $u = \pm\alpha$. So, the kernel $K(u)$ is analytic in the strip $|\text{Im} u| < 1/2$. Using the asymptotic formula for the Gamma function at large arguments,

$$\ln \Gamma(z) = z \ln z - z - \frac{1}{2} \ln z + O(1),$$

we find that at $u \to \infty$

$$K(u)\Big|_{u\to\infty} = 1.$$

Indeed, we have

$$\ln K(u) = \overbrace{2 \ln u + \ln\left[\frac{\Gamma(\frac{1}{2} - iu)}{\Gamma[1 - \frac{i}{2}(u+\alpha)]\Gamma[1 - \frac{i}{2}(u-\alpha)]}\right]}^{\ln K_+(u)}$$
$$+ \underbrace{\ln\left[\frac{\Gamma(\frac{1}{2} + iu)}{\Gamma[1 - \frac{i}{2}(u+\alpha)]\Gamma[1 + \frac{i}{2}(u+\alpha)]}\right]}_{\ln K_-(u)}$$

and

$$\ln K(u)|_{u\to\infty} \to \underbrace{\left[-iu\ln 2 + \ln u\right]}_{\ln K_+(u)|_{u\to\infty}} + \underbrace{\left[iu\ln 2 - \ln u\right]}_{\ln K_-(u)|_{u\to\infty}} = 0.$$

To provide the asymptotic behavior at large arguments, we redefine $K_+(u)$ and $K_-(u)$ as follows

$$K_+(u) = \frac{2^{iu}(u^2-\alpha^2)\,\Gamma(\frac{1}{2}-iu)}{\Gamma[1-\frac{i}{2}(u+\alpha)]\Gamma[1-\frac{i}{2}(u-\alpha)]},$$

$$K_-(u) = \frac{\pi}{2}\,\frac{2^{-iu}\,\Gamma(\frac{1}{2}+iu)}{\Gamma[1+\frac{i}{2}(u+\alpha)]\Gamma[1+\frac{i}{2}(u-\alpha)]},$$

so that the product $K_+(u)K_-(u)$ is still the same. Returning to the Wiener-Hopf equation, we have

$$K_+(u)F_+(u) = -\frac{G_-(u)}{K_-(u)} = L\,,$$

where L is some constant. Hence using the inversion formula, we have

$$f(x) = \frac{1}{2\pi}\int_P e^{-iux}F_+(u)\,du = \frac{L}{2\pi}\int_P e^{-iux}\,\frac{\Gamma[-\frac{i}{2}(u+\alpha)]\Gamma[-\frac{i}{2}(u-\alpha)]}{2^{iu}\Gamma(\frac{1}{2}-iu)}\,du,$$

where the contour P lies above all singularities (and zeros) of $K_+(u)$. At $x>0$ the integrand $|e^{-iux}| = e^{u_2x}$ ($u_2 = \mathrm{Im}u$), so that we need to close the integration contour in the lower half-plane. [At $x<0$ $|e^{-iux}| = e^{-u_2x}$ the contour must be closed in the upper half-plane—the analyticity region of $K_+(u)$, so that $f(x)|_{x<0} = 0$, as required.] Here the integrand has the poles at the points $u = u_n^\pm = -2in \mp \alpha$ ($n = 0, 1, 2, ...$). Since for small ζ

$$\Gamma[-\frac{i}{2}(u_n^+ + \zeta + \alpha)] = \Gamma[-n - \frac{i\zeta}{2}]\Big|_{\zeta\to 0} = \frac{2i\,(-1)^n}{\zeta n!}$$

and similarly for u_n^-, by the residue theorem we have

$$f(x) \sim \sum_{n=0}^{\infty}\frac{(-1)^n}{n!}2^{-2n}e^{-2nx}\left[\frac{2^{i\alpha}e^{i\alpha x}\Gamma(-n+i\alpha)}{\Gamma(\frac{1}{2}+i\alpha-2n)} + \frac{2^{-i\alpha}e^{-i\alpha x}\Gamma(-n-i\alpha)}{\Gamma(\frac{1}{2}-i\alpha-2n)}\right].$$

[Note: because we are dealing with a homogeneous problem, f is defined up to a multiplicative constant.] To make further progress we use the complement formula (2.6) in order to transform the argument of the Gamma function (effectively we achieve $n \to -n$), then

$$\frac{\Gamma(-n+i\alpha)}{\Gamma(\frac{1}{2}+i\alpha-2n)} = -i\,(-1)^n \text{ctg}(\pi\alpha)\frac{\Gamma(\frac{1}{2}-i\alpha+2n)}{\Gamma(1-i\alpha+n)}\,.$$

In the next step we would like to 'half' the argument of the numerator in the rhs using the duplication formula (2.12),

$$\Gamma\left(\frac{1}{2}-i\alpha+2n\right) = \frac{2^{2n-i\alpha-1/2}}{\sqrt{\pi}}\,\Gamma\left(n+\frac{1}{4}-\frac{i\alpha}{2}\right)\Gamma\left(n+\frac{3}{4}-\frac{i\alpha}{2}\right).$$

Gathering everything we obtain

$$f(x) \sim e^{i\alpha x}\sum_{n=0}^{\infty}\frac{\Gamma(n+\frac{1}{4}-\frac{i\alpha}{2})\,\Gamma(n+\frac{3}{4}-\frac{i\alpha}{2})}{\Gamma(1-i\alpha+n)\,n!}\left(e^{-2x}\right)^n - \left(\alpha\to-\alpha\right).$$

It is easy to see that the obtained series represents the conventional hypergeometric function (2.34) and the required solution has the form

$$f(x) = C\Bigg(e^{i\alpha x}\frac{\Gamma(\frac{1}{4}-\frac{i\alpha}{2})\Gamma(\frac{3}{4}-\frac{i\alpha}{2})}{\Gamma(1-i\alpha)}F\left(\tfrac{1}{4}-\tfrac{i\alpha}{2},\,\tfrac{3}{4}-\tfrac{i\alpha}{2};\,1-i\alpha;\,e^{-2x}\right)$$
$$-e^{-i\alpha x}\frac{\Gamma(\frac{1}{4}+\frac{i\alpha}{2})\Gamma(\frac{3}{4}+\frac{i\alpha}{2})}{\Gamma(1+i\alpha)}F\left(\tfrac{1}{4}+\tfrac{i\alpha}{2},\,\tfrac{3}{4}+\tfrac{i\alpha}{2};\,1+i\alpha;\,e^{-2x}\right)\Bigg),$$

where C is an arbitrary constant.

Problem 3.5
(a): First we insert the kernel into the equation and shift both t and t' by τ:

$$D(t+\tau,t'+\tau) = D_0(t-t') - \Gamma e^{-i\Delta(t+\tau)}\int_0^t dt_1\,e^{i\Delta t_1}\,D(t_1,t'+\tau)\,.$$

Now we replace t_1 by $t_1+\tau$ and obtain an equation for $D(t+\tau,t'+\tau)$, which is completely identical to the one for $D(t,t')$. From this kind of translational invariance in the variable t immediately follows $D(t,t') = D(t-t')$.
(b): First we observe that $D(t-t')$ can possess a $\Theta(t-t')$ prefactor. So we make the following substitution,

$$D(t-t') = -i\Theta(t-t')\,e^{-i\Delta(t-t')}\,f(t-t')\,,$$

where $f(t-t')$ is the new unknown function, an equation for which has the following form

$$f(t-t') = 1 - \Gamma\int_{t'}^{t} dt_1\,f(t_1-t')\,.$$

An iterative solution of this equation immediately leads to the final result

$$f(t-t') = e^{-\Gamma(t-t')} .$$

(c): On the other hand we can simplify the Eq. (3.5) further and write

$$f(t) = 1 - \Gamma \int_0^t dt_1 \, f(t_1) .$$

The integral on the rhs is then nothing but a convolution of functions 1 and $f(t_1)$. After the Laplace transformation this convolution just becomes a product of the transforms [see, for instance, Eq. (1.47)] of 1, equal to $1/p$, and of $F(p)$. Then we obtain

$$F(p) = 1/p - \Gamma \, F(p)/p \quad \text{and thus} \quad F(p) = 1/(p+\Gamma) .$$

The inverse Laplace transformation then immediately leads to the result $f(t) = e^{-\Gamma t}$.

Problem 3.6

After the Fourier transformation the equation reads

$$K(s)F_+(s) + G_-(s) = P(s)$$

with

$$K(s) = \frac{s^2+9}{s^2+1}, \qquad P(s) = \frac{1}{1-is} \equiv P_+(s),$$

and the strip of common analyticity being: $-1 < \text{Im}(s) < 1$. An obvious factorization of $K(s)$ consists of taking, for example,

$$K_+(s) = \frac{s+3i}{s+i}, \quad K_-(s) = \frac{s-3i}{s-i},$$

which leads to the following sum decomposition:

$$\frac{P_+(s)}{K_-(s)} = \frac{s-i}{(1-is)(s-3i)} = \underbrace{\frac{1}{2(1-is)}}_{\text{`plus'–function}} + \underbrace{\frac{i}{2(s-3i)}}_{\text{`minus'–function}} .$$

We then have

$$K_+(s)F_+(s) - \frac{1}{2(1-is)} = -\frac{G_-(s)}{K_-(s)} + \frac{i}{2(s-3i)} = E(s) ,$$

therefore

$$F_+(s) = \frac{(s+i)E(s)}{s+3i} + \frac{i}{2(s+3i)}, \quad G_-(s) = -\frac{(s-3i)E(s)}{s-i} + \frac{i}{2(s-i)}.$$

From the asymptotic form of $F_+(s)$ we immediately conclude that $E(s) = 0$. $F_+(s)$ is bounded for $s \neq -3i$ and tends to zero at infinity, and we obtain

$$E(s) = \frac{s+3i}{s+i} F_+(s) - \frac{1}{2(1-is)}.$$

We conclude that the entire function $E(s)$ is bounded and tends to zero at infinity. Liouville's theorem then implies that $E(s) = 0$. So, applying the inversion formula we find the unique solution:

$$f(x) = \frac{1}{2} e^{-3x}.$$

Problem 3.7

In order to solve the equation we use the formula (3.63) with $g(t) = 1 - t^2$. The only integral to solve is then the following one:

$$I(x) = \mathrm{P} \int_{-1}^{1} \frac{(1-t^2)^{3/2}}{t-x} dt\,.$$

As before we define an auxiliary function of the variable $|z| > 1$ by the prescription

$$G(z) = \frac{1}{2\pi i} \int_{-1}^{1} dt\, \frac{(1-t^2)^{3/2}}{t-z}\,.$$

Here we perform the same trigonometric substitution, $t = \cos\theta$ and obtain

$$G(z) = \frac{1}{2\pi i} \int_{0}^{\pi} \frac{\sin^4\theta}{\cos\theta - z} d\theta\,.$$

This integral is very close to the one we calculated on page 172. Using basically the same method we obtain

$$I(z) = \frac{1}{2i} \left[(z^3 - 3z/2) - (z^2 - 1)^{3/2} \right].$$

Performing the analytic continuation we obtain for the principal value

$$I(x) = \pi \left(x^3 - 3x/2 \right).$$

Therefore for the solution with the help of (3.63) we obtain

$$f(x) = \frac{3x/2 - x^3 + A}{\sqrt{1-x^2}}\,.$$

Problem 3.8

We first represent the function $f(x)$ by a sum: $f(x) = f_+(x) + f_-(x)$, where $f_-(x) = 0$ for $x > 0$ and $f_+(x) = 0$ for $x < 0$. The original equation then can be rewritten in the form

$$\int_{-\infty}^{\infty} k_1(x-t) f_+(t)dt + \int_{-\infty}^{\infty} k_2(x-t) f_-(t)dt = k_1(x).$$

In the next step we apply the Fourier transformation to both sides of the above equation. One then obtains

$$K_1(s)F_+(s) + K_2(s)F_-(s) = K_1(s),$$

where

$$K_1(s) = \frac{1}{(s-2i)(s-3i)}, \quad K_2(s) = \frac{1}{s+2i}.$$

We observe that $K_1(s)$ is analytic in the lower half–plane for Im$s < 2$, $K_1(s) = K_-(s)$, whereas $K_2(s)$ is analytic in the upper half–plane at Im$s > -2$, $K_2(s) = K_+(s)$. So we have

$$\frac{F_+(s)}{K_+(s)} = -\frac{F_-(s)}{K_-(s)} + \frac{1}{K_+(s)}.$$

In the class of functions that vanish at infinity the solution is

$$F_+(s) = \frac{C}{s+2i}, \quad F_-(s) = -\frac{C-2i-s}{(s-2i)(s-3i)},$$

where C is an arbitrary constant. Finally the solution $f(x)$ is given by the inversion formula,

$$f_\pm(x) = \frac{1}{2\pi} \int_{-\infty}^{\infty} F_\pm(s) e^{-isx} ds.$$

Closing the integration contour in the lower (for $x > 0$) or the upper (for $x < 0$) half–plane by the semicircle C_R, the residue theorem (in both cases the C_R-integral vanishes in the limit $R \to \infty$) leads to the answer.

Problem 3.9

We rewrite the respective Wiener-Hopf equation

$$F_+(s) + k(s)F_+(s) + G_-(s) = P_+(s)$$

in the form

$$K(s)F_+(s) - P_+(s) = -G_-(s),$$

with $K(s) = 1 + k(s)$. For the Fourier transform of the kernel one obtains

$$k(s) = \frac{2a}{s^2+1} - \frac{2b(s^2-1)}{(s^2+1)^2} \quad \rightarrow \quad K(s) = \frac{M(s)}{(s^2+1)^2},$$

where the numerator of $K(s)$ is a polynomial

$$M(u) = u^4 + 2(a-b+1)u^2 + (2a+2b+1).$$

To exclude the zeros of $K(s)$ on the real axis, we impose additional requirements on the constants a and b assuming that the polynomial $M(s)$ has no real roots. Then if $s_1 = \alpha + i\beta$ (with $\alpha > 0$ and $\beta > 0$) is a root of the biquadratic equation $M(s) = 0$, the other three roots are: $s_2 = -\alpha - i\beta$, $s_3 = \overline{s_1} = \alpha - i\beta$, and $s_4 = \overline{s_2} = -\alpha + i\beta$, where $\overline{s_i}$ is complex conjugate of s_i. Therefore we can factorise the function $K(s) = K_+(s)K_-(s)$ by:

$$K_+(s) = \frac{(s-s_2)(s-\overline{s_1})}{(s+i)^2}, \quad K_-(s) = \frac{(s-s_1)(s-\overline{s_2})}{(s-i)^2}.$$

Note that the functions $K_\pm(s) \rightarrow 1$ at $|s| \rightarrow \infty$. Applying this result, the Wiener-Hopf equation reads

$$F_+(s)K_+(s) - \frac{P_+(s)}{K_-(s)} = -\frac{G_-(s)}{K_-(s)}.$$

On the basis of the analytic continuation theorem both sides of this relation are equal to

$$\frac{C_1}{s-s_1} + \frac{C_2}{s-\overline{s_2}},$$

where the constants C_1 and C_2 are still to be determined. So we obtain

$$F_+(s) = \frac{1}{K_+(s)}\left(\frac{(s-i)^2P_+(s)}{(s-s_1)(s-\overline{s_2})} + \frac{C_1}{s-s_1} + \frac{C_2}{s-\overline{s_2}}\right).$$

For $F_+(s)$, the poles s_1 and $\overline{s_2}$ in the upper half-plane may be avoided by a special choice of the constants C_1 and C_2, namely

$$C_1 = -\frac{(s_1-i)^2P_+(s_1)}{2\alpha}, \quad C_2 = \frac{(\overline{s_2}-i)^2P_+(\overline{s_2})}{2\alpha}.$$

These poles in this case are effectively removed. For simplicity, we now consider two terms in the above equation for $F_+(s)$ separately. Let us denote the first and the second summands by

$$F_1(s) = \frac{1}{K_+(s)} \frac{(s-i)^2 P_+(s)}{(s-s_1)(s-\overline{s_2})}, \quad F_2(s) = \frac{1}{K_+(s)} \left(\frac{C_1}{s-s_1} + \frac{C_2}{s-\overline{s_2}} \right),$$

respectively. Rewriting the first term in the form

$$F_1(s) = P_+(s) + R(s)P_+(s),$$

the corresponding inverse transform is

$$f_1(x) = p(x) + \int_0^\infty R(x-t)p(t)dt.$$

Evidently, the function $R(s)$ plays here a role of the resolvent for the considered integral equation but on the entire axis. For $R(s)$, after some algebra, we find

$$R(s) = \frac{\gamma}{s^2 - s_1^2} + \frac{\overline{\gamma}}{s^2 - \overline{s_1}^2}, \quad \gamma = i\frac{s_1^2(a-b) + a + b}{2\alpha\beta}.$$

Its inverse transform can be calculated with the help of the residue theorem (remember that $x > 0$ and then two poles at $s = -s_1$ and $s = +\overline{s_1}$ in the lower half-plane contribute) and is equal to

$$R(x) = \frac{\gamma e^{-(\beta - i\alpha)|x|}}{2(\beta - i\alpha)} + \frac{\overline{\gamma} e^{-(\beta + i\alpha)|x|}}{2(\beta + i\alpha)} = \frac{\rho}{2}\, e^{-\beta|x|} \cos(\theta + \alpha|x|),$$

where $\rho e^{i\theta} = \gamma/(\beta - i\alpha)$. Hence

$$f_1(x) = p(x) + \rho \int_0^\infty e^{-\beta|x-t|} \cos(\theta + \alpha|x-t|)p(t)dt.$$

The second part of $F_+(s)$ is given by

$$F_2(s) = C_1\Big[\frac{(s+i)^2}{(s-s_2)(s-\overline{s_1})(s-s_1)}\Big] + C_2\Big[\frac{(s+i)^2}{(s-s_2)(s-\overline{s_1})(s-\overline{s_2})}\Big].$$

Its inverse transform can also be calculated by residue theorem, the contributing poles at $s = s_2$ and $s = \overline{s_1}$ are the same for both summands above. After tedious but simple algebra, one obtains for $x > 0$

$$f_2(x) = \frac{e^{-\beta x}}{2}\Big[A\Big(P_+(s_1)e^{-i\alpha x} + P_+(\overline{s_2})e^{i\alpha x}\Big) + B\Big(P_+(s_1)e^{i\alpha x}e^{i\psi} + P_+(\overline{s_2})e^{-i\alpha x}e^{-i\psi}\Big)\Big],$$

with

$$A = \frac{[\alpha^2 + (\beta-1)^2]^2}{4\alpha^2\beta}, \quad B = \frac{R}{4\alpha^2}, \text{ and } Re^{i\psi} = \frac{[\alpha + i(\beta-1)]^4}{\alpha + i\beta}.$$

Taking into account that

$$P_+(s_1) = \int_0^\infty e^{i(\alpha+i\beta)t}p(t)dt,$$

and similarly for $P_+(\overline{s_2})$, one obtains

$$f_2(x) = \int_0^\infty e^{-\beta(x+t)}\Big[A\cos[\alpha(x-t)] + B\cos[\psi + \alpha(x+t)]\Big]p(t)dt.$$

The desired solution $f(x)$ is the sum of the functions $f_1(x)$ and $f_2(x)$.

Problem 3.10
We take the Laplace transforms of both sides of the equation

$$f(x) = \int_0^x \frac{u(t)}{(x-t)^\mu}\,dt \equiv \int_0^x u(t)v(x-t)\,dt, \qquad v(\tau) = \tau^{-\mu}.$$

Since the rhs is the convolution of $u(t)$ and $v(t)$ after the transformation we obtain[2]

$$f_L(s) = u_L(s)v_L(s) = u_L(s)\frac{\Gamma(1-\mu)}{s^{1-\mu}},$$

valid for $\mu < 1$, where Γ is the Gamma function. Consequently,

[2] For the power function t^ν, the Laplace transform is given by

$$\int_0^\infty e^{-st}t^\nu dt = \frac{\Gamma(\nu+1)}{s^{\nu+1}}, \qquad \nu > -1,$$

which follows from definition (2.1) of the Gamma function.

$$u_L(s) = \frac{s^{1-\mu}}{\Gamma(1-\mu)} f_L(s)$$

or ($\mu > 0$ must hold as well, of course)

$$u_L(s) = \frac{s}{\pi}\Big[\frac{\pi}{\Gamma(1-\mu)\Gamma(\mu)} \frac{\Gamma(\mu)}{s^\mu} f_L(s)\Big] = \frac{\sin(\pi\mu)}{\pi} s \left[\frac{\Gamma(\mu)}{s^\mu} f_L(s)\right].$$

In the above expression we used the complement formula (2.6). We now observe that the last factor in the rhs brackets is the Laplace transform of the function [see the Laplace transformation property (1.50)]

$$h(x) = \int_0^x f(t)(x-t)^{\mu-1}\, dt \quad \leftrightarrow \quad h_L(s) = f_L(s)\frac{\Gamma(\mu)}{s^\mu}$$

and since $sh_L(s) = h'_L(s)$ [property (1.42), also note: $h(0) = 0$] we have

$$u_L(s) = \frac{\sin(\mu\pi)}{\pi} h'_L(s).$$

Applying the inverse transformation to both sides of the above equality yields the answer:

$$u(x) = \frac{\sin(\mu\pi)}{\pi} \frac{d}{dx} \int_0^x \frac{f(t)\, dt}{(x-t)^{1-\mu}}.$$

In the particular case of $f(t) = t^\nu$ we have

$$u(x) = \frac{\sin(\mu\pi)}{\pi} \frac{d}{dx} \int_0^x \frac{t^\nu}{(x-t)^{1-\mu}}\, dt.$$

By the substitution $t = xy$ the integral in the rhs reduces to the Beta function (2.10):

$$\int_0^x \frac{t^\nu}{(x-t)^{1-\mu}}\, dt = \frac{x^{\nu+1}}{x^{1-\mu}} \int_0^1 y^\nu (1-y)^{\mu-1}\, dy \equiv x^{\nu+\mu} B(\nu+1, \mu).$$

An ensuing differentiation gives the answer.

5.4 Chapter 4

Problem 4.1
First we bring the equation to the form (4.1) and identify

$$p_0 = 6,\ p_1 = 0,\ p_2 = -6,\ \ q_0 = -5,\ q_1 = 1\,.$$

From (4.3) we then immediately see that $\lambda_n = n(6n+5)$. Using Eq. (4.4) we then obtain

$$a_{n-1} = a_n \frac{5n}{12n-1}\,,$$

while (4.2) yields

$$a_r = \frac{5a_{r+1}(r+1) - 6a_{r+2}(r+2)(r+1)}{n(6n+5) - r(6r+5)}\,.$$

Therefore for the lowest order $n = 0$ we obviously obtain $y_0(x) = y_0 = 1$. For $n = 1$ then follows

$$a_0 = \frac{5}{11}\,a_1,$$

so that

$$y_1(x) = a_0 + a_1 x = \frac{a_1}{11}\,(5 + 11x):\ \ y_1(x)\ \rightarrow\ 11x + 5.$$

For $n = 2$, $\lambda_2 = 34$,

$$a_1 = \frac{10}{23}\,a_2 \ \text{ and } \ a_0 = -\frac{113}{391}\,a_2,$$

so that

$$y(x) = a_0 + a_1 x + a_2 x^2 = \frac{a_2}{391}(-113 + 170x + 391x^2) : y_2 \ \rightarrow 391x^2 + 170x - 113.$$

For the Rodrigues' formula we first need the weight function. Since

$$\frac{q(x)}{p(x)} = \frac{5-x}{6(x^2-1)}$$

the weight is simply

$$w(x) = \exp\Big(\int \frac{q(x)}{p(x)}\,dx\Big) = (1-x)^{1/3}\,(x+1)^{-1/2}.$$

With $p(x) = 6(1-x^2)$ Rodrigues' formula states:

$$y_n(x) = \frac{1}{w(x)} \frac{d^n}{dx^n}(wp^n)$$
$$= \frac{6^n}{(1-x)^{1/3}(x+1)^{-1/2}} \frac{d^n}{dx^n}\Big[(1-x)^{n+1/3}(x+1)^{n-1/2}\Big].$$

In this normalisation,

$$y_0(x) = 1,$$
$$y_1(x) = -\frac{6}{(x-1)^{1/3}(x+1)^{-1/2}} \frac{d}{dx}\Big((x-1)^{4/3}(x+1)^{1/2}\Big)$$
$$= -(11x+5),$$
$$y_2(x) = \frac{36}{(x-1)^{1/3}(x+1)^{-1/2}} \frac{d^2}{dx^2}\Big((x-1)^{7/3}(x+1)^{3/2}\Big)$$
$$= 36\Big[\frac{7}{3}\frac{4}{3}(x+1)^2 + 2\frac{7}{3}\frac{3}{2}(x^2-1) + \frac{3}{2}\frac{1}{2}(x-1)^2\Big]$$
$$= 391x^2 + 170x - 113.$$

Problem 4.2

According to (4.23), the Jacobi polynomial of degree n is defined by the nth derivative of the function $(1-x)^{\alpha+n}(1+x)^{\beta+n}$: $y_n(x) = (1-x)^{-\alpha}(1+x)^{-\beta}\frac{d^n}{dx^n}[(1-x)^{\alpha+n}(1+x)^{\beta+n}]$. The more widespread form of the Rodrigues' formula is

$$P_n^{(\alpha,\beta)}(x) = \frac{(-1)^n}{2^n n!}(1-x)^{-\alpha}(1+x)^{-\beta}\frac{d^n}{dx^n}[(1-x)^{n+\alpha}(1+x)^{n+\beta}],$$

which differs from our notation by an additional factor $(-1)^n/2^n n!$ (we set $1/K_n = (-1)^n/2^n n!$). It is similar to Legendre polynomials $P_n(x)$, which are a special case of the Jacobi polynomials $P_n^{(\alpha,\beta)}(x)$ with $\alpha = \beta = 0$.
Let us now evaluate the generation function $\Psi(x,t) = \sum_{n=0}^{\infty} P_n^{(\alpha,\beta)}(x)\,t^n$ following Example 4.10. Then we have [recall: factor $(-1)^n/2^n$ leads to a replacement $p(x) = (1-x^2) \to (x^2-1)/2$]

$$\Psi(x,t) = \frac{w(z)}{w(x)}\frac{\partial z}{\partial x} = \frac{1}{R}\frac{(1-z_1)^\alpha(1+z_1)^\beta}{(1-x)^\alpha(1+x)^\beta}, \qquad z_1 = \frac{1-R}{t},$$

where $R = \sqrt{1-2xt+t^2}$. It is easy to show, using the equation $z = x + tp(z) \to z = x + \frac{t}{2}(z^2-1)$, that

$$(1-z_1) = 2\,\frac{1-x}{1-t+R} \quad \text{and} \quad (1+z_1) = 2\,\frac{1+x}{1+t+R},$$

so that the generating function is

$$2^{\alpha+\beta} R^{-1}(1-t+R)^{-\alpha}(1+t+R)^{-\beta} = \sum_{n=0}^{\infty} P_n^{(\alpha,\beta)}(x)\, t^n.$$

In particular, at $\alpha = \beta = 0$ this result immediately reduces to the expression (4.36) for the generating function of the (normalized) Legendre polynomials. To obtain the polynomials at $x = 1$, we expand the generating function about $t = 0$ and obtain

$$\Psi(1,t) = \sum_{n=0}^{\infty} P_n^{(\alpha,\beta)}(1)\, t^n = \frac{1}{(1-t)^{1+\alpha}} = \sum_{n=0}^{\infty} \frac{\Gamma(n+1+\alpha)}{n!\,\Gamma(1+\alpha)}\, t^n.$$

Comparing the coefficients at t^n yields

$$P_n^{(\alpha,\beta)}(1) = \frac{\Gamma(n+1+\alpha)}{n!\,\Gamma(1+\alpha)}.$$

According to the Rodrigues' formula, the following relation

$$P_n^{(\alpha,\beta)}(x) = (-1)^n P_n^{(\beta,\alpha)}(-x)$$

obviously holds and therefore at $x = -1$ we get

$$P_n^{(\alpha,\beta)}(-1) = (-1)^n P_n^{(\beta,\alpha)}(1) = (-1)^n \frac{\Gamma(n+1+\beta)}{n!\,\Gamma(1+\beta)}.$$

Problem 4.3

(a): First we observe that for $C_n^{\lambda}(x) \sim P_n^{(\lambda-\frac{1}{2},\lambda-\frac{1}{2})}(x)$ the Rodrigues' formula acquires the form (see Problem 4.2):

$$C_n^{\lambda}(x) = \frac{(-1)^n}{2^n} \frac{\Gamma(\lambda+\frac{1}{2})\Gamma(n+2\lambda)}{\Gamma(2\lambda)\Gamma(n+\lambda+\frac{1}{2})} \frac{(1-x^2)^{\frac{1}{2}-\lambda}}{n!} \frac{d^n}{dx^n}\left[(1-x^2)^{\lambda+n-\frac{1}{2}}\right],\ \lambda > -\frac{1}{2}.$$

Then the normalization coefficient can be written down in the following way:

$$N_n = \int_{-1}^{1} [C_n^{\lambda}(x)]^2 (1-x^2)^{\lambda-\frac{1}{2}} dx = A(n,\lambda) \int_{-1}^{1} C_n^{\lambda}(x) \frac{d^n}{dx^n}\left[(1-x^2)^{\lambda+n-\frac{1}{2}}\right] dx,$$

where

$$A(n,\lambda) = \frac{(-1)^n}{2^n n!} \frac{\Gamma(\lambda+\frac{1}{2})\Gamma(n+2\lambda)}{\Gamma(2\lambda)\Gamma(n+\lambda+\frac{1}{2})}.$$

Integrating by parts yields (the boundary terms vanish)

$$N_n = A(n,\lambda)(-1)^n \int_{-1}^{1} (1-x^2)^{\lambda+n-\frac{1}{2}} \frac{d^n}{dx^n} C_n^\lambda(x)dx$$

$$= c_n A(n,\lambda)(-1)^n n! \int_{-1}^{1} (1-x^2)^{\lambda+n-\frac{1}{2}} dx$$

$$= c_n A(n,\lambda)(-1)^n n!\, B\Big(\lambda+n+\frac{1}{2},\frac{1}{2}\Big),$$

where B is the Beta function and

$$c_n = \frac{2^n\Gamma(n+\lambda)}{n!\Gamma(\lambda)}$$

is the coefficient in front of x^n in the series representation of $C_n^\lambda(x)$. This leads to:

$$N_n = \frac{\sqrt{\pi}}{n!}\,\frac{\Gamma(n+2\lambda)}{(\lambda+n)\Gamma(\lambda)}\,\frac{\Gamma(\lambda+\frac{1}{2})}{\Gamma(2\lambda)} = \frac{2^{1-2\lambda}\pi\Gamma(n+2\lambda)}{(\lambda+n)[\Gamma(\lambda)]^2}.$$

(b): To obtain an explicit form of the three–term recurrence relation (4.39)

$$y_{n+1}(x) = \alpha_n x y_n(x) + a_{n,n} y_n(x) + a_{n,n-1} y_{n-1}(x)$$

for $y_n(x) = C_n^\lambda(x)$, we need to calculate the involved coefficients

$$\alpha_n = \frac{1}{b_{n,n+1}},\ a_{n,r} = -\frac{b_{n,r}}{b_{n,n+1}},\ b_{n,r} = \frac{1}{N_r}\int_{-1}^{1} x C_n^\lambda(x) C_r^\lambda(x)(1-x^2)^{\lambda-\frac{1}{2}}dx.$$

These integrals can be evaluated in the same way as above. First, at $r = n$ the integrand is an odd function of x [note that $C_n^\lambda(x) = (-1)^n C_n^\lambda(-x)$] and therefore the coefficient $b_{n,n} = 0$. At $r = n \pm 1$ one obtains

$$b_{n,n+1} = \frac{\sqrt{\pi}}{2N_{n+1}}\,\frac{\Gamma(2\lambda+n+1)}{n!(\lambda+n)(\lambda+n+1)}\,\frac{\Gamma(\lambda+\frac{1}{2})}{\Gamma(\lambda)\Gamma(2\lambda)} = \frac{n+1}{2(\lambda+n)},$$

$$b_{n,n-1} = \frac{\sqrt{\pi}}{2N_{n-1}}\,\frac{\Gamma(2\lambda+n)}{(n-1)!(\lambda+n-1)(\lambda+n)}\,\frac{\Gamma(\lambda+\frac{1}{2})}{\Gamma(\lambda)\Gamma(2\lambda)} = \frac{2\lambda+n-1}{2(\lambda+n)}.$$

Finally,

$$\alpha_n = \frac{2(\lambda+n)}{n+1},\quad a_{n,n} = 0,\quad a_{n,n-1} = -\frac{2\lambda+n-1}{n+1}.$$

With the above coefficients the three–term recurrence relation for C_n^λ takes the form (we substituted $n+1 \to n$):

$$nC_n^\lambda(x) - 2(n-1+\lambda)xC_{n-1}^\lambda(x) + (n+2\lambda-2)C_{n-2}^\lambda(x) = 0.$$

(c): Let us denote the generating function in the following way

$$\sum_{n=0}^{\infty} C_n^\lambda(x)t^n = h(t).$$

Then the derivative is

$$h'(t) = \sum_{n=0}^{\infty} n\, C_n^\lambda(x)\, t^{n-1} \equiv \sum_{n=1}^{\infty} n\, C_n^\lambda(x)\, t^{n-1}\,.$$

Using for $C_n^\lambda(x)$ the above recurrence relation, one can show that the function $h(t)$ satisfies the differential equation

$$h'(t)(1-2xt+t^2) = h(t)2\lambda(x-t).$$

Since by definition $h(0) = C_0^\lambda(x) = 1$, the solution is

$$h(t) = \frac{1}{(1-2xt+t^2)^\lambda}.$$

Problem 4.4
From the given Rodrigues' formula we read off:

$$w(x) = e^{-x}x^\alpha, \quad \text{and} \quad p(x) = x.$$

In order to compute the generating function we use (4.30),

$$z = x + y\,p(z) \quad \to z = \frac{x}{1-y}.$$

Then the generating function immediately follows from (4.31),

$$G^\alpha(x,y) = \frac{w(z)}{w(x)}\,\frac{dz}{dx} = \frac{e^{-x/(1-y)}\,x^\alpha}{e^{-x}x^\alpha(1-y)^\alpha}\,\frac{1}{1-y} = (1-y)^{-\alpha-1}e^{-xy/(1-y)}\,.$$

In order to obtain the polynomials we expand

$$G^\alpha(x,y) = \sum_{n=0}^{\infty} \frac{L_n^\alpha(x)}{n!}\,y^n$$

for small y and extract the coefficients:

$$\begin{aligned}
L_0^\alpha(x) &= 1\,,\\
L_1^\alpha(x) &= (1+\alpha-x),\\
\frac{L_2^\alpha(x)}{2} &= \frac{1}{2}[(1+\alpha)(2+\alpha)-2(2+\alpha)x+x^2],\\
\frac{L_3^\alpha(x)}{6} &= \frac{1}{6}[(1+\alpha)(2+\alpha)(3+\alpha)-3(2+\alpha)(3+\alpha)x+3(3+\alpha)x^2-x^3].
\end{aligned}$$

The ODE for $G^\alpha(x,y)$ can be checked directly. For the derivative we have

$$\begin{aligned}
\frac{\partial G^\alpha(x,y)}{\partial y} &= \frac{\alpha+1}{(1-y)^{\alpha+2}}\,e^{-xy/(1-y)} - \frac{x}{(1-y)^{\alpha+3}}\,e^{-xy/(1-y)}\\
&= \frac{\alpha+1}{(1-y)}\,G^\alpha(x,y) - \frac{x}{(1-y)^2}\,G^\alpha(x,y).
\end{aligned}$$

That immediately leads to the ODE.

In order to obtain the recursion relation we substitute into the ODE the expansion $G^\alpha(x,y)=\sum_{n=0}^{\infty} L_n^\alpha(x)y^n/n!$. That procedure yields

$$(1-2y+y^2)\sum_{n=1}^{\infty}\frac{nL_n^\alpha(x)}{n!}y^{n-1} = [\alpha+1-x-(\alpha+1)y]\sum_{n=0}^{\infty}\frac{L_n^\alpha(x)}{n!}y^n .$$

Matching coefficients at equal powers of y gives the recurrence relation

$$L_{n+1}^\alpha(x) + (x-\alpha-1-2n)\;L_n^\alpha(x) + n\;(n+\alpha)\;L_{n-1}^\alpha(x) = 0.$$

Note: for polynomials which are normalized differently, for example $\overline{L_n^\alpha(x)} = L_n^\alpha(x)/n!$ the above recurrence relation reads

$$(n+1)\;\overline{L_{n+1}^\alpha(x)} + (x-\alpha-1-2n)\;\overline{L_n^\alpha(x)} + (n+\alpha)\;\overline{L_{n-1}^\alpha(x)} = 0.$$

Evidently, the Rodrigues' formula for the polynomials $\overline{L_n^\alpha(x)}$ also differs from that for $L_n^\alpha(x)$ by the factorial $n!$:

$$\overline{L_n^\alpha(x)} = \frac{1}{n!}e^x x^{-\alpha}\frac{d^n}{dx^n}\left(e^{-x}x^{n+\alpha}\right).$$

The above generated function reads

$$\frac{e^{-xy/(1-y)}}{(1-y)^{\alpha+1}} = \sum_{n=0}^{\infty}\overline{L_n^\alpha(x)}y^n\,.$$

Problem 4.5

(a): First we expand the generating function in powers of $(2xt - t^2)$ and then expand in powers of t. One obtains

$$\Psi(x,t) = \frac{1}{\sqrt{1-2xt+t^2}} = \sum_{s=0}^{\infty} \frac{(2s)!}{2^{2s}(s!)^2}\,(2xt-t^2)^s$$

$$= \sum_{s=0}^{\infty} \frac{(2s)!}{2^{2s}(s!)^2}\,t^s \sum_{r=0}^{s}(-1)^r \frac{s!}{r!(s-r)!}\,(2x)^{s-r}t^r .$$

Now we introduce $n = s + r$ which we keep fixed, that is $s = n - r$, and take into account that $1/(n-2r)! \equiv 0$ for $r > [n/2]$, where $[n/2] = n/2$ for even n and $[n/2] = (n-1)/2$ for odd n. Then we have

$$\Psi(x,t) = \sum_{n=0}^{\infty}\sum_{r=0}^{[\frac{n}{2}]} \frac{(2n-2r)!}{2^n r!(n-r)!(n-2r)!}(-1)^r\,x^{n-2r}\,t^n .$$

Hence the series expansion of Legendre polynomials has the form

$$P_n(x) = x^n \sum_{r=0}^{[\frac{n}{2}]} \frac{(2n-2r)!}{2^n r!(n-r)!(n-2r)!}(-1)^r\,x^{-2r} .$$

From this relation follows, in particular, that $P_n(-x) = (-1)^n\,P_n(x)$.

(b): We differentiate the generating function once and perform a series expansion with respect to t. One obtains

$$\frac{\partial\Psi}{\partial t} = \sum_{n=0}^{\infty} nP_n(x)t^{n-1} = \frac{x-t}{1-2xt-t^2}\,\Psi(x,t) .$$

Inserting the series expansion of $\Psi(x,t)$ into the rhs, after some algebra we have three separate sums:

$$\sum_{n=0}^{\infty} nP_n(x)t^{n-1} - \sum_{n=0}^{\infty}(2n+1)xP_n(x)t^n + \sum_{n=0}^{\infty}(n+1)P_n(x)t^{n+1} = 0.$$

Reindexing of the first and the last sums (to make all powers of t the same)

$$\sum_{n=0}^{\infty} n P_n(x) t^{n-1} = \sum_{n=1}^{\infty} n P_n(x) t^{n-1} = \sum_{k=0}^{\infty} (k+1) P_{k+1}(x) t^k,$$

$$\sum_{n=0}^{\infty} (n+1) P_n(x) t^{n+1} = \sum_{k=1}^{\infty} k P_{k-1}(x) t^k = \sum_{k=0}^{\infty} k P_{k-1}(x) t^k$$

leads to a single sum

$$\sum_{k=0}^{\infty} \left[(k+1) P_{k+1}(x) - (2k+1) x P_k(x) + k P_{k-1}(x)\right] t^k = 0.$$

The above equality is valid for all t and therefore the coefficients of the t^k are zero. Hence for Legendre polynomials the three-term recurrence relation has the form

$$(k+1) P_{k+1}(x) - (2k+1) x P_k(x) + k P_{k-1}(x) = 0.$$

(c): Squaring the generating function, we have

$$\frac{1}{1 - 2xt + t^2} = \left[\sum_{n=0}^{\infty} P_n(x) t^n\right]^2 = \sum_{n=0}^{\infty} \sum_{m=0}^{\infty} P_n(x) P_m(x) t^{n+m}.$$

Integrating now in the interval $[-1, 1]$ and using the orthogonality of the Legendre polynomials, we get

$$\int_{-1}^{1} \frac{dx}{1 - 2xt + t^2} = \sum_{n=0}^{\infty} t^{2n} \int_{-1}^{1} P_n^2(x)\, dx.$$

The integral in the lhs easily evaluates, that is

$$\int_{-1}^{1} \frac{dx}{1 - 2xt + t^2} = -\frac{1}{2t} \int_{(1+t)^2}^{(1-t)^2} \frac{dy}{y} = -\frac{1}{2t} \ln\left[\frac{(1-t)^2}{(1+t)^2}\right] = \frac{1}{t} \ln\left(\frac{1+t}{1-t}\right).$$

Expanding this result around $t = 0$, we obtain

$$\frac{1}{t} \ln\left(\frac{1+t}{1-t}\right) = \sum_{n=0}^{\infty} \frac{2}{2n+1} t^{2n} = \sum_{n=0}^{\infty} t^{2n} \int_{-1}^{1} P_n^2(x)\, dx\,.$$

Therefore for the normalization constant we find

$$N_n = \int_{-1}^{1} P_n^2(x)\, dx = \frac{2}{2n+1}.$$

Problem 4.6

The problem can be solved in two different ways.

First method: Write $K(x) = (1-x)M(x) + K(1)$, where $M(x)$ is polynomial of degree strictly less than n. Then

$$\int_{-1}^{1} S_n(x)K(x)dx = \int_{-1}^{1} S_n(x)\Big[(1-x)M(x) + K(1)\Big]dx.$$

Now expand $S_n(x)$ and $M(x)$ as a sum of Legendre polynomials,

$$S_n(x) = \sum_{r=0}^{n} a_r P_r(x), \quad M(x) = \sum_{m=0}^{n-1} b_m P_m(x).$$

Ultimately we have to show that

$$\int_{-1}^{1} \sum_{r=0}^{n} a_r P_r(x)\Big[(1-x)\sum_{m=0}^{n-1} b_m P_m(x) + K(1)\Big]dx = K(1).$$

This simplifies further, since the x-independent term vanishes by orthogonality, while the x-dependent terms are eliminated using the recursion relation,

$$x P_r(x) = \frac{(r+1)P_{r+1}(x) + r P_{r-1}(x)}{2r+1}.$$

Thus we get

$$\sum_{r=0}^{n-1} b_r a_r \frac{2}{2r+1} - \sum_{r=0}^{n-1} b_r a_{r+1} \frac{2(r+1)}{(2r+1)(2r+3)}$$

$$-\sum_{r=0}^{n-1} b_r a_{r-1} \frac{2r}{(2r+1)(2r-1)} + 2a_0\, K(1) = K(1)\,,$$

which must hold for all b_r. Hence $a_0 = 1/2$, and the coefficient of b_r must vanish:

$$a_r = \frac{r}{2r-1}\, a_{r-1} + \frac{r+1}{2r+3}\, a_{r+1} \quad \text{for } r = 1, ..., n-1 \text{ and } a_0 = a_1/3.$$

This recurrence relation has the solution

$$a_r = \frac{2r+1}{2},$$

as is easily checked. Therefore $S_n(x)$ is indeed a polynomial.
Second method: Expand $K(x)$ as a sum of the Legendre polynomials:

$$K(x) = \sum_{r=0}^{n} k_r P_r(x) \quad \text{with} \quad K(1) = \sum_{r=0}^{n} k_r, \quad \text{as} \quad P_n(1) = 1.$$

Expand $S_n(x)$ in the same way as before:

$$S_n(x) = \sum_{m=0}^{n} a_m P_m(x).$$

Therefore

$$\int_{-1}^{1} S_n(x)K(x)dx = \sum_{r=0}^{n}\sum_{m=0}^{n} k_r a_m \int_{-1}^{1} P_r(x)P_m(x)dx$$
$$= \sum_{r=0}^{n} k_r \underbrace{a_r \frac{2}{2r+1}}_{=1} = \sum_{r=0}^{n} k_r = K(1).$$

Hence again $a_r = (2r+1)/2$.

Problem 4.7
(a): Hermite polynomials are defined with the weight $w(x) = e^{-x^2}$. Using the orthogonality relations we have to deal with the integrals of the type

$$\int_{-\infty}^{\infty} dx e^{-x^2} x^k = \begin{cases} 0, & k = 2n+1 \\ \Gamma(n+\frac{1}{2}), & k = 2n, \end{cases}$$

where the Gamma function $\Gamma(n+1/2) = \sqrt{\pi}(2n)!/2^{2n}n!$.
We choose $H_0 = 1$ and $H_1(x) = x + a$. Then a is determined from

$$\int_{-\infty}^{\infty} dx e^{-x^2} H_0 H_1(x) = \int_{-\infty}^{\infty} dx e^{-x^2}(x+a) = 0,$$

therefore $a = 0$ and thus $H_1(x) = x$. For the next polynomial we set $H_2(x) = x^2 + ax + b$, where a, b are determined from the relations

$$\int_{-\infty}^{\infty} dx e^{-x^2} \underbrace{(x^2 + ax + b)}_{H_0 H_2} = 0, \quad \int_{-\infty}^{\infty} dx e^{-x^2} \underbrace{x(x^2 + ax + b)}_{H_1 H_2} = 0.$$

From the first condition we find $b = -1/2$, whereas the second one leads to $a = 0$, so that $H_2(x) = x^2 - 1/2$. For the fourth polynomial we substitute $H_3(x) = x^3 + ax^2 + bx + c$, where the coefficients are obtained from the three orthogonality relations

$$\int_{-\infty}^{\infty} dx e^{-x^2} \overbrace{(x^3 + ax^2 + bx + c)}^{H_0 H_3} = 0, \quad \int_{-\infty}^{\infty} dx e^{-x^2} \overbrace{x(x^3 + ax^2 + bx + c)}^{H_1 H_3} = 0,$$

$$\int_{-\infty}^{\infty} dx e^{-x^2} \underbrace{\left(x^2 - \frac{1}{2}\right)(x^3 + ax^2 + bx + c)}_{H_2 H_3} = 0.$$

The solution is: $a = c = 0$ and $b = -3/2$. Note that for the obtained polynomials the coefficient in front of highest exponent term equals to 1.

(b): Using the Rodrigues' formula (4.28) the contour integral representation of Hermite polynomials is

$$H_n(x) = (-1)^n e^{x^2} \frac{n!}{2\pi i} \int_C \frac{e^{-z^2}}{(z - x)^{n+1}} dz,$$

where the contour C encloses the point $z = x$. Substituting $t = x - z$ leads to a Schläfli integral in the form

$$H_n(x) = \frac{n!}{2\pi i} \int_\gamma \frac{e^{2xt - t^2}}{t^{n+1}} dt,$$

where the contour γ encloses the point $t = 0$. Expanding the numerator of the integrand in series in x and t yields

$$H_n(x) = \frac{n!}{2\pi i} \sum_{k=0}^{\infty} \frac{(2x)^k}{k!} \sum_{l=0}^{\infty} \frac{(-1)^l}{l!} \int_\gamma t^{2l+k-n-1} dt.$$

Note that the contour integral in the rhs is zero unless $2l + k - n = 0$, otherwise it is equal to $2\pi i$, so that we have

$$H_n(x) = n! \sum_{k=0}^{\infty} (-1)^{\frac{n-k}{2}} \frac{(2x)^k}{k!(\frac{n-k}{2})!},$$

where $k - n$ must be even. Finally, letting $l = (n - k)/2$ yields the answer:

$$H_n(x) = n! \sum_{l=0}^{[\frac{n}{2}]} (-1)^l \frac{(2x)^{n-2l}}{l!(n-2l)!}.$$

(c): To prove the real axis integral representation we can expand the binomial series in the integrand arriving back to the series representation of $H_n(x)$. Indeed,

$$H_n(x) = \frac{2^n}{\sqrt{\pi}} \int_{-\infty}^{\infty} dt e^{-t^2} (x + it)^n = \frac{2^n}{\sqrt{\pi}} \sum_{k=0}^{n} \frac{n!\, i^{n-k}}{k!(n-k)!} x^k \int_{-\infty}^{\infty} dt e^{-t^2} t^{n-k}$$

$$= \frac{2^n n!}{\sqrt{\pi}} \sum_{l=0}^{[\frac{n}{2}]} \frac{(-1)^l x^{n-2l}}{(2l)!(n-2l)!} \Gamma\left(l + \frac{1}{2}\right) = n! \sum_{l=0}^{[\frac{n}{2}]} (-1)^l \frac{(2x)^{n-2l}}{l!(n-2l)!}.$$

Problem 4.8

We set $z = w^{-1}$ ($\frac{d}{dz} = -w^2 \frac{d}{dw}$) and rewrite the original Legendre equation as

$$\frac{d}{dw}\left[(1 - w^2)\frac{du}{dw}\right] - \frac{\nu(\nu+1)}{w^2} u = 0.$$

In the next step we introduce $u(w) = w^\alpha v(w)$, where the exponent α is (for now) an arbitrary number. The respective derivatives are

$$u' = w^\alpha \left(v' + \frac{\alpha}{w} v\right), \quad u'' = w^\alpha \left[v'' + \frac{2\alpha}{w} v' + \frac{\alpha(\alpha-1)}{w^2} v\right],$$

so that the new function $v(w)$ satisfies the equation

$$(1 - w^2)\frac{d^2 v}{dw^2} + \left[\frac{2\alpha}{w} - 2(\alpha+1)w\right]\frac{dv}{dw} + \left[\frac{\alpha(\alpha-1) - \nu(\nu+1)}{w^2} - \alpha(\alpha+1)\right]v = 0.$$

Let us now assume that the parameter α satisfies the identity: $\alpha(\alpha - 1) = \nu(\nu + 1)$, that is $\alpha = -\nu$ or $\alpha = \nu + 1$. The above equation then reduces to:

$$(1 - w^2)\frac{d^2 v}{dw^2} + \left[\frac{2\alpha}{w} - 2(\alpha+1)w\right]\frac{dv}{dw} - \alpha(\alpha+1)v = 0.$$

We now suppose that $v(w)$ is given as a hypergeometric series of the variable w^2 with the parameters a, b, and c still to be determined,

$$v(w) = \sum_{n=0}^{\infty} A_n w^{2n}, \qquad \frac{A_{n+1}}{A_n} = \frac{(a+n)(b+n)}{(n+1)(c+n)}.$$

Substituting the derivatives

$$v' = \sum_{n=0}^{\infty} 2n A_n w^{2n-1} = \sum_{n=0}^{\infty} (2n+2) A_{n+1} w^{2n+1},$$

$$v'' = \sum_{n=0}^{\infty} 2n(2n-1) A_n w^{2n-2} = \sum_{n=0}^{\infty} (2n+2)(2n+1) A_{n+1} w^{2n}$$

into the above equation we obtain the recurrence relation for A_n:

$$(2n+2)\Big(2n+2\alpha+1\Big) A_{n+1} - \Big[2n(2n-1+2\alpha+2) + \alpha(\alpha+1)\Big] A_n = 0.$$

Remember that $\alpha = -\nu$ or $\alpha = \nu + 1$.

- For $\alpha = -\nu$ this reduces to

$$\frac{A_{n+1}}{A_n} = \frac{(n-\frac{\nu}{2})(n-\frac{\nu}{2}+\frac{1}{2})}{(n+1)(n+\frac{1}{2}-\nu)}.$$

Hence $a = -\nu/2,\ b = (1-\nu)/2,\ c = 1/2 - \nu$, and the respective solution is

$$u_1(z) = z^{\nu} F\left(-\frac{\nu}{2}, \frac{1-\nu}{2}; \frac{1}{2} - \nu; \frac{1}{z^2}\right),$$

up to some constant prefactor.

- Similarly, for $\alpha = \nu + 1$ we obtain

$$\frac{A_{n+1}}{A_n} = \frac{(n+\frac{\nu+1}{2})(n+\frac{\nu+2}{2})}{(n+1)(n+\nu+\frac{3}{2})},$$

so that now $a = (\nu+1)/2,\ b = (\nu+2)/2,\ c = \nu + 3/2$, and the second solution is

$$u_2(z) = z^{-\nu-1} F\left(\frac{\nu+1}{2}, \frac{\nu+2}{2}; \nu + \frac{3}{2}; \frac{1}{z^2}\right),$$

again with an accuracy of a constant prefactor. Finally, the sum of these two solutions, each with its own coefficient, leads to the answer.

For integer values of $\nu = n = 0, 1, \ldots$, the polynomial solution of Legendre equation is unique and regular at $z = 0$. Hence only the first solution survives in this case and for the Legendre polynomials on the real axis we have

$$P_n(x) = A_n\, x^n F\left(-\frac{n}{2}, \frac{1-n}{2}; \frac{1}{2} - n; \frac{1}{x^2}\right), \quad n = 0, 1, 2, \ldots.$$

To determine A_n we write the lhs and the rhs as the series in the variable x. For the rhs we have

$$F\left(-\frac{n}{2}, \frac{1-n}{2}; \frac{1}{2}-n; \frac{1}{x^2}\right) = \sum_{k=0}^{[\frac{n}{2}]} \frac{(-\frac{n}{2})_k(\frac{1-n}{2})_k}{k!\,(\frac{1}{2}-n)_k} x^{-2k},$$

where $[\frac{n}{2}] = \frac{n}{2}$ for n-even and $[\frac{n}{2}] = \frac{1}{2}(n-1)$ for n-odd due to the first and second Pochhammer symbol, respectively. By definition, Pochhammer symbol is $(a)_k = a(a+1)...(a+k-1)$, so that the numerator is given by

$$\left(-\frac{n}{2}\right)_k \left(\frac{1-n}{2}\right)_k = 2^{-2k}\frac{n!}{(n-2k)!}.$$

For the denominator we can write the Pochhammer symbol in terms of the Gamma functions, $(a)_k = \Gamma(a+k)/\Gamma(a)$, and using the duplication formula for Γ yields

$$\frac{1}{(\frac{1}{2}-n)_k} = \frac{(-1)^k 2^{2k} n!(2n-2k)!}{(2n)!(n-k)!}.$$

Collecting the results, one obtains

$$\text{rhs} = A_n x^n \frac{(n!)^2}{(2n)!} \sum_{k=0}^{[\frac{n}{2}]} (-1)^k \frac{(2n-2k)!}{k!(n-k)!(n-2k)!} x^{-2k}.$$

For the lhs one can use, for instance, the series expansion of the Rodrigues' formula for Legendre polynomial:

$$P_n(x) = \frac{1}{2^n n!}\frac{d^n}{dx^n}(x^2-1)^n = \frac{1}{2^n n!}\frac{d^n}{dx^n}\sum_{k=0}^{n}\frac{(-1)^k n!}{k!(n-k)!}x^{2(n-k)}.$$

Performing the differentiation, we find [see also Problem 4.5 (a)]

$$\text{lhs} = \frac{x^n}{2^n}\sum_{k=0}^{[\frac{n}{2}]}(-1)^k \frac{(2n-2k)!}{k!(n-k)!(n-2k)!}x^{-2k}.$$

Comparing these results, we identify

$$A_n = \frac{(2n)!}{2^n (n!)^2},$$

whence the answer.

Problem 4.9

Let us consider the confluent hypergeometric Eq. (2.57) and make the substitution $z \to \xi^2$ to obtain

$$F'' - \left(\frac{1}{\xi} - \frac{2c}{\xi} + 2\xi\right)F' - 4aF = 0.$$

At $c = 1/2$ and $a = -m/2$ this equation reduces to Hermite Eq. (4.17). Hence, generally, we have

$$H_m(x) = A_m\, F\left(-\frac{m}{2}, \frac{1}{2}; x^2\right) + B_m\, x\, F\left(-\frac{m}{2} + \frac{1}{2}, \frac{3}{2}; x^2\right).$$

[Remember that $F(a, c; z)$ and $z^{1-c}F(a-c+1, 2-c; z)$ are two linearly independent solutions of (2.57).] In order to determine A_m and B_m we consider the even and odd m separately.

• At $m = 2n$ the second term does not survive due to the symmetry property $H_{2n}(x) = H_{2n}(-x)$, that is $B_{2n} = 0$, so that

$$H_{2n}(x) = H_{2n}(0)F\left(-n, \frac{1}{2}; x^2\right).$$

The Hermite polynomials at the origin $x = 0$ can be evaluated by means of the generating function

$$G(x, -t) = e^{-2xt+t^2} = \sum_{k=0}^{\infty} H_k(x)\frac{t^k}{k!},$$

see Example 4.11. Expanding $G(0, -t)$ in powers of t, the equality

$$H_{2n}(0) = (-1)^n \frac{(2n)!}{n!}$$

readily follows.

• At $m = 2n + 1$, on the contrary, only the second term survives, since now $H_{2n+1}(x) = -H_{2n+1}(-x)$, that is $A_{2n+1} = 0$ and hence

$$H_{2n+1}(x) = B_{2n+1}\, x\, F\left(-n, \frac{3}{2}; x^2\right).$$

The coefficient B_{2n+1} can be determined as follows. We write out the recurrence relation (2.60) for the confluent hypergeometric series,

$$2(2n+1)F\left(-n, \frac{3}{2}; x^2\right) - 2F\left(-n, \frac{1}{2}; x^2\right) - 4nF\left(-n+1, \frac{3}{2}; x^2\right) = 0,$$

and compare it to the relation

$$B_{2n+1}F\Big(-n,\frac{3}{2};x^2\Big)-2H_{2n}(0)F\Big(-n,\frac{1}{2};x^2\Big)+4nB_{2n-1}F\Big(-n+1,\frac{3}{2};x^2\Big)=0,$$

which follows from the identity $H_{2n+1}(x) - 2xH_{2n}(x) + 4nH_{2n-1}(x) = 0$ [see recurrence relation (4.43)]. Then we find

$$B_{2n+1} = 2(2n+1)(-1)^n\frac{(2n)!}{n!}.$$

Note, that one also obtains that $B_{2n-1} = -H_{2n}(0)$ and hence

$$H_{2n-1}(x) = -xH_{2n}(0)F\Big(-n+1,\frac{3}{2};x^2\Big).$$

This result, however, is not independent, but is a consequence of the identity $H_{2n}(x) = H_{2n}(0)F(-n,\frac{1}{2};x^2)$: $H'_{2n}(x) = 4nH_{2n-1}(x) = H_{2n}(0)F'(-n,\frac{1}{2};x^2) = H_{2n}(0)\,(-4nx)\,F(-n+1,\frac{3}{2};x^2)$.

We see that the Eq. (2.57) is converted to (4.14) at $a = -n$ and $c = \alpha+1$, that is $L_n^\alpha(x) = CF(-n,\alpha+1;x)$, where, evidently $C = L_n^\alpha(0)$. The Laguerre polynomials at the origin $x = 0$ can be evaluated from the Rodrigues' formula (see Problem 4.4)

$$L_n^\alpha(x) = \frac{x^{-\alpha}e^x}{n!}\frac{d^n}{dx^n}\big(e^{-x}x^{n+\alpha}\big).$$

In the limit $x \to 0$ we have by the Leibniz rule

$$\frac{d^n}{dx^n}\big(e^{-x}x^{n+\alpha}\big) = \sum_{k=0}^{n}\frac{n!}{k!(n-k)!}\frac{d^k}{dx^k}x^{n+\alpha}\frac{d^{n-k}}{dx^{n-k}}e^{-x} \;\to\; x^\alpha e^{-x}\frac{(n+\alpha)!}{\alpha!},$$

so that we get

$$L_n^\alpha(0) = \frac{(n+\alpha)!}{n!\alpha!}.$$

Using the above results for the Hermite polynomials, the relation between $H_{2n}(x)$ and $L_n^{-\frac{1}{2}}(x^2)$, as well as between $H_{2n+1}(x)$ and $L_n^{\frac{1}{2}}(x^2)$, immediately follows. [Remember that $\Gamma(n+1/2) = \sqrt{\pi}(2n)!/4^n n!$.]

Problem 4.10

(a): By the substitution $x = \cos t$ the Chebyshev equation is converted to the following simple form

$$\frac{d^2y}{dt^2} + n^2y = 0.$$

The general solution of this equation is written as

$$y(t) = C \cos(nt + \alpha),$$

where C and α are arbitrary real numbers. The condition $y(x)|_{x\to 1} = C\cos(n \arccos x + \alpha)|_{x\to 1} = 1$ is satisfied at $\alpha = 0$ and $C = 1$, so that

$$T_n(x) = \cos(n \arccos x).$$

Setting again $x = \cos t$, we observe that

$$T_n = \cos\bigl(n \arccos(\cos t)\bigr) = \cos(nt) = \mathrm{Re}\, e^{int}.$$

On the other hand, according de Moivre's theorem

$$e^{\pm int} = (\cos t \pm i \sin t)^n = (x \pm i\sqrt{1-x^2})^n,$$

which immediately leads to the relation

$$T_n(x) = \frac{1}{2}\Bigl[(x + i\sqrt{1-x^2})^n + (x - i\sqrt{1-x^2})^n\Bigr].$$

From this relation follows the normalization: $T_n(1) = 1$.
(b): The three–term recurrence relation for $T_n = \cos(nt)$ $(\cos t = x)$, $T_{n-1} = \cos\bigl((n-1)t\bigr)$, and $T_{n+1} = \cos\bigl((n+1)t\bigr)$ directly follows from the summation formula

$$\cos\bigl[(n-1)t\bigr] + \cos\bigl[(n+1)t\bigr] = 2\cos(nt)\cos t.$$

(c): The generating functions can be easily obtained with the trigonometric representation of T_n. So

$$\begin{aligned}\sum_{n=0}^{\infty} T_n s^n &= \frac{1}{2}\Bigl(\sum_{n=0}^{\infty} e^{int} s^n + \sum_{n=0}^{\infty} e^{-int} s^n\Bigr) = \frac{1}{2}\Bigl(\frac{1}{1-se^{it}} + \frac{1}{1-se^{-it}}\Bigr)\\ &= \frac{1}{2}\,\frac{2 - s(e^{it}+e^{-it})}{1 - s\underbrace{(e^{it}+e^{-it})}_{2x} + s^2} = \frac{1-xs}{1-2xs+s^2}.\end{aligned}$$

Similarly,

$$\begin{aligned}\sum_{n=0}^{\infty} \frac{T_n}{n!} s^n &= \frac{1}{2}\Bigl(\sum_{n=0}^{\infty} \frac{e^{int}}{n!} s^n + \sum_{n=0}^{\infty} \frac{e^{-int}}{n!} s^n\Bigr) = \frac{1}{2}\Bigl(e^{se^{it}} + e^{se^{-it}}\Bigr)\\ &= \frac{1}{2}\Bigl[e^{s(x-\sqrt{x^2-1})} + e^{s(x+\sqrt{x^2-1})}\Bigr].\end{aligned}$$

Problem 4.11
Let us consider the equation

$$\int_{-1}^{1} \frac{y(t)dt}{\sqrt{1+x^2-2xt}} = f(x)$$

with the original kernel but the rhs being an arbitrary polynomial function of x. First we observe that the kernel is a generating function for the Legendre polynomials:

$$\frac{1}{\sqrt{1+x^2-2xt}} = \sum_{m=0}^{\infty} P_m(t)\, x^m.$$

Therefore we can seek a solution in the form

$$y(x) = \sum_{n=0}^{\infty} A_n P_n(x).$$

Taking into account the orthogonality relations (see also Problem 4.5)

$$\int_{-1}^{1} P_n^2(x)dx = \frac{2}{2n+1}, \quad \int_{-1}^{1} P_n(x)P_m(x)(x)dx = 0 \quad (n \neq m)$$

and expanding the function $f(x)$ in the rhs of the equation into a Maclaurin series, for constants A_n we find

$$A_n = \frac{2n+1}{2n!} \frac{d^n}{dx^n} f(x)\big|_{x=0}.$$

Therefore the general solution of the above equation is

$$y(x) = \frac{1}{2} \sum_{n=0}^{\infty} \frac{2n+1}{n!} f^{(n)}(0) P_n(x).$$

For a special case $f(x) = x^m$, the nth derivative is clearly zero unless $n = m$, when it is equal to $n!$, so the answer.

Bibliography

1. Marsden, J., Hoffman, M.: Basic Complex Analysis. W. H Freeman, New York (1998)
2. Markushevich, A., Silverman, R.: Theory of Functions of a Complex Variable. AMS Chelsea Publishing Series. AMS Chelsea (2005)
3. Whittaker, E.T., Watson, G.N.: A Course of Modern Analysis. Cambridge University Press, Cambridge (1927)
4. Morse, P., Feshbach, H.: Methods of Theoretical Physics. McGraw-Hill, New York (1953)
5. Hardy, G.H.: Divergent Series. Oxford University Press, Oxford (1963)
6. Bateman, H., Erdélyi, A.: Tables of Integral Transforms, Bateman Manuscript project. McGraw-Hill, New York (1954)
7. Stalker, J.: Complex Analysis: Fundamentals of the Classical Theory of Functions. Birkhäuser, Boston (1998)
8. Bateman, H., Erdélyi, A.: Higher Transcendental Functions, Higher Transcendental Functions, bateman manuscript project ed. vol. 1, McGraw-Hill, New York (1955)
9. Watson, G.: A Treatise on the Theory of Bessel Functions, Cambridge Mathematical Library. Cambridge University Press, Cambridge (1995)
10. Abrikosov, A., Gorkov, L., Dzyaloshinskii, I.: Quantum Field Theoretical Methods in Statistical Physics, International Series of Monographs in Natural Philosophy. Dover (1975)
11. Landau, L., Lifshitz, E.: Quantum Mechanics: Non-relativistic Theory, Course of Theoretical Physics. vol. 3, Butterworth-Heinemann, (1977)
12. Gogolin, A.O., Mora, C., Egger, R.: Phys. Rev. Lett. **100**, 140404 (2008)
13. Smirnov, V.I.: A Course of Higher Mathematics: Integral Equations and Partial Differential Equations, A Course of Higher Mathematics. vol. 4, Pergamon Press, Oxford (1964)
14. Courant, R., Hilbert, D.: Methods of Mathematical Physics, Wiley classics library. vol. 1, Wiley, New York (1989)
15. Titchmarsh, E.C.: Introduction to the Theory of Fourier Integrals. Clarendon Press, Oxford (1948)
16. Hewson, A.C.: The Kondo Problem to Heavy Fermions, Cambridge Studies in Magnetism. Cambridge University Press, Cambridge (1997)
17. Hopf, E.: Mathematical Problems of Radiative Equilibrium. Buchanan Press, (2007)
18. Davison, B.: Neutron transport theory, International series of monographs on physics. Clarendon PressClarendon Press, Oxford (1958)
19. Noble, B.: Methods Based on the Wiener-Hopf Technique for the Solution of Partial Differential Equations, Chelsea Publishing Series. Chelsea Publishing Company, New York (1988)
20. Muskhelishvili, N.: Singular Integral Equations P. Noordhoff, Holland (1953)
21. Spitzer, F.: Duke Math. J. **24**(3), 327 (1957)
22. Widom, H.: Illinois J. Math. **2**(2), 261 (1958)

A. O. Gogolin (edited by E. G. Tsitsishvili and A. Komnik), *Lectures on Complex Integration*, 281
Undergraduate Lecture Notes in Physics, DOI: 10.1007/978-3-319-00212-5,

23. Titchmarsh, E.C.: Eigenfunction Expansions Associated with Second-Order Differential Equations. Clarendon Press, Oxford (1946)
24. Glauert, H.: The Elements of Aerofoil and Airscrew Theory, Cambridge Science Classics. Cambridge University Press, Cambridge (1983)
25. Langreth, D.C., Nordlander, P.: Phys. Rev. B **43**, 2541 (1991)
26. Schmidt, T.L., Werner, P., Mühlbacher, L., Komnik, A.: Phys. Rev. B **78**, 235110 (2008)
27. Jackson, J.D.: Classical Electrodynamics, Wiley, New York (1998)

Index

A. O. Gogolin (edited by E. G. Tsitsishvili and A. Komnik), *Lectures on Complex Integration*, 283
Undergraduate Lecture Notes in Physics, DOI: 10.1007/978-3-319-00212-5,